Cysteine Proteases of Pathogenic Organisms

ADVANCES IN EXPERIMENTAL MEDICINE AND BIOLOGY

Recent Volumes in this Series

Volume 704
TRANSIENT RECEPTOR POTENTIAL CHANNELS
Md. Shahidul Islam

Volume 705
THE MOLECULAR IMMUNOLOGY OF COMPLEX CARBOHYDRATES-3
Albert M. Wu

Volume 706
ADHESION-GPCRS: STRUCTURE TO FUNCTION
Simon Yona and Martin Stacey

Volume 707
HORMONAL AND GENETIC BASIS OF SEXUAL DIFFERENTIATION DISORDERS AND HOT TOPICS IN ENDOCRINOLOGY
Maria I. New and Joe Leigh Simpson

Volume 708
INVERTEBRATE IMMUNITY
Kenneth Söderhäll

Volume 709
HISTAMINE IN INFLAMMATION
Robin L. Thurmond

Volume 710
RECENT ADVANCES ON MODEL HOSTS
Eleftherios Mylonakis

Volume 711
EPIGENETIC CONTRIBUTIONS IN AUTOIMMUNE DISEASE
Esteban Ballestar

Volume 712
CYSTEINE PROTEASES OF PATHOGENIC ORGANISMS
Mark W. Robinson and John P. Dalton

A Continuation Order Plan is available for this series. A continuation order will bring delivery of each new volume immediately upon publication. Volumes are billed only upon actual shipment. For further information please contact the publisher.

Cysteine Proteases of Pathogenic Organisms

Edited by

Mark W. Robinson, PhD
Infection, Immunity and Innovation (i3) Institute, University of Technology Sydney, Ultimo, Sydney, Australia

John P. Dalton, PhD
Institute of Parasitology, McDonald Campus, McGill University, St. Anne de Bellevue, Quebec, Canada

Springer Science+Business Media, LLC

Landes Bioscience

Springer Science+Business Media, LLC
Landes Bioscience

Printed in the USA.

Springer Science+Business Media, LLC, 233 Spring Street, New York, New York 10013, USA
http://www.springer.com

Please address all inquiries to the publishers:
Landes Bioscience, 1806 Rio Grande, Austin, Texas 78701, USA
Phone: 512/ 637 6050; FAX: 512/ 637 6079
http://www.landesbioscience.com

The chapters in this book are available in the Madame Curie Bioscience Database.
http://www.landesbioscience.com/curie

Cysteine Proteases of Pathogenic Organisms, edited by Mark W. Robinson and John P. Dalton. Landes Bioscience / Springer Science+Business Media, LLC dual imprint / Springer series: Advances in Experimental Medicine and Biology.

ISBN: 978-1-4419-8413-5

While the authors, editors and publisher believe that drug selection and dosage and the specifications and usage of equipment and devices, as set forth in this book, are in accord with current recommendations and practice at the time of publication, they make no warranty, expressed or implied, with respect to material described in this book. In view of the ongoing research, equipment development, changes in governmental regulations and the rapid accumulation of information relating to the biomedical sciences, the reader is urged to carefully review and evaluate the information provided herein.

Library of Congress Cataloging-in-Publication Data

Cysteine proteases of pathogenic organisms / edited by Mark W. Robinson, PhD, Infection, Immunity and Innovation (i3) Institute, University of Technology Sydney, Ultimo, Sydney, Australia, John P. Dalton, PhD, Institute of Parasitology, McDonald Campus, McGill University, St. Anne de Bellevue, Quebec, Canada.
p. ; cm. -- (Advances in experimental medicine and biology ; v. 712)
Includes bibliographical references and index.
ISBN 978-1-4419-8413-5 (hardback)
1. Cysteine proteinases--Pathophysiology. 2. Microbial enzymes. 3. Pathogenic microorganisms. I. Robinson, Mark W., 1977- editor. II. Dalton, John P. (John Pius), 1958- editor. III. Series: Advances in experimental medicine and biology ; v. 712.
[DNLM: 1. Cysteine Proteases--chemistry. 2. Bacteria--enzymology. 3. Cysteine Proteases--immunology. 4. Eukaryota--enzymology. 5. Viruses--enzymology. W1 AD559 v.712 2011 / QU 136]
QP609.C94.C965 2011
616.9'041--dc22

2010050245

PREFACE

Cysteine proteases expressed by pathogenic organisms play key roles in virulence including host entry, feeding and suppression of host immune responses. This volume gives comprehensive coverage to all aspects of pathogen cysteine proteases and brings together numerous scientific advances which have been made over many years. Thus, the biochemistry, molecular biology and structure-function relationships of these important pathogen enzymes are covered in detail.

Chapters 1 and 2 review recent developments in bacterial cysteine proteases. Chapter 1 highlights the role of staphopains as virulence factors from *Staphylococcus aureus* which is timely given the increasing prevalence of antibiotic-resistant isolates of this pathogen. Chapter 2 covers the central role of *Porphyromonas gingivalis* gingipains in the development of periodontal disease. With emphasis on structure-function relationships, this chapter probes the potential for bacterial cysteine proteases as novel targets for anti-infective therapy.

Chapters 3 to 7 focus on protozoan parasites of medical importance. In Chapter 3, the cysteine proteases, falcipains, expressed by malaria (*Plasmodium spp.*) are described with emphasis on their involvement in the digestion of haemoglobin by the intra-erythrcocytic stages of this parasite. Falcipains are promising drug targets and efforts to optimize falcipain inhibitors as antimalarials are discussed. Chapter 4 describes how cysteine proteases are employed by the apicomplexan parasite *T. gondii* for host cell invasion, replication, and nutrient acquisition. Chapter 5 provides an overview of how cysteine proteases from the intestinal parasite *Entamoeba histolytica* manipulate and destroy host defences to facilitate nutrient acquisition and colonisation of new hosts. Chapter 6 details advances in our understanding of the cysteine proteases of the kinetoplastid parasites *Leishmania* and *Trypanosoma.* The crucial role of cysteine proteases in kinetoplastid metabolism and virulence has made these enzymes leading targets for antiparasite therapy. This topic is covered in detail in Chapter 7 which chronicles the development of new drugs that target *Trypanosome* cysteine proteases.

Chapters 8 to 10 cover cysteine proteases from helminths of medical and veterinary importance. With emphasis on the cathepsin L family, Chapter 8 highlights how recent advances in proteomics, gene silencing and crystallography are allowing researchers to probe the molecular functions of these important enzymes from trematodes of the genus *Fasciola* and *Schistosoma* and others. Building on this, Chapter 9 details how cysteine proteases play a central role in the disease caused by *Trichobilharzia*, a helminth parasite of birds that is emerging as a significant human infection. Chapter 10 covers how cysteine proteases play a crucial role in the acquisition of nutrients in blood-feeding nematodes and reviews the latest developments in their potential use as antinematode vaccines.

Chapter 11 highlights how cysteine proteases facilitate digestion and embryogenesis in several important parasitic arthropods. Chapter 12 covers the various mechanisms by which pathogen-derived cysteine proteases modulate host immune responses to promote their survival. Chapter 13 concludes the book with a review of pathogen cysteine protease inhibitors (cystatins) thus providing a comprehensive view of cysteine proteases from both proteolytic and inhibitory standpoints. Written by leading researchers from Europe, Australia and North America, *Cysteine Proteases of Pathogenic Organisms* is essential reading for students and professionals interested in human medicine and infectious disease research.

Mark W. Robinson, PhD
Infection, Immunity and Innovation (i3) Institute
University of Technology Sydney, Ultimo, Sydney, Australia

ABOUT THE EDITORS...

MARK W. ROBINSON is a Chancellor's Postdoctoral Research Fellow at the Infection, Immunity and Innovation (i3) Institute at the University of Technology, Sydney (UTS), Australia. Robinson received a BSc (Hons) in Zoology (1999) and a PhD in molecular parasitology (2003) from Queen's University, Belfast (Northern Ireland). As a postdoctoral researcher, he worked on nematode proteomics and molecular biology at the University of Aberdeen (Scotland) until 2007 when he was awarded a prestigious Wain International Fellowship from the British Biotechnology and Biological Sciences Research Council (BBSRC) to continue his proteomics research at UTS (Australia). Robinson was awarded a UTS Chancellor's Postdoctoral Fellowship in 2008 which he currently holds as an independent researcher. His research interests include proteomics of host-pathogen interactions, structure-function relationships of parasite immuno-modulators and the molecular biology of helminth proteases.

ABOUT THE EDITORS...

JOHN P. DALTON is a Canada Research Chair in Infectious Diseases at the Institute of Parasitology, McGill University, Montreal, Canada. He is also a Professor in Biotechnology and Director of McGill's Graduate Program in Biotechnology. Before joining McGill, Dalton was Director of the Institute for the Biotechnology of Infectious Diseases (IBID), University of Technology Sydney (UTS), Australia. He was awarded the New South Wales Government BioFirst Award in Biotechnology from 2004-2008. Dalton received his BSc (Hons) in Zoology/Biochemistry (1980) and a PhD in Parasitology (1984) at University College Dublin. He spent postdoctoral training periods in the Johns Hopkins School of Medicine, (Baltimore, USA) and at the Laboratories of Parasitic Diseases, National Institute of Allergies and Infectious Diseases (NIAID) at the National Institutes of Health (Bethesda, MD, USA). He has also spent several years at the Queensland Institute for Medical Research (QIMR), Brisbane, Australia, where he is currently an adjunct Professor. He was a lecturer and researcher in biotechnology at Dublin City University (DCU) from 1997-2003. His interests include vaccine and drug design against helminth parasites, antimalaria drug design and the development of novel parasite-derived immunotherapeutics.

PARTICIPANTS

Linda S. Brinen
Sandler Center for Drug Discovery
and
Department of Cellular and Molecular Pharmacology
University of California–San Francisco
San Francisco, California
USA

Conor R. Caffrey
Sandler Center for Drug Discovery
California Institute for Quantitative Biosciences
University of California San Francisco
San Francisco, California
USA

Eric Calvo
Laboratory of Malaria and Vector Research
National Institute of Allergy and Infectious Diseases
National Institutes of Health
Rockville, Maryland
USA

Vern B. Carruthers
Department of Microbiology and Immunology
University of Michigan School of Medicine
Ann Arbor, Michigan
USA

Kris Chadee
Gastrointestinal Research Group
University of Calgary
Calgary, Alberta
Canada

John P. Dalton
Institute of Parasitology
McGill University
St. Anne de Bellevue, Quebec
Canada

Emilia Daniłowicz-Luebert
Department of Molecular Parasitology
Humboldt University
Berlin
Germany

Kateřina Dolečková
Department of Parasitology
Faculty of Science
Charles University
Prague
Czech Republic

Sheila Donnelly
Infection, Immunity and Innovation (i3) Institute
University of Technology
Sydney, Ultimo
Australia

Zhicheng Dou
Department of Microbiology
and Immunology
University of Michigan School of Medicine
Ann Arbor, Michigan
USA

Jan Dvořák
Institute of Parasitology
Biology Centre of the Academy of Sciences
of the Czech Republic
Budějovice
Czech Republic

Ivo M.B. Francischetti
Laboratory of Malaria and Vector Research
National Institute of Allergy
and Infectious Diseases
National Institutes of Health
Rockville, Maryland
USA

Vladimír Hampl
Department of Parasitology
Faculty of Science
Charles University
Prague
Czech Republic

Susanne Hartmann
Department of Molecular Parasitology
Humboldt University
Berlin
Germany

Petr Horák
Department of Parasitology
Faculty of Science
Charles University
Prague
Czech Republic

Tomasz Kantyka
Department of Microbiology
Faculty of Biochemistry
Biophysics and Biotechnology
Jagiellonian University
Krakow
Poland

Martin Kašný
Department of Parasitology
Faculty of Science
Charles University
Prague
Czech Republic

Vanessa Kissoon-Singh
Gastrointestinal Research Group
University of Calgary
Calgary, Alberta
Canada

Christian Klotz
Department of Molecular Parasitology
Humboldt University
Berlin
Germany

David Knox
Moredun Research Institute
Pentlands Science Park
Bush Loan, Penicuik
Midlothian, Scotland
UK

Michalis Kotsyfakis
Laboratory of Genomics and Proteomics
of Disease Vectors
Institute of Parasitology
Biology Centre of the Academy of Sciences
of the Czech Republic
Ceske Budejovice
Czech Republic

Ana-Paula Lima
Instituto de Biofísica Carlos Chagas Filho
Universidade Federal do Rio de Janeiro
Cidade Universitária
Ilha do Fundão
Rio de Janeiro
Brazil

James H. McKerrow
Sandler Center for Drug Discovery
University of California–San Francisco
San Francisco, California
USA

Libor Mikeš
Department of Parasitology
Faculty of Science
Charles University
Prague
Czech Republic

Leanne Mortimer
Gastrointestinal Research Group
University of Calgary
Calgary, Alberta
Canada

Marian Novotný
Department of Cell Biology
Faculty of Science
Charles University in Prague
Prague
Czech Republic

Robert N. Pike
Department of Biochemistry and Molecular Biology
and
CRC for Oral Health Sciences
Monash University
Clayton, Victoria
Australia

Jan Potempa
Department of Microbiology
Faculty of Biochemistry
Biophysics and Biotechnology
Jagiellonian University
Krakow
Poland
and
Oral Health and Systemic Disease Research Facility
University of Louisville School of Dentistry
Louisville, Kentucky
USA

Stephanie A. Robertson
Sandler Center for Drug Discovery
University of California–San Francisco
San Francisco, California
USA

Mark W. Robinson
Infection, Immunity and Innovation (i3) Institute
University of Technology
Sydney, Ultimo
Australia

Philip J. Rosenthal
Department of Medicine
San Francisco General Hospital
University of California
San Francisco, California
USA

Mohammed Sajid
Leiden University Medical Center
Afd. Parasitologie
Leiden
The Netherlands

Lindsey N. Shaw
Department of Cell Biology
Microbiology and Molecular Biology
University of South Florida
Tampa, Florida
USA

Daniel Sojka
Laboratory of Vector Immunology
Institute of Parasitology
Biology Centre of the Academy of Sciences of the Czech Republic
Ceske Budejovice
Czech Republic

Colin Stack
School of Biomedical and Health Sciences
University of Western Sydney (UWS)
Narellan Road
Campbelltown, NSW
Australia

Dietmar Steverding
BioMedical Research Centre
School of Medicine
Health Policy and Practice
University of East Anglia
Norwich
UK

Tang Yongqing
Department of Biochemistry and Molecular Biology
and
CRC for Oral Health Sciences
Monash University
Clayton, Victoria
Australia

Lakshmi C. Wijeyewickrema
Department of Microbiology
Faculty of Biochemistry, Biophysics and Biotechnology
Jagiellonian University
Krakow
Poland

Thomas Ziegler
Department of Molecular Parasitology
Humboldt University
Berlin
Germany

CONTENTS

ACKNOWLEDGEMENTS

The authors wish to thank Cynthia Conomos, Erin O'Brien, Brett Aycock, Celeste Carlton and all at Landes Bioscience for their invaluable assistance in preparing this book.

CHAPTER 1

PAPAIN-LIKE PROTEASES OF *STAPHYLOCOCCUS AUREUS*

Tomasz Kantyka,[1] Lindsey N. Shaw[2] and Jan Potempa*,[1,3]

[1]*Department of Microbiology, Faculty of Biochemistry, Biophysics and Biotechnology, Jagiellonian University, Poland;* [2]*Department of Cell Biology, Microbiology and Molecular Biology, University of South Florida, Tampa, Florida, USA;* [3]*Oral Health and Systemic Disease Research Facility, University of Louisville School of Dentistry, Louisville, Kentucky, USA*

**Corresponding Author: Jan Potempa—Email: jspote01@louisville.edu*

Abstract: *Staphylococcus aureus* remains one of the major human pathogens, causing a number of diverse infections. The growing antibiotic resistance, including vancomycin and methicilin-resistant strains raises the special interest in virulence mechanism of this pathogen. Among a number of extracellular virulence factors, *S. aureus* secretes several proteases of three catalytic classes—metallo, serine and papain-like cysteine proteases. The expression of proteolytic enzymes is strictly controlled by global regulators of virulence factors expression *agr* and *sar* and proteases take a role in a phenotype change in postlogarithmic phase of growth. The staphylococcal proteases are secreted as proenzymes and undergo activation in a cascade manner.

Staphopains, two cysteine, papain-like proteases of *S. aureus* are both ~20 kDa proteins that have almost identical three-dimensional structures, despite sharing limited primary sequence identity. Although staphopain A displays activity similar to cathepsins, recognising hydrophobic residues at P2 position and large charged residues at P1, staphopain B differs significantly, showing significant preference towards β-branched residues at P2 and accepting only small, neutral residues at the P1 position. There is limited data available on the virulence potential of staphopains in in vivo models. However, in vitro experiments have demonstrated a very broad activity of these enzymes, including destruction of connective tissue, disturbance of clotting and kinin systems and direct interaction with host immune cells. Staphopain genes in various staphylococci species are regularly followed by a gene encoding an extremely specific inhibitor of the respective staphopain. This pattern is conserved across species and it is believed that inhibitors (staphostatins) protect the cytoplasm of the cell from premature activation of staphopains during protein folding. Notably, production and activity of staphopains is controlled on each level, from gene expression, through presence of specific inhibitors in cytoplasm,

Cysteine Proteases of Pathogenic Organisms, edited by Mark W. Robinson and John P. Dalton.

to the cascade-like activation in extracellular environment. Since these systems are highly conserved, this points to the importance of these proteases in the survival and/or pathogenicity of *S. aureus*.

INTRODUCTION

All cysteine peptidases belonging to clan CA have a common fold motif, consisting of an aminoterminal domain that is mostly α-helical and a carboxy-terminal domain featuring an antiparallel β-sheet, with the Cys and His catalytic residues forming a thiolate-imidazolium dyad. The clan is divided into 24 families, out of which papain-like proteases constitute the major, most populous family (C1) exemplified by papain and mammalian cysteine cathepsins. It is an evolutionary paradox that this major family of cysteine peptidases encompassing more than 3,300 sequences listed in the MEROPS Data Base (http://merops.sanger.ac.uk) have relatively few representatives in bacteria. All together, circa 300 homologues of papain have been identified in prokaryotes and a gene encoding papain-like protein were found in 121 out of 764 bacteria with completely sequenced genomes. In this context, it is interesting to note that all *Lactococci* spp. and *Streptococci* spp. with sequenced genomes contain at least one C1 family representative, aminopeptidase C. Up to date there are no reports that this peptidase or its homologues are involved in any aspect of bacterial pathogenicity. In stark contrast, cysteine proteases grouped in four different families, including family C10, C47, C58, and C66 and exemplified by streptopain (SpeB) of *Streptococcus pyogenes*, staphopains of *Staphylococcus aureus*, YopT peptidase of *Yersinia pestis*, and IdeS peptidase *S. pyogenes* all contribute to the bacterial pathogenicity. Apparently, despite preserving the typical fold of papain-like proteases these enzymes evolved to function as virulence factors and this is discussed in the following sections of this chapter taking as the example staphopains.

THE PATHOGENIC POTENTIAL OF *S. AUREUS*

Despite the progress in the disciplines of infectious medicine and pharmacology, *S. aureus* continues to be a major human pathogen. Infections caused by this micro-organism range from superficial lesions, such as wound infections and abscesses, to life-threatening syndromes, including bacteremia, endocarditis, meningitis and osteomyelitis. The increasing prevalence of antibiotic-resistant strains, especially methicillin resistant *S. aureus* (MRSA) and the emergence of vancomycin resistant *S. aureus* (VRSA), has rendered staphylococcal infections a serious public health concern.

S. aureus expresses more than 50 extracellular virulence factors, ranging from toxins (hemolysin, enterotoxins, toxic shock syndrome toxin), through to surface proteins (Protein A, FnBP and other MSCRAMM's), to secreted enzymes, such as proteases, nucleases and lipase.[1,2] As these factors are employed at different stages of infection, they are strictly and co-ordinately regulated. The major control of expression and activity of these factors, is exerted by a quorum sensing system, referred to as the accessory gene regulator (Agr, see Fig. 1).[3-5] This complex regulatory mechanism ultimately functions via the RNAIII product, generated from one of the two divergent promoters of the *agr* operon, P3. The other promoter, P2, encodes the *agr* structural genes, *agrBCDA*. AgrB

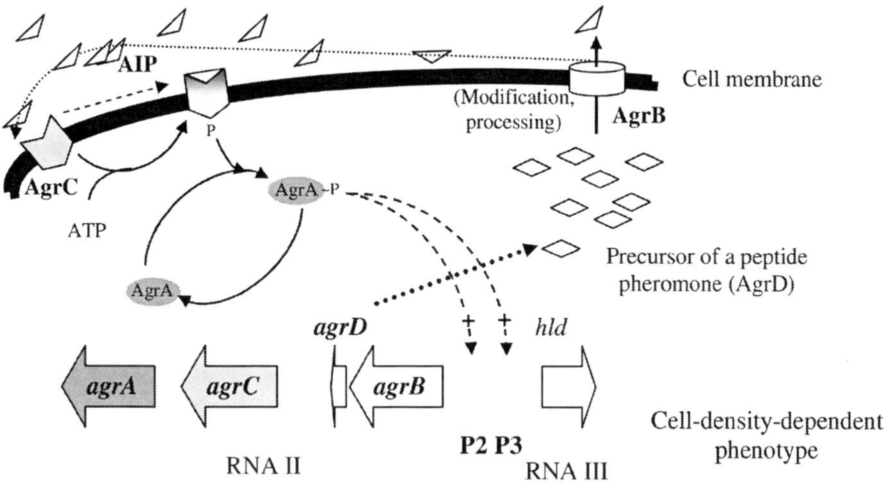

Figure 1. A schematic representation of the Staphylococcus aureus quorum sensing system.

is a transmembrane transport protein with endopeptidase activity and facilitates AgrD cleavage, followed by secretion of the resulting octapeptide,[6] subsequently termed the auto-inducing peptide (AIP). AgrC forms a membrane-bound histidine kinase, which senses external levels of AIP. This binding event results in auto-phosphorylation of AgrC, followed by transfer of this phosphate to the intracellular AgrA protein. AgrA is a DNA-binding protein which leads to activation of the P2 and P3 promoters, as well as direct activation of specific toxin genes[7,8] (Fig. 1). This process is required to diversify the composition of cell surface proteins and to transform the bacterial phenotype from one of adhesion, expressing a variety of surface proteins responsible for binding to host tissues, into one of invasion, producing factors which degrade tissues and allow for the spread of infection from initial colonisation foci.[9-11] An additional regulatory system, mediated by the DNA-binding protein SarA (staphylococcal accessory regulator) acts in a somewhat opposing manner.[12-14] While the *agr* system promotes transformation of *S. aureus* cells into a migratory phenotype, SarA promotes the adhesive properties of the bacterial cell. Additionally, *sarA* regulates virulence factor synthesis, both directly or indirectly, by modulating the transcription of other virulence determinant regulators.[15] Furthermore, analysis of the phenotype of *sarA* mutants reveals intensive overexpression of extracellular proteases,[16] demonstrating the restricting effect of SarA on protease production. Along with these intracellular regulating factors, this pathway of phenotypic switching, which is a hallmark of *S. aureus* infection, requires proteolytic enzymes to cleave the surface proteins, which decorate the exterior of the cell.

It has been proposed that a core assembly of secreted proteins, commonly identified among clinical isolates of *S. aureus*, constitute its basic virulence unit; a minimal adaptation required to cause the majority of infections.[17] However, as significant diversity of secreted proteins between strains has been described,[18] the importance of particular proteins and expression patterns for successful pathogenesis and/or commensalism remains poorly understood.[19]

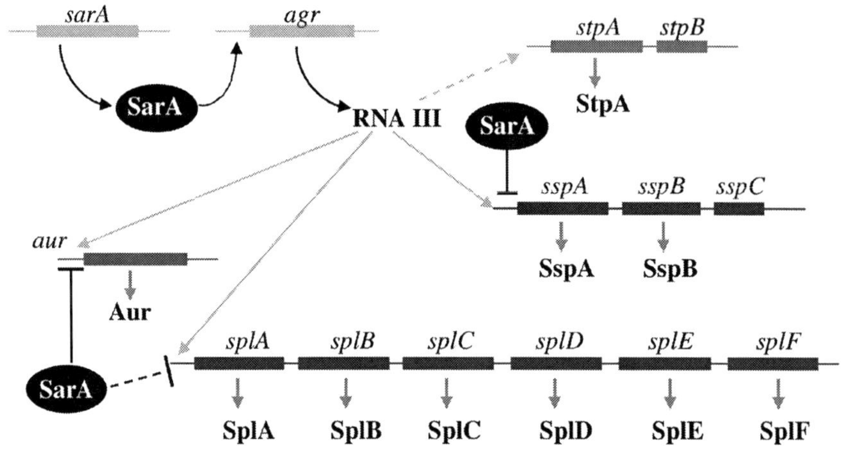

Figure 2. Protease operons under control of the *agr* and *sar* global regulatory systems of virulence factor synthesis. (arrows—stimulation, bars—repression)

THE PROTEOLYTIC ENZYMES OF *S. AUREUS*

Amongst the proteins secreted by *S. aureus* are proteolytic enzymes belonging to three distinct catalytic classes, namely metallo- (aureolysin), serine (V8, Spl proteases, epidermiolytic toxins) and cysteine (staphopain A, StpA and staphopain B, StpB) proteases.[20-24] Based on in vitro studies, staphylococcal proteases are considered potentially important virulence factors. They possess the ability to degrade critical components of the host defence system, such as elements of the complement system, cytokines and receptors on host immune cells;[25-29] thus providing defence against the host immune response.[17,30] They can also contribute to virulence directly, or indirectly, by interception of host enzymes and degrading endogenous proteinase inhibitors, resulting in tissue damage.[25,31] Besides host-derived proteins, *S. aureus* proteases cleave a score of self-derived surface proteins, affecting its adhesive phenotype, thus contributing to bacterial detachment.[9-11] Taken together, this implies the need for strict control of their expression and regulation of their activity. Therefore, it is not surprising that the transcription of each protease encoding operon is tightly controlled in a growth phase-dependent manner. Specifically, their transcription is positively regulated by the *agr* quorum-sensing system and repressed by the pleiotropic virulence determinant regulator, SarA (Fig. 2).[16,32-34] The interplay between these two major global regulators, alongside several other modulators of gene expression, results in enhanced synthesis of these extracellular proteases in the postlogarithmic phase of growth, similarly to the majority of secreted virulence factors in *S. aureus*.[15,16,32,33,35]

GENETIC ORGANISATION OF THE STAPHYLOCOCCAL PROTEOLYTIC ENZYMES

The genes encoding extracellular proteolytic enzymes in *S. aureus* are clustered into four distinct loci on the chromosome (Fig. 3). The staphylococcal serine protease (*ssp*) operon consists of three genes: *sspA* encoding the serine V8 protease

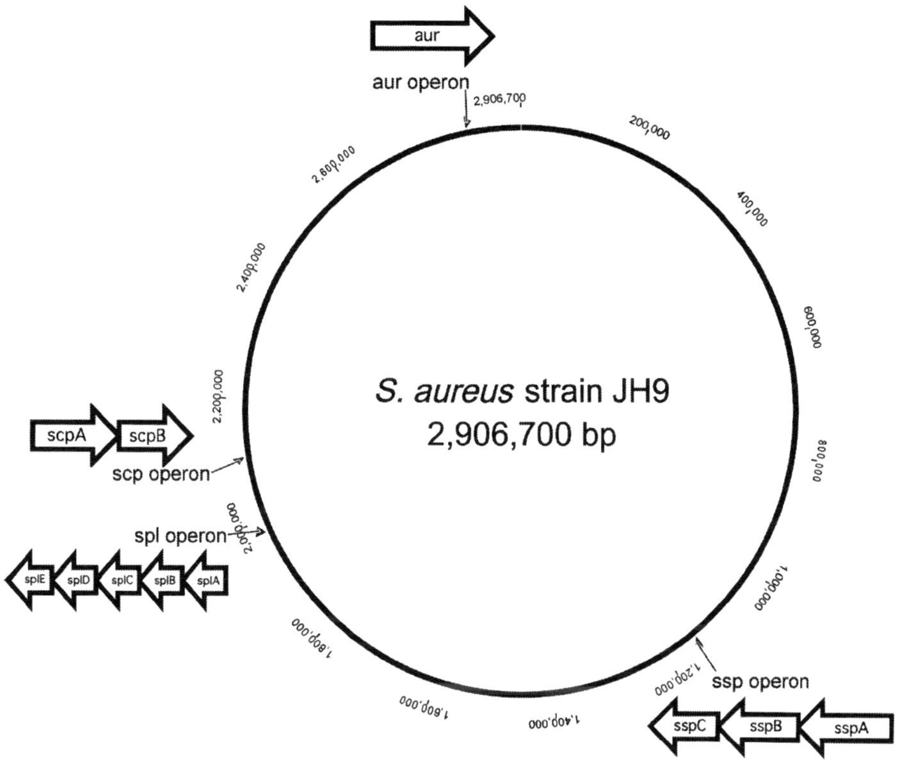

Figure 3. The organisation of protease genes within the *S. aureus* genome. Positions of protease operons in the *S. aureus* JH9 genome[53] are marked with respective distances. The schematics of protease operons are not drawn according to scale.

(glutamylendopeptidase I), *sspB*, encoding the cysteine protease staphopain B (StpB) and *sspC* encoding staphostatin B (StpinB), which is a specific inhibitor of StpB activity.[19,24,36-38] The serine protease-like operon (*spl*) is located 0.8 Mb downstream of the *ssp* operon, grouping together up to six genes that are homologous to *sspA*. This operon is closely followed by the staphylococcal cysteine protease operon (*scp*), consisting of two genes: *scpA* encoding staphopain A (StpA) and *scpB* encoding staphostatin A (StpinA), a specific, intracellular inhibitor of StpA, homologous to StpinB.[34,38-41] Further downstream of these loci, a single gene, *aur*, encodes the metalloprotease aureolysin.

EXPRESSION AND ACTIVATION OF *S. AUREUS* EXTRACELLULAR PROTEASES

Staphylococcal proteolytic enzymes are expressed as preproproteins, with a large propeptide domain. Firstly, signal sequences target the polypeptide chains for translocation across the cell membrane. Subsequently, in the extracellular environment, the proproteins are folded into enzymatically inert zymogens, most likely in the vicinity of the bacterial cell surface.

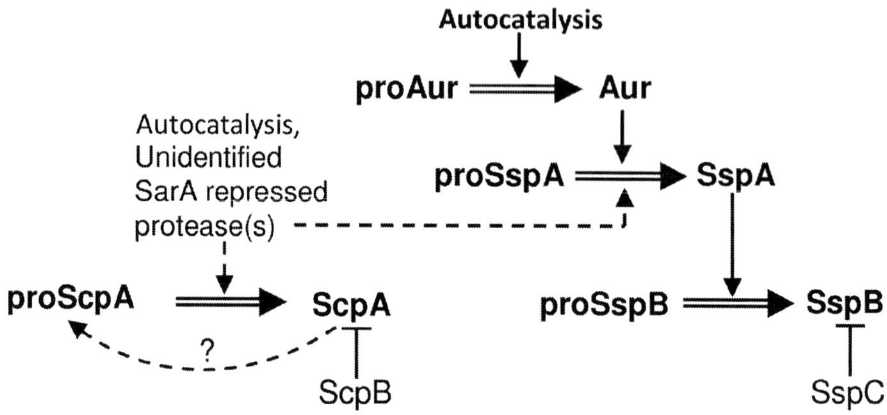

Figure 4. Diagrammatic representation of the posttranslational steps of activation of extracellular proteases of *S. aureus*. Broken lines represent putative proteolytic steps in the cascade, arrows with lines represent activation events and bars represent inhibition.

The zymogens are activated in an interdependent, cascade-like manner. The staphylococcal metalloprotease, aureolysin, activates SspA (V8 protease) by limited proteolysis, which in turns activates proStpB, releasing the mature active form of the enzyme.[24,32,34,36] Aureolysin appears to be activated via autocatalysis;[42] which is in agreement with the self-processing of other thermolysin-like enzymes.[43] Interestingly, prostaphopain A (pro-StpA) seems to be located outside this activation cascade, although it has been suggested that the zymogen is proteolytically activated by an as yet unknown enzyme (Fig. 4). In an attempt to explain prostaphopain A activation, it has recently been suggested that its maturation involves autocatalytic processing, as some structural adaptations for autocatalytic processing have been identified in the proenzyme.[17,18,44,45] However, these conclusions were drawn based on in silico modelling of the pro-StpA structure and thus require experimental verification.[45] Nevertheless, if confirmed, the suggested mechanism may provide a back-up system to the proteolytic cascade, as activation of proSspA and proStpB has been observed in *aur*-negative mutant strains.[34]

The mature forms of both the staphopain enzymes are ~20 kDa proteins. Despite negligible similarity at the level of amino acid sequence to papain, or cathepsins, the staphopains are folded in a papain-like manner.[37,44] Interestingly, although StpA and StpB share almost identical 3-dimensional structures (Fig. 5), they differ significantly in their specificity. While both enzymes prefer hydrophobic amino acid residues at the S2 specificity subsite, StpB strongly favours β-branched side chains, such as found with isoleucine. Even more significantly, the staphopains differ with respect to S1 subsite specificity. In stark contrast to the StpA S1 subsite, which accepts large, charged residues, StpB prefers small, neutral ones, such as glycine and alanine. This makes StpB a unique papain-like protease with regards to its specificity. Interestingly, the observed differences in specificity cannot be explained by a comparison of StpA and StpB structures.[37] Apparently, the minor difference in topology of substrate-binding clefts dictates this remarkable disparity in staphopain specificity.

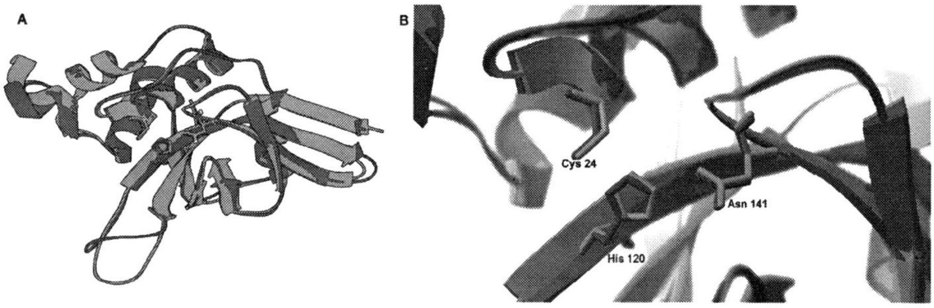

Figure 5. Comparison of the 3-dimensional structure of the staphopains. The superimposed structures indicate high similarity of the staphopain molecules and their fold. StpA (red) (accession code 1CV8,[44]) and StpB (green) (accession code 1Y4H,[37]) differ mainly in the exposed loop region (panel A). The structure of the staphopain active sites (panel B) reveals that the topology of their catalytic residues (displayed as sticks) is basically identical. A color version of this image is available at www.landesbioscience.com/curie.

STAPHOPAIN ACTIVITY AND ITS RELEVANCE TO VIRULENCE

The elaborate and unusual mechanism of controlling proteolytic activity in *S. aureus* strongly suggests that extracellular proteases and staphopains in particular, are very important factors for growth and survival. Each step in releasing staphopain activity is strictly regulated, starting at the level of gene transcription, through to preproprotein secretion and folding and finally zymogen activation and control of mature enzyme activity by specific inhibitors. Such complicated mechanisms have to be energetically expensive, yet their conservation among *S. aureus* strains emphasizes the importance of these enzymes for the pathophysiology of *S. aureus*. Results from numerous in vitro investigations support this conclusion, but there is still a lack of convincing in vivo data demonstrating that the staphopains are essential virulence factors.

Initial results have shown significant reductions in virulence in murine abscess, bacteraemia and wound infection models when analyzing polar *sspA* mutants.[46] With that said, other research appears to contradict this, as reviewed elsewhere.[47] To date, only the importance of StpB in a mouse abscess model of infection has been documented.[34] However, it must be kept in mind that mice serum, in stark contrast to human serum, inhibits StpB activity. In addition, this inhibitory capacity for staphopains is at much higher levels than seen in humans, due to enzyme inhibition by α_2-macrogobulin and macroglobulin-related proteins described as murinoglobulins (our unpublished results). Taking this fact into consideration, murine, or more generally rodent models, may not be entirely suitable for investigating the role of staphylococcal proteases in infection.

In contrast to these inconclusive in vivo studies, the results of in vitro experiments strongly implicate *S. aureus*-derived proteases as important pathogenic determinants (Fig. 6). Specifically, these enzymes are able to inactivate α-1-protease inhibitor and α-1-antichymotrypsin, two important endogenous protease inhibitors that are essential for controlling neutrophil serine proteases.[25,26] In addition, they have been shown to degrade a variety of pathogenically-relevant substrates, including fibrinogen, fibronectin and high molecular mass kinininigen.[19,25,48,49] From the perspective of staphylococcal lung infections it is important to underline that StpA possesses very potent elastinolytic activity (48). Furthermore, prothrombin activation by aureolysin[50] and the action of the

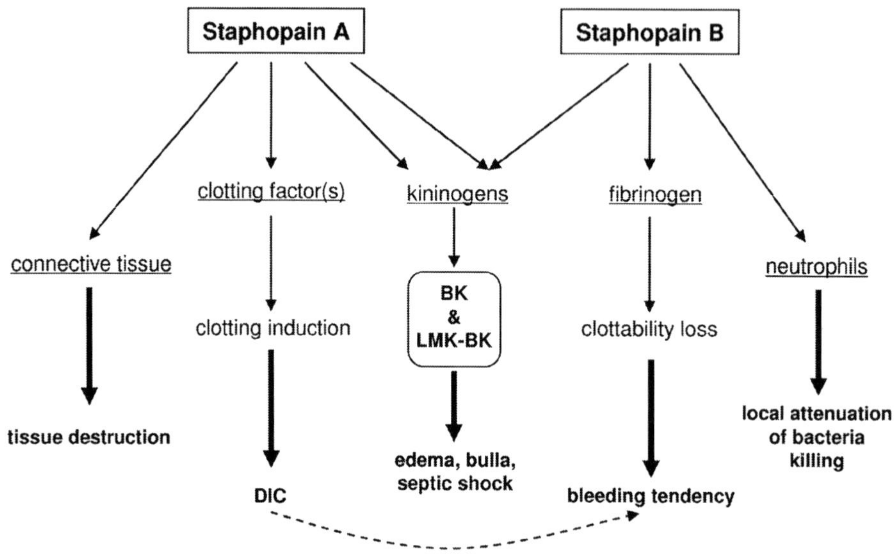

Figure 6. Staphopains as potential virulence factors.

staphopains on kininogen, may contribute to the pathological outcomes of septic *S. aureus* infections. To this end, studies using a guinea pig model of infection demonstrated that synergistic release of bradykinin by the staphopains leads to an increase in capillary blood vessel leakage into surrounding tissues.[29] Finally, staphopain B has been shown to interact with cells of the host immune system. By shedding CD31 from the surface of neutrophils, this protease may affect apoptotic neutrophil clearance in inflamed infection loci in host organisms.[30] These results, alongside the finding that the staphopains are the most intensively secreted proteolytic enzymes of all staphylococcal proteases,[51] and their conservation among clinical *S. aureus* isolates, strongly argues for the important role of these enzymes in staphylococcal infections.

STAPHOSTATINS ARE EFFECTIVE REGULATORS OF STAPHOPAIN ACTIVITY

As briefly described above, the activity of the staphopains is strictly regulated posttranslationally. This regulation is partially enforced by their grouping in the same operon as the gene encoding their specific inhibitors, the staphostatins. The first protein belonging to this group, staphostatin B ($StpinB_{aur}$), encoded by the *sspC* gene, was described in 2002,[19] where its role in the maturation of staphopain B ($StpB_{aur}$) was suggested. However, the finding that $StpinB_{aur}$ is located intracellularly and lacks a signal peptide,[38] led to the proposition that it actually plays a protective role as a cytoplasmic inhibitor of prematurely unleashed proteolytic activity of preproStpB. The second inhibitor of the staphostatin family identified was staphostatin A ($StpinA_{aur}$), encoded by the *scpB* gene. This protein is found in the *scpAB* operon, akin to that of the staphopain B-staphostatin B arrangement and constitutes a highly efficient inhibitor of $StpA_{aur}$ activity.[38]

Table 1. Amino acid sequence identity (%) among staphopains (Stps and staphostatins (Stpins)

	Sequence Identity					
Inhibitor	StpinA$_{aur}$	StpinA$_{epi}$	StpinA$_{car}$	StpinA$_{aurCH\text{-}91}$	StpinA$_{warn}$*	StpinB$_{aur}$
StpinA$_{aur}$	–					
StpinA$_{epi}$	54	–				
StpinA$_{car}$	21	20	–			
StpinA$_{aurCH\text{-}91}$	99	53	21	–		
StpinA$_{warn}$*	24	26	23	22	–	
StpinB$_{aur}$	19	26	26	19	15	–
StpinB$_{war}$	20	21	24	20	0	56
Protease	StpA$_{aur}$	StpA$_{epi}$	StpA$_{car}$*	StpA$_{aurCH\text{-}91}$	StpA$_{warn}$	StpB$_{aur}$
StpA$_{aur}$	–					
StpA$_{epi}$	75	–				
StpA$_{car}$*	59	59	–			
StpA$_{aurCH\text{-}91}$	78	79	61	–		
StpA$_{warn}$	68	68	65	69	–	
StpB$_{aur}$	49	48	49	49	47	–
StpB$_{warn}$	49	49	50	51	49	82

All sequences of the respective genes were obtained from GeneBank. Translated protein sequences were used for comparison. (*) denote truncated and nonfunctional pseudogenes.

Seemingly, the clustering of staphopain and staphostatin genes into operons is conserved, although to varying degrees. In addition to the *S. aureus* examples, staphopain followed by staphostatin in one transcriptional unit has been found in *S. warneri* (StpB$_{war}$ and StpinB$_{war}$), *S. carnosus* (StpA$_{car}$ and StpinA$_{car}$) and *S. epidermidis* (StpA$_{epi}$ and StpinA$_{epi}$). Interestingly, this arrangement is absent in *S. capitis*, *S. haemolyticus*, *S. hominis*, *S. lugdunensis* and *S. saphrophyticus*. Moreover, in the *S. aureus* poultry isolate CH-91, there is a third StpA-like (StpA$_{aurCH\text{-}91}$) enzyme, in addition to the classical StpA$_{aur}$ and StpB$_{aur}$; with all three protease genes followed by a specific inhibitor (StpinA$_{aur}$, StpinB$_{aur}$ and StpinA$_{aurCH\text{-}91}$, respectively). A comparison of amino acid sequences clearly shows that conserved StpAs are clustered with conserved StpinAs, while StpBs are found together with StpinBs (Table 1). Cumulatively, this pattern suggests that the staphopains have evolved alongside their inhibitors and argues that the staphostatins are indispensable for regulating and/or folding of their cysteine protease counterparts.

A perhaps more compelling line of evidence indicating the coevolution of staphopains and staphostatins comes from an analysis of the inhibitory interaction of these proteins in different staphylococcal species. As summarized in Table 2, StpAs, regardless of the species of origin, are inhibited by StpinAs, while StpBs interact only with StpinBs, without any cross-reactivity between the groups. This unique specificity of inhibition is especially intriguing in light of the high conservation of staphopain A and staphopain B structure.

Table 2. Inhibition pattern of staphopains (Stps) and staphostatins (Stpins)

	Protease				
Inhibitor	$StpA_{aur}$	$StpA_{epi}$	$StpA_{aurCH\text{-}91}$	$StpB_{aur}$	$StpB_{warn}$
$StpinA_{aur}$	+++	+++	++	–	–[a]
$StpinA_{epi}$	–	+++	++	–	–
$StpinA_{aurCH\text{-}91}$	+	+++	+++	–	–
$StpinB_{aur}$	–	–	–	+++	–
$StpinB_{war}$	–	–	–	+++	(+)[b]

Inhibition was classified according to the K_i order of magnitude. (+++) $K_i \leq 3$ nM, (++) $K_i \leq 30$ nM, + $K_i > 30$ nM. (a)—no inhibition detected, (b)—complex formation detected using gel filtration; no kinetic data available.[55]

Taken together, this analysis strongly suggests that groups A and B arose by duplication of an ancestral operon in a common progenitor of modern staphylococci and have been subject to independent evolution ever since. In each subgroup, a change in protease specificity requires parallel adjustments in the specificity of the co-expressed inhibitor. Such continuous and subtle adjustment in the structure of the two interacting proteins represents an intriguing example of molecular co-evolution. The lack of cross-group inhibition indicates that from the time of initial operon duplication, the staphopains in each subgroup have diverged to fulfill different functions.

Despite the limited degree of identity at the level of amino acid sequence (Table 1), staphostatin A and staphostatin B have very similar 3D structures, which resemble that of the lipocalins (Fig. 7).[41] Each polypeptide chain is folded into 8-stranded, antiparallel β-barrels, with a short α-helix extending from one side. The reactive-site, constituted by the Gly100-Thr101 and Gly98-Thr99 peptide bonds (P1 and P1′ positions), in staphopain A and staphopain B, respectively, is located on a loop near the C-terminus of the molecule.

Analysis of the structure of the StpB-Stpin inhibitory complex reveals that staphostatin binds to the protease in a substrate-like manner, without any proteolysis occurring. Apparently, this is due to the presence of a glycine residue at the P1 position of the inhibitor. From the analysis of Ramachandran plots it is clear that Gly98 takes up a sterically unique conformation within the enzymatic pocket of the protease. This makes the Gly-Thr peptide bond inaccessible for nucleophilic attack by the catalytic machinery of the protease, thus rendering the complex incredibly stable (Fig. 8). Indeed, mutagenesis experiments which change the P1-Gly to any other residue, convert the inhibitor into a very good substrate for the target enzyme.

It is clear that the staphostatins are powerful intracellular inhibitors of staphopain proteases and constitute a novel and distinct family of protein inhibitors, bearing a characteristic fold and unique mechanism of inhibition. They inhibit their respective staphopain with K_i values in the nanomolar range, constituting a shield against accidental proteolytic activity of prematurely folded and processed staphopains in the cytoplasm. Although an additional chaperone role for the staphostatins was suggested,[45] it appears their primary role is as inhibitors. This is confirmed by the lethal effect of expressing staphopain A without its respective inhibitor, staphostatin A, in *E. coli.*[52] Significantly and conversely, active-site mutated enzyme can be expressed in very large amounts

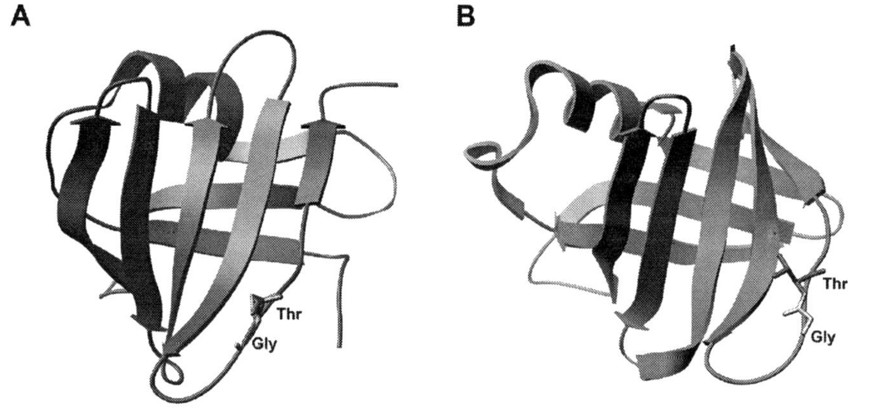

Figure 7. A comparison of 3-dimensional structures of staphostatin A and B. Shown are ribbon representations of NMR-derived staphostatin A (red, pdb accession code 1OH1[40]) and the crystal structure of staphostatin B (green, pdb accession code 1NYC[41]) of *S. aureus*. The P1 Gly and P1′ Thr are displayed as sticks. The overall structure of both proteins is highly similar, nevertheless noticeable differences in location of helices and differential folds of RLS can be observed. A color version of this image is available at www.landesbioscience.com/curie.

without any affect on the host, or requirement for inhibitor. The widespread nature of this regulation system, of cysteine protease followed by inhibitor in the same operon in many staphylococcal *sp.*, indicates the indispensability of controlling staphopain activity.

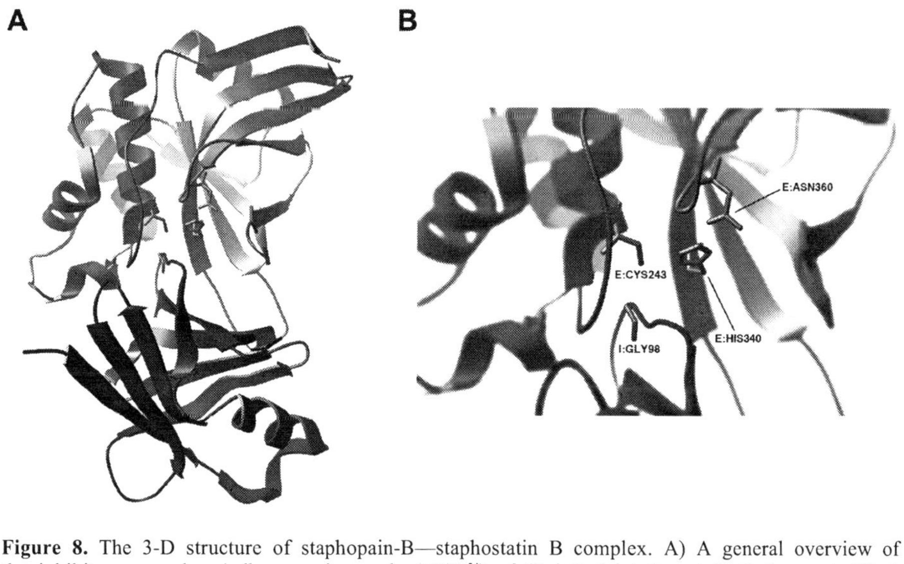

Figure 8. The 3-D structure of staphopain-B—staphostatin B complex. A) A general overview of the inhibitory complex (pdb accession code 1Y4H[54]) of StpinB (violet) and StpB (orange). B) A close-up view of the active site of the enzyme with an inserted inhibitory reactive loop. The catalytic residues, together with the P1 Gly residue of StpinB, are displayed as sticks and are labeled. Due to the characteristic fold of the inhibitor reactive site loop, its P1 residue clearly points out of the active site, rendering hydrolysis impossible.

CONCLUSION

In many respects, the staphopains are a very interesting family (C47) of clan CA of cysteine proteases. The tight, multilevel regulation and conservation of staphopain-staphostatin arrangements, along with the unique specificity of staphopain B and the spectrum of human proteins degraded by these enzymes, argues that they are important virulence factors of *S. aureus*. This aspect of staphopain function deserves intensive investigation, since if they are proven to be essential factors in human staphylococcal infections they may prove to be important targets for the development of therapeutic inhibitors. The clear difference in proteolytic specificity when compared to the cathepsins, which is apparently related to unique features of the substrate binding cleft, is attractive for the design of staphopain-specific inhibitors, which would likely have low or no cross-reactivity with similar human enzymes.

ACKNOWLEDGEMENTS

This work was supported by the grants from the Polish Ministry of Science and Education (N N301 309237 and N N301 461438). The Faculty of Biochemistry, Biophysics and Biotechnology at the Jagiellonian University is a beneficiary of structural funds from the European Union (grant No: POIG.02.01.00-12-064/08—"Molecular biotechnology for health").

REFERENCES

1. Foster TJ. The Staphylococcus aureus "superbug". J Clin Invest 2004; 114(12):1693-1696.
2. Ferry T, Perpoint T, Vandenesch F et al. Virulence determinants in Staphylococcus aureus and their involvement in clinical syndromes. Curr Infect Dis Rep 2005; 7(6):420-428.
3. Abdelnour A, Arvidson S, Bremell T et al. The accessory gene regulator (agr) controls Staphylococcus aureus virulence in a murine arthritis model. Infect Immun 1993; 61(9):3879-3885.
4. Cheung AL, Eberhardt KJ, Chung E et al. Diminished virulence of a sar-/agr- mutant of Staphylococcus aureus in the rabbit model of endocarditis. J Clin Invest 1994; 94(5):1815-1822.
5. Vuong C, Gotz F, Otto M. Construction and characterization of an agr deletion mutant of Staphylococcus epidermidis. Infect Immun 2000; 68(3):1048-1053.
6. Kavanaugh JS, Thoendel M, Horswill AR. A role for type I signal peptidase in Staphylococcus aureus quorum sensing. Mol Microbiol 2007; 65(3):780-798.
7. Queck SY, Jameson-Lee M, Villaruz AE et al. RNAIII-independent target gene control by the agr quorum-sensing system: insight into the evolution of virulence regulation in Staphylococcus aureus. Mol Cell 2008; 32(1):150-158.
8. Novick RP. Autoinduction and signal transduction in the regulation of staphylococcal virulence. Mol Microbiol 2003; 48(6):1429-1449.
9. McGavin MJ, Zahradka C, Rice K et al. Modification of the Staphylococcus aureus fibronectin binding phenotype by V8 protease. Infect Immun 1997; 65(7):2621-2628.
10. Karlsson A, Saravia-Otten P, Tegmark K et al. Decreased amounts of cell wall-associated protein A and fibronectin-binding proteins in Staphylococcus aureus sarA mutants due to up-regulation of extracellular proteases. Infect Immun 2001; 69(8):4742-4748.
11. McAleese FM, Walsh EJ, Sieprawska M et al. Loss of clumping factor B fibrinogen binding activity by Staphylococcus aureus involves cessation of transcription, shedding and cleavage by metalloprotease. J Biol Chem 2001; 276(32):29969-29978.
12. Cheung AL, Ying P. Regulation of alpha- and beta-hemolysins by the sar locus of Staphylococcus aureus. J Bacteriol 1994; 176(3):580-585.

13. Chien Y, Manna AC, Projan SJ et al. SarA, a global regulator of virulence determinants in Staphylococcus aureus, binds to a conserved motif essential for sar-dependent gene regulation. J Biol Chem 1999; 274(52):37169-37176.
14. Rechtin TM, Gillaspy AF, Schumacher MA et al. Characterization of the SarA virulence gene regulator of Staphylococcus aureus. Mol Microbiol 1999; 33(2):307-316.
15. Novick RP, Ross HF, Projan SJ et al. Synthesis of staphylococcal virulence factors is controlled by a regulatory RNA molecule. EMBO J 1993; 12(10):3967-3975.
16. Chan PF, Foster SJ. Role of SarA in virulence determinant production and environmental signal transduction in Staphylococcus aureus. J Bacteriol 1998; 180(23):6232-6241.
17. Arvidson S. Extracellular Enzymes. In: Fischetti VA, Novick RP, ferretti JJ, Potrnoy DA, Rood JIe, editors. Gram-positive pathogens. Washington D.C.: American Society for Microbiology, 2000: 379-385.
18. Ziebandt AK, Weber H, Rudolph J et al. Extracellular proteins of Staphylococcus aureus and the role of SarA and sigma B. Proteomics 2001; 1(4):480-493.
19. Massimi I, Park E, Rice K et al. Identification of a novel maturation mechanism and restricted substrate specificity for the SspB cysteine protease of Staphylococcus aureus. J Biol Chem 2002; 277(44):41770-41777.
20. Drapeau GR, Boily Y, Houmard J. Purification and properties of an extracellular protease of Staphylococcus aureus. J Biol Chem 1972; 247(20):6720-6726.
21. Arvidson S, Holme T, Lindholm B. Studies on extracellular proteolytic enzymes from Staphylococcus aureus. I. Purification and characterization of one neutral and one alkaline protease. Biochim Biophys Acta 1973; 302(1):135-148.
22. Arvidson S. Studies on extracellular proteolytic enzymes from Staphylococcus aureus. II. Isolation and characterization of an EDTA-sensitive protease. Biochim Biophys Acta 1973; 302(1):149-157.
23. Reed SB, Wesson CA, Liou LE et al. Molecular characterization of a novel Staphylococcus aureus serine protease operon. Infect Immun 2001; 69(3):1521-1527.
24. Rice K, Peralta R, Bast D et al. Description of staphylococcus serine protease (ssp) operon in Staphylococcus aureus and nonpolar inactivation of sspA-encoded serine protease. Infect Immun 2001; 69(1):159-169.
25. Potempa J, Watorek W, Travis J. The inactivation of human plasma alpha 1-proteinase inhibitor by proteinases from Staphylococcus aureus. J Biol Chem 1986; 261(30):14330-14334.
26. Potempa J, Fedak D, Dubin A et al. Proteolytic inactivation of alpha-1-anti-chymotrypsin. Sites of cleavage and generation of chemotactic activity. J Biol Chem 1991; 266(32):21482-21487.
27. Prokesova L, Potuznikova B, Potempa J et al. Cleavage of human immunoglobulins by proteinase from Staphylococcus aureus. Adv Exp Med Biol 1995; 371A:613-616.
28. Potempa J, Banbula A, Travis J. Role of bacterial proteinases in matrix destruction and modulation of host responses. Periodontol 2000; 24:153-192.
29. Imamura T, Tanase S, Szmyd G et al. Induction of vascular leakage through release of bradykinin and a novel kinin by cysteine proteinases from Staphylococcus aureus. J Exp Med 2005; 201(10):1669-1676.
30. Smagur J, Guzik K, Bzowska M et al. Staphylococcal cysteine protease staphopain B (SspB) induces rapid engulfment of human neutrophils and monocytes by macrophages. Biol Chem 2009; 390(4):361-371.
31. Maeda H, Yamamoto T. Pathogenic mechanisms induced by microbial proteases in microbial infections. Biol Chem Hoppe Seyler 1996; 377(4):217-226.
32. Lindsay JA, Foster SJ. Interactive regulatory pathways control virulence determinant production and stability in response to environmental conditions in Staphylococcus aureus. Mol Gen Genet 1999; 262(2):323-331.
33. Dunman PM, Murphy E, Haney S et al. Transcription profiling-based identification of Staphylococcus aureus genes regulated by the agr and/or sarA loci. J Bacteriol 2001; 183(24):7341-7353.
34. Shaw L, Golonka E, Potempa J et al. The role and regulation of the extracellular proteases of Staphylococcus aureus. Microbiology 2004; 150(Pt 1):217-228.
35. Tegmark K, Morfeldt E, Arvidson S. Regulation of agr-dependent virulence genes in Staphylococcus aureus by RNAIII from coagulase-negative staphylococci. J Bacteriol 1998; 180(12):3181-3186.
36. Drapeau GR. Role of metalloprotease in activation of the precursor of staphylococcal protease. J Bacteriol 1978; 136(2):607-613.
37. Filipek R, Rzychon M, Oleksy A et al. The Staphostatin-staphopain complex: a forward binding inhibitor in complex with its target cysteine protease. J Biol Chem 2003; 278(42):40959-40966.
38. Rzychon M, Sabat A, Kosowska K et al. Staphostatins: an expanding new group of proteinase inhibitors with a unique specificity for the regulation of staphopains, Staphylococcus spp. cysteine proteinases. Mol Microbiol 2003; 49(4):1051-1066.
39. Karlsson A, Arvidson S. Variation in extracellular protease production among clinical isolates of Staphylococcus aureus due to different levels of expression of the protease repressor sarA. Infect Immun 2002; 70(8):4239-4246.
40. Dubin G, Krajewski M, Popowicz G et al. A novel class of cysteine protease inhibitors: solution structure of staphostatin A from Staphylococcus aureus. Biochemistry 2003; 42(46):13449-13456.

41. Rzychon M, Filipek R, Sabat A et al. Staphostatins resemble lipocalins, not cystatins in fold. Protein Sci 2003; 12(10):2252-2256.
42. Nickerson NN, Joag V, McGavin MJ. Rapid autocatalytic activation of the M4 metalloprotease aureolysin is controlled by a conserved N-terminal fungalysin-thermolysin-propeptide domain. Mol Microbiol 2008; 69(6):1530-1543.
43. Marie-Claire C, Roques BP, Beaumont A. Intramolecular processing of prothermolysin. J Biol Chem 1998; 273(10):5697-5701.
44. Hoffman B, Schomburg D, Hecht HJ. Crystal structure of a thiol proteinase from Staphylococcus aureus V8 in the E-64 inhibitor complex. Acta Crystallographica 1993;(49):102.
45. Nickerson N, Ip J, Passos DT et al. Comparison of Staphopain A (ScpA) and B (SspB) precursor activation mechanisms reveals unique secretion kinetics of proSspB (Staphopain B) and a different interaction with its cognate Staphostatin, SspC. Mol Microbiol 2010; 75(1):161-177.
46. Coulter SN, Schwan WR, Ng EY et al. Staphylococcus aureus genetic loci impacting growth and survival in multiple infection environments. Mol Microbiol 1998; 30(2):393-404.
47. Potempa J, Pike RN. Bacterial peptidases. Contrib Microbiol 2005; 12:132-180.
48. Potempa J, Dubin A, Korzus G et al. Degradation of elastin by a cysteine proteinase from Staphylococcus aureus. J Biol Chem 1988; 263(6):2664-2667.
49. Dubin G. Extracellular proteases of Staphylococcus spp. Biol Chem 2002; 383(7-8):1075-1086.
50. Wegrzynowicz Z, Heczko PB, Drapeau GR et al. Prothrombin activation by a metalloprotease from Staphylococcus aureus. J Clin Microbiol 1980; 12(2):138-139.
51. Jones RC, Deck J, Edmondson RD et al. Relative quantitative comparisons of the extracellular protein profiles of Staphylococcus aureus UAMS-1 and its sarA, agr and sarA agr regulatory mutants using one-dimensional polyacrylamide gel electrophoresis and nanocapillary liquid chromatography coupled with tandem mass spectrometry. J Bacteriol 2008; 190(15):5265-5278.
52. Wladyka B, Puzia K, Dubin A. Efficient co-expression of a recombinant staphopain A and its inhibitor staphostatin A in Escherichia coli. Biochem J 2005; 385(Pt 1):181-187.
53. Copeland A, Lucas S, Lapidus A et al. Complete sequence of chromosome of Staphylococcus aureus subsp.aureus JH9. 2007.
54. Filipek R, Potempa J, Bochtler M. A comparison of staphostatin B with standard mechanism serine protease inhibitors. J Biol Chem 2005; 280(15):14669-14674.
55. Dubin G, Wladyka B, Stec-Niemczyk J et al. The staphostatin family of cysteine protease inhibitors in the genus Staphylococcus as an example of parallel evolution of protease and inhibitor specificity. Biol Chem 2007; 388(2):227-235.

CHAPTER 2

THE LYSINE-SPECIFIC GINGIPAIN OF *PORPHYROMONAS GINGIVALIS*

Importance to Pathogenicity and Potential Strategies for Inhibition

Tang Yongqing,[1] Jan Potempa,[2] Robert N. Pike*,[1] and Lakshmi C. Wijeyewickrema[3]

[1]*Department of Biochemistry and Molecular Biology and CRC for Oral Health Sciences, Monash University, Clayton, Victoria, Australia;* [2]*Oral Health and Systemic Disease Research Facility, University of Louisville School of Dentistry, Louisville, Kentucky, USA;* [3]*Department of Microbiology, Faculty of Biochemistry, Biophysics and Biotechnology, Jagiellonian University, Krakow, Poland*

**Corresponding Author: Robert N. Pike— Email: rob.pike@monash.edu*

Abstract: Periodontitis is a disease affecting the supporting structures of the teeth. The most severe forms of the disease result in tooth loss and have recently been strongly associated with systemic diseases, including cardiovascular and lung diseases and cancer. The disease is caused by biofilms of predominantly anaerobic bacteria. A major pathogen associated with severe, adult forms of the disease is *Porphyromonas gingivalis*. This organism produces potent cysteine proteases known as gingipains, which have specificity for cleavage after arginine or lysine residues. The lysine-specific gingipain, Kgp, appears to be the major virulence factor of this organism and here we describe its structure and function. We also discuss the inhibitors of the enzyme produced to date and the potential pathways to newer versions of such molecules that will be required to combat periodontitis.

INTRODUCTION: PERIODONTAL DISEASE—SIGNIFICANCE AND AETIOLOGY

Periodontal disease is a complex disorder involving Gram-negative anaerobic bacteria interacting with host cells with the combined effect leading to the destruction of the supporting structures of teeth. The supporting structures include the gingiva

Cysteine Proteases of Pathogenic Organisms, edited by Mark W. Robinson and John P. Dalton.

(gums), the periodontal ligament between the cementum and the junctional epithelium and the alveolar bone. Periodontal disease affects approximately 30% of the adult population and is amongst the most prevalent oral diseases contributing to tooth loss in adults.[1,2] Furthermore, epidemiological evidence from the past few decades has pointed to a correlation between periodontal disease and several systemic disorders, including circulatory diseases, respiratory infections, premature delivery and low birth weight of infants, neuronal damage, renal malfunction, certain cancers and autoimmune diseases.[3-10] Although the mechanisms underlying this correlation may vary among systemic conditions, insult by periodontopathic bacteria and their toxic products and/or the hyper-inflammatory effects of the host immune response are likely to be involved.[11]

Periodontal disease is an inflammatory disorder instigated and mediated by multiple bacterial species, which usually change dynamically during different states of disease and differ among individuals.[12] The disease starts in the marginal gingiva as a reversible inflammatory state, called gingivitis, resulting from the excessive accumulation of bacterial dental plaque reaching deep into the gingival sulcus.[13-15] This generates a local environment prone to colonization by and the proliferation of anaerobic periodontopathogenic bacteria. As the infection progresses, some of the harmful bacterial species induce damage to the sub-gingival tissues such as the periodontal ligament and the junctional epithelium, leading to loss of attachment and the formation of periodontal pockets. These pathogenic changes are accompanied by massive destruction of the soft tissues and permanent alveolar bone resorption.[14,16] This more advanced stage is defined as periodontitis,[15] which eventually leads to tooth loss due to the lack of periodontium support.[13,17] Periodontal disease is a multifactorial disease, in which the onset and outcome are affected by a variety of risk factors involving the host, bacteria and environment.[14] Of the 300~400 bacterial species residing in the periodontal environment, only 10~20 may be periodontopathic,[12] including, but not limited to the cluster of 'red complex' bacteria comprising *Tannarella forsythia*, *Treponema denticola* and *Porphyromonas gingivalis,* often found in large numbers in the biofilms of adult periodontitis patients.[18,19]

Porphyromonas gingivalis

Among the putative periodontopathic bacteria, *P. gingivalis* is believed to be one of the major pathogens involved in the progression of periodontal disease, based on the observation that increased *P. gingivalis* levels were associated with an increased severity of periodontal disease.[20] The results of experimental oral infection of primates with *P. gingivalis* strongly support this hypothesis, by showing that the sub-gingival inoculation of the bacterium into *P. gingivalis*-free monkeys resulted in the development of periodontitis. Significantly, in this model, the levels of the bacterium in the sub-gingival area closely correlated with the aggravation of periodontitis, marked by an exacerbated loss of alveolar bone.[21]

P. gingivalis is a nonmotile, Gram-negative obligatorily anaerobic coccobacillus, typically 0.5~3.5 μm in diameter[22] that is a late colonizer of the dental bacterial plaque. It utilizes bacteria already resident in the oral cavity for initial settlement.[23] The bacterium has a hemolytic nature and exhibits a characteristic black pigment that contains the metabolic derivatives of heme, extracted from hemoglobin or other heme-containing proteins, on the cell surface and in the centre of aged colonies grown on blood agar.[24,25] Metabolically, *P. gingivalis* is an asaccharolytic species which generates energy and obtains carbon by fermentation of peptides and amino acids.[24,26] Different *P. gingivalis* strains can be categorised according to their pathogenicity, which is evaluated by the severity of the inflammatory responses in mice following subcutaneous injection of bacteria.[27] Virulent strains such as W50 can cause severe spreading abscesses and induce fatal sepsis in murine models after 2~4 days, while less-virulent strains such as W186 may only result in localized pustules.[27]

P. gingivalis exerts its pathogenic effects via the production of a variety of virulence factors, including the capsule, outer membrane vesicles and abundant adhesive structures, such as fimbriae, hemagglutinin/adhesins and lipopolysaccharide (LPS).[24,28-32] More importantly, the virulence factors of *P. gingivalis* also comprise a group of proteolytic endopeptidases, of which cysteine proteases with trypsin-like activity are closely associated with tissue damage and host immunity disruption in periodontitis.[23,33-35] The enzymes, referred to as gingipains,[36] are responsible for the vast majority of the extracellular proteolytic (85%) and trypsin-like amidolytic activities (99%) of *P. gingivalis* and hence are believed to be critical for the virulence of this pathogen.[37] Indeed, a few spontaneous mutants of the virulent *P. gingivalis* strain W50 with lowered trypsin-like amidolytic activities have been found to be avirulent in murine models.[38,39]

GINGIPAINS

Members of gingipains (family C25) strictly cleave peptide bonds with either Arg-Xaa or Lys-Xaa at the cleavage site and the responsible enzymes are referred to as Arg-gingipain and Lys-gingipain, respectively.[37,40,41] Arg-gingipains include two members, Arg-gingipain A (RgpA) and Arg-gingipain B (RgpB), encoded by two closely related genes, *rgpA* and *rgpB*, respectively. Lys-gingipain (Kgp) is encoded by a single gene, *kgp*.[42,43] A number of studies have shown that *P. gingivalis* mutants with a deficiency of one or more of the gingipain genes exhibited lower pathogenic potential in mice than wild type bacteria, hence confirming that the gingipains are the main contributors to the virulence of *P. gingivalis*.[44-47] Furthermore, O'Brien-Simpson et al[47] compared the virulence of *P. gingivalis* isogenic mutants defective in the individual gingipain gene in a murine lesion model, to estimate the extent of contribution of each of the three gingipain proteases to the pathogenic potential of the bacterium. They reported that Kgp contributed more to pathogenicity than the two Arg-gingipains and thus this gingipain can be considered to be the major virulence factor of *P. gingivalis*.[47] This finding has been further confirmed by Pathirana et al who conducted a similar experiment with a recently developed murine periodontitis model and found that Kgp contributed more to pathogenicity than RgpB and much more than RgpA. Therefore, Kgp represents one of the major determinants of the virulence of *P. gingivalis*.[48]

THE BIOLOGICAL FUNCTIONS OF KGP

Kgp participates in many *P. gingivalis*-mediated pathogenic processes by binding various targets and importantly, cleaving multiple proteins (Fig. 1).[49] Kgp can bind red blood cells and many heme-containing proteins, including hemoglobin and acts as a major hemolytic enzyme to produce and store iron/heme, which is a vital growth factor essential for the survival and function of *P. gingivalis*.[50-52] The importance of Kgp in providing iron/heme for *P. gingivalis* has been demonstrated by the observation that

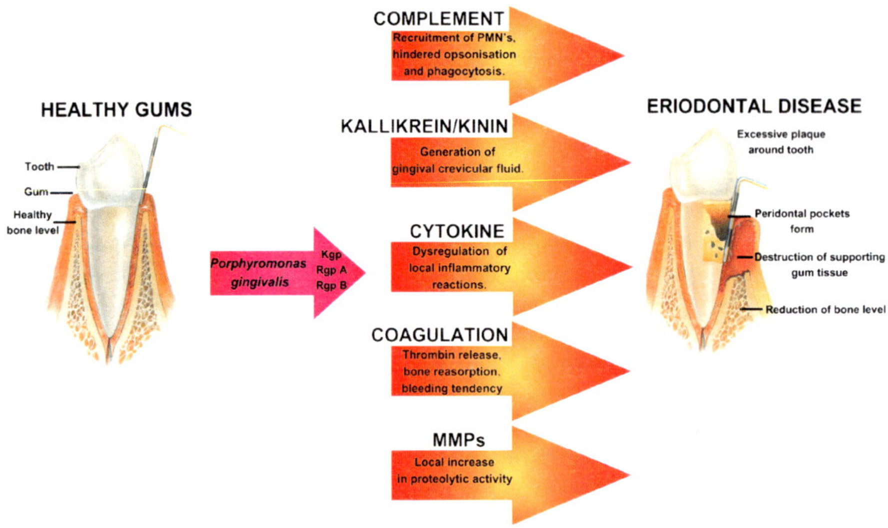

Figure 1. Diagrammatic overview of the postulated contributions of the gingipain enzymes from *Porphyromonas gingivalis* to the progression of periodontitis. A healthy tooth and surrounding oral tissue, is converted to one exhibiting the symptoms of periodontitis as shown by the summarised clinical indicators on the right of the diagram. Sites of periodontitis show destruction of supporting tissue, lowered levels of the alveolar bone socket and enhanced depth of probing (visualized by the metal probe on the right). The gingipains of *P. gingivalis* are postulated to contribute to the progression of the disease by affecting the indicated systems to dysregulate normal responses to the bacterium and thus enhance the clinical indications and contribute to disease.

Kgp-null *P. gingivalis* strains lacked the black pigment characteristic for this species and were less virulent.[46,53] In concert with the two Rgp members, Kgp also directly binds and cleaves fibrinogen and indirectly breaks down collagens via the activation of the matrix metalloproteinase system, resulting in an increased bleeding tendency and tissue damage at the infected periodontal site.[54-57] Whilst the proteolytic activity of Kgp is important for generating peptides/amino acids as the energy and carbon source suitable for the asaccharolytic nature of *P. gingivalis*, it is also utilized by *P. gingivalis* to manipulate and evade the host immune response via complex mechanisms,[58-62] mainly involving the degradation of surface molecules of immune cells [63-66] and interference with the cytokine system of the host.[61,67] In addition, Kgp can also bind host epithelial cells and other bacteria, providing a biological basis for the attachment and colonization of *P. gingivalis*.[49,68] Cumulatively, Kgp is critical for the survival and pathogenicity of *P. gingivalis* and hence is a promising target for inhibitors to control periodontal disease.

THE STRUCTURAL CHEMISTRY OF KGP

The precursor of Kgp is a polyprotein consisting of domain components evolutionarily related to those of RgpA and RgpB[17,37,49,69] (Table 1). The RgpB polyprotein does not have HA domains[70] and is 72%, 99%, 52% and 51% identical in the sequences of the pro-domain, catalytic sub-domain, IgSF sub-domain and C-terminal domain, respectively,

Table 1. The domain structure of RgpA-, RgpB- and Kgp-polyproteins

Polyprotein	Domain Structure (from N- to C-terminus)	Descriptions	References
RgpA (1706 residues)	Signal peptide	23 residues	17, 37, 49, 69
	Pro-domain	204 residues	
	Arg-specific catalytic subdomain	Together forming a 45 kDa Arg-specific protease domain ($RgpA_{cat}$)	
	IgSF* subdomain		
	$RgpA_{A1}$	RgpA hemagglutinin/adhesin (HA) domain 1, formerly rHGP-44*	
	$RgpA_{A2}$	RgpA HA domain 2, formerly rHGP-15	
	$RgpA_{A3}$	RgpA HA domain 3, formerly rHGP-17	
	$RgpA_{A4}$	RgpA HA domain 4, formerly rHGP-27	
	C-terminal domain	70 residues	
RgpB (736 residues)	Signal peptide	24 residues	49
	Pro-domain	205 residues	
	Arg-specific catalytic subdomain	Together forming a 45 kDa Arg-specific protease domain ($RgpB_{cat}$)	
	IgSF subdomain		
	C-terminal domain	71 residues	
Kgp (1723 or1732 residues)	Signal peptide	19 residues	37, 40, 49, 76-78
	Pro-domain	209 residues	
	Lys-specific catalytic subdomain	Together forming a 48 kDa Lys-specific protease domain (Kgp_{cat})	
	IgSF subdomain		
	$Kgp_{A1, 2, 3...}$(see Table 2)	Kgp HA domains, formerly kHGP*	
	C-terminal domain	70 residues	

* IgSF = immunoglobin superfamily; rHGP = Arg-specific high molecular weight gingipain; kHGP = Lys-specific high molecular weight gingipain. The HA domains of RgpA and Kgp were formerly named by their molecular weights.

to those of the RgpA polyprotein.[71] However, Kgp_{cat} only shares 27% sequence identity to $RgpA_{cat}$ and RgpB.[72] The C-terminal HA domains of Kgp polyprotein are highly similar in sequence to those of RgpA polyprotein, in particular Kgp_{A2}, which, apart from two residues, is identical to $RgpA_{A2}$.[72,73] Several repetitive Adhesin Binding Motifs (ABMs) were identified in the HA domains of Kgp and RgpA. These ABMs connect the HA domains in terms of their sequence similarity and link the HA domains to other proteins,[74] for example, the ABM2 (SYTYTVYRDGTKIKEGLTATTFEEDGVAA) is responsible for the interaction between Kgp and extracellular matrix proteins,[49,75] while ABM3

Table 2. The HA domains of Kgp-polyproteins from different *P. gingivalis* strains

P. gingivalis Strain	HA Domain Structure	Former Name	References
381	Kgp_{A1}	kHGP-39	
	Kgp_{A2}	kHGP-15	78
	Kgp_{A3}	kHGP-44	
W50	Kgp_{A1}	kHGP-39	
	Kgp_{A2}	kHGP-15	
	Kgp_{A3}		37
	Kgp_{A4}	Collectively kHGP-44	
	Kgp_{A5}		
HG66 and ATCC33277	Kgp_{A1}		
	Kgp_{A2}	Collectively kHGP-44	
	Kgp_{A3}	kHGP-15	40
	Kgp_{A4}	kHGP-17	
	Kgp_{A5}	kHGP-27	
W12 and W83	Kgp_{A1}	kHGP-44	
	Kgp_{A2}	kHGP-15	
	Kgp_{A3}		76,77
	Kgp_{A4}	Collectively kHGP-44	

(VTLKWDAPNGTPNPNPNPNPNPNPGTTTLSESF) is critical for attachment of Kgp to hemoglobin and red blood cells.[49,75] These HA domains of the Kgp polyprotein vary in number and order among *P. gingivalis* strains. Four Kgp variants have been identified to date[37,40,76-78] (Table 2). It is interesting to note that recently Li et al[79] have postulated a different set of protein boundaries for the HA domains of Kgp and RgpA and proceeded to successfully express and purify a so-called K2 domain which is largely equivalent to the domain termed Kgp_{A2} or kHGP-15 previously (Table 2). Bioinformatic analysis suggests that the domains belong to the *Cleaved_adhesin* domain family, with affiliation to the galactose-binding domain-like superfamily. They also solved the structure of this protein (Fig. 2) and showed that it was a β-strand rich structure in a "jelly-roll" fold, with numerous loops connecting the β-strands, resembling carbohydrate binding domains and so-called MAM domains found in receptor-type tyrosine-protein phosphatases and ephrin type A/B receptors. They demonstrated that the domain was able to weakly bind to galactose containing carbohydrates and that it could act as a hemolysin in its uncleaved form. This study most likely will constitute a breakthrough in our understanding of the structure and function of the HA domains of the gingipains and further work in this regard will be of great interest.

Similar to other gingipains, the mature Kgp exists extracellularly in multiple isoforms, including monomeric soluble forms, membrane-associated forms of monomers and RgpA-Kgp complexes.[71,73,80-87] The processes of secretion and maturation of Kgp are complex and not fully understood. Generally glycosylation at the C-terminal domain

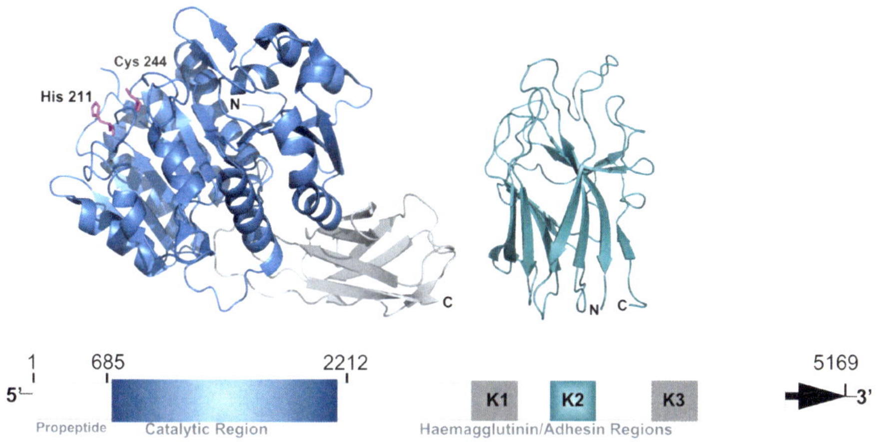

Figure 2. The domain structure of Kgp and known three-dimensional structures for the domains. The gene sequence for Kgp is shown as a grey bar overlaid with the translated domain structure as recently elucidated by Li et al.[79] The structure of the catalytic region of Kgp (blue bar) is postulated to be similar to that solved for RgpB,[92] the three-dimensional structure of which is shown as a ribbon diagram. The catalytic residues, Cys_{244} and His_{211}, are shown in pink ball and stick to indicate the position of the active site. The recently determined structure of the K2 haemagglutinin/adhesin domain of Kgp is shown as a ribbon diagram in light green.

is believed to provide the anchor for the membrane-associated Kgp isoforms,[81,88] and the maturation of Kgp requires N- and C-terminal modifications in the intra-domains of Kgp polyprotein. Interestingly, these intra-domain modifications were found to require the activities of both Rgp and Kgp.[71] Furthermore, the Kgp polyprotein in an Rgp-null *P. gingivalis* mutant was found in the fully mature form,[89] and the introduction of RgpB into a *P. gingivalis* mutant lacking all three gingipain members was found to result in mature RgpB.[49] These phenomena of auto-processing of the gingipains may explain the initial source of active gingipains, which participate in the proteolytic modification of subsequently produced Kgp polyprotein.

The crystal structure of Kgp is yet to be determined, but despite the fact that the identity between the catalytic domains of Kgp and RgpA/RgpB is only around 27%, there is a strong likelihood that Kgp_{cat} will have a similar conformation to RgpB, for which a crystal structure is available.[90,91] Topologically, the crystal structure of RgpB displays an N-terminal domain and a C-terminal domain (Fig. 2).[92] The N-terminal domain is comprised of A- and B-sub-domains, each of which has a characteristic α/β motif that is made up of a central β-sheet sandwiched by α-helices. This α/β-sandwich structure can also be seen in caspase-1 and -3.[92] The B-sub-domain encompasses the catalytic residues of the Cys-His dyad and hence the N-terminal domain represents the catalytic domain. The C-terminal IgSF domain is composed entirely of β-sheets, which is a topological homologue to an IgG domain (Fig. 2).[92]

As mentioned, Kgp cleaves Lys-Xaa bonds and RgpB cleaves Arg-Xaa bonds exclusively, meaning a strict preference for Lys and Arg as the P1-residue (note that the nomenclature for proteases[93] indicates that cleavage occurs in substrates between the P1 and P1′ residues, with substrate residues N-terminal to P1 labelled P2, P3 etc and those

C-terminal to P1′ labelled P2′, P3′ etc; subsites binding the substrate residues in the enzymes are correspondingly labelled, eg. the S1 subsite binds the P1 residue etc.) for Kgp and RgpB, respectively and a broad spectrum of residues at the P1′ position for both gingipains.[37,93] The structural basis for this substrate specificity of Kgp can be inferred based on that of RgpB from its crystal structure. On the surface of the B-sub-domain of RgpB, adjacent to the catalytic dyad (His_{211} and Cys_{244}), a deep S1 pocket is formed by the peptide segments: Tyr_{283}~Met_{288}, Thr_{209}~His_{211} and Val_{242}~Cys_{244}. Of these, Met_{288} and Val_{242} were hypothesized to be important in determining the P1-Arg specificity of RgpB by participating in lining the side and bottom of the S1 pocket, respectively.[92] This S1 pocket is not only an optimal conformational fit for the side chain of the P1-Arg, but also contains a negatively-charged Asp_{163} at its bottom, contributing an electrostatic stabilisation for the bound P1-Arg.[92] Compared to this, in the modelled S1-pocket of Kgp, Asp_{288} and Glu_{287} form a small negatively-charged patch,[40] which may be important for binding the positively-charged Lys; Phe_{242} in Kgp is a bigger residue than Val_{242} in RgpB and thus may contribute to altering the shape of the S1 pocket to exclusively accommodate Lys residues. On the other hand, outside the S1 pocket of RgpB, around the catalytic dyad, there is a relatively flat and open surface, which can accommodate different types of amino acids.[92] Based on the topological and functional similarities and molecular modelling experiments, a similar flat and open area is also expected to exist outside the S1-pocket of Kgp, around its catalytic dyad (His_{217} and Cys_{249}).[40] Therefore, for all members of the gingipain family, in contrast to the high specificity for the P1-residue of the substrate, less specificity is predicted for the residues closely flanking the P1-residue, including the P3-, P2-, P1′- and P2′-residues.[92]

THE CATALYTIC MECHANISM OF KGP

A detailed understanding of the mechanism by which Kgp interacts and cleaves its substrates is critical for the development of Kgp inhibitors. Since members of the gingipain family (C25) share the characteristic catalytic dyad motif, His-Gly-X-Ala-Cys, with all other familial members of the CD clan of cysteine proteases, including family 14 (caspases), family 11 (bacterial clostripains), family 13 (plant and animal legumains) and family 50 (separase), the principal catalytic mechanism is therefore likely to be shared by all members of cysteine proteases in the CD clan, including Kgp.[49,94] This mechanism, in general, is similar to that exerted by papain-like (Clan CA) cysteine proteases.

In this catalytic mechanism (Fig. 3), under activated conditions, the negatively-charged Sγ atom of the catalytic Cys first attacks the carbonyl carbon of the scissile bond of the substrate, forming an enzyme-substrate complex.[95,96] The resulting proximity of the substrate and the catalytic dyad leads to a nucleophilic interaction between the N atom of the scissile bond and the Nδ atom of the catalytic His. Consequently, the scissile bond breaks, releasing the prime side of the substrate in the form of an amide.[97] The remaining acyl-enzyme intermediate is the subject of nucleophillic attack by a water molecule, which attacks the carbonyl carbon from the previous scissile bond, resulting in the release of the remaining N-terminus of the substrate and the functional protease molecule.[95,96]

Based on the catalytic mechanism and the P1 specificity of Kgp, it is possible to attenuate the proteolytic activity of Kgp via the inhibition of the catalytic Sγ atom, by introducing a peptidyl inhibitor containing a P1-Lys and close to it, one or more functional groups such as an aldehyde group, which can tightly bind to the Sγ atom.[98] In this scheme, the interaction of the peptide chain with the catalytic site of Kgp, including the binding of

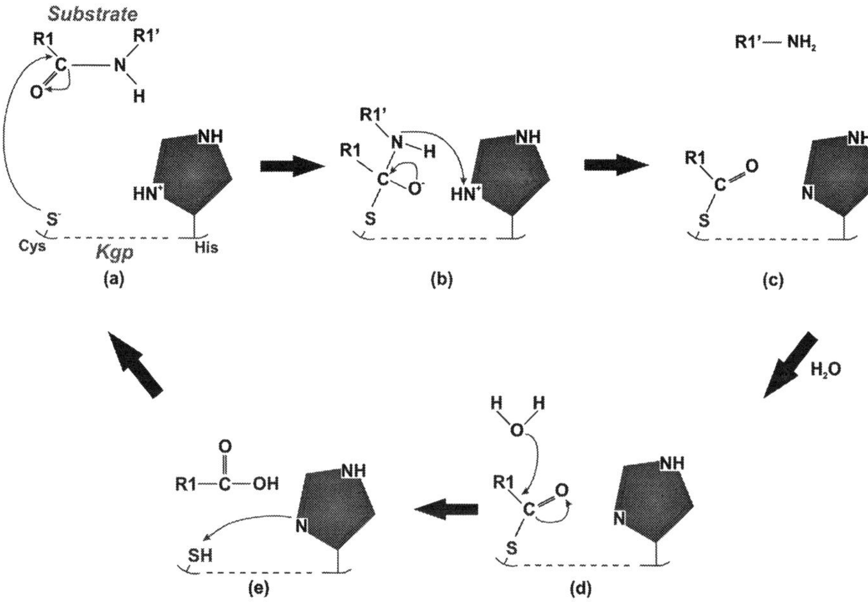

Figure 3. The catalytic mechanism of cysteine proteases. The catalytic dyad is illustrated in blue. The red arrows indicate the movement of protons. A) The catalytic Sγ atom attacks the carbonyl carbon of the scissile bond of a substrate, forming (B) an enzyme-substrate complex. In this complex, the N atom of the scissile bond attacks the Nδ atom of the catalytic His residue, resulting in the breakage of the scissile bond and the release of the prime side of the substrate in the form of an amine. C) An acyl-enzyme intermediate is formed. D) A water molecule attacks the acyl-enzyme complex, leading to (E) the release of the N-terminus of the substrate in the form of a carboxylic acid and a functional enzyme molecule. Under activated conditions, the catalytic dyad of the cysteine protease switches into the active or charged form (A).

P1-Lys to the S1-pocket, brings the functional group into close proximity to the catalytic Sγ atom. Consequently, a covalent Sγ-functional group bond is formed, interrupting the catalytic function of Sγ.[97,98] Furthermore, it can be postulated that the efficacy of a peptidyl Kgp inhibitor would be greatly dependent on the effectiveness of the interaction between the peptide chain and the Kgp catalytic site. This interaction is not only affected by the binding of the P1-Lys to the S1 pocket, but also by the interaction amongst the residues flanking P1-Lys, including the P3-, P2-, P1′- and P2′-residues, with their corresponding sub-sites on Kgp. Supporting this hypothesis, Abe et al found that there was a decrease in cleavage of Lys-Xaa bonds by Kgp when the P2-residue is an Arg or Lys, although the amino acid specificity for P2-residue is theoretically broad.[99] Moreover, in another study, Abe et al found that Kgp was able to cleave the peptidyl substrate Z-His-Glu-Lys-MCA nearly 20-fold more effectively than Boc-Val-Leu-Lys-MCA, indicating that there may be preferences for the P3- and/or P2-position of substrates cleaved by Kgp.[100] Thus, it is possible that peptides of certain amino acid sequence may be more preferred by Kgp for interaction and cleavage. These peptides therefore would constitute optimal scaffolds for developing effective Kgp inhibitors. However, the overall specificity of Kgp at the S3, S2, S1′ and S2′ sub-sites, where the P3-, P2-, P1′- and P2′-residues of the substrates or inhibitors bind, has never been fully profiled.

PREVIOUSLY DEVELOPED KGP INHIBITORS

A few Kgp inhibitors from natural sources have been tested (Table 3)[101-105], however, they only showed limited inhibitory activity. Furthermore, a number of enzyme-substrate-based Kgp inhibitors have been developed (Table 4)[41,106-109]. For example, the ketopeptide inhibitor KYT-36 (carbobenzoxy-Glu(NHN(CH_3)Ph)-Lys-CO-NHCH_2Ph) was designed based on the sequence of histatin 5, a natural gingipain inhibitor in human saliva.[108] It displayed modest inhibitory effects in keeping with the weak inhibition of Kgp by histatin 5. In stark contrast, an acyloxymethyl ketone inhibitor (carbobenzoxy-Phe-Lys-CH_2OCO-2,4,6-Me_3Ph) and an aza-peptide Michael acceptor inhibitor (PhCH_2CH_2CO-Leu-ALys-CH = CHCOOEt) for Kgp have shown much higher inhibitory potency for Kgp, with k_{ass} values (the second-order rate constant of an inhibitory reaction, which indicates how fast an effective inhibition occurs) in the range of 10^6 $M^{-1}s^{-1}$.[41] As mentioned, since the amino acid specificity of Kgp at the positions closely flanking the S1-pocket has not been fully profiled, there may be some optimal

Table 3. Kgp inhibitors from natural sources

Inhibitor	Source	Inhibition or Association Rate Constants***	Reference
Histatin 5	Human saliva	$IC_{50} = 1.4 \times 10^{-5}$ M	101
CrmA (Asp>Lys)	Cowpox virus	$k_{ass} = 2.1 \times 10^5$ $M^{-1}s^{-1}$	102
p35	Baculovirus	$Ki = 2 \times 10^{-10}$ M	102
Cranberry nondialysable fraction	Cranberry	not determined	103
Cranberry polyphenol fraction	Cranberry	not determined	104
Pancreatic trypsin inhibitor	Pancreas	$k_{ass} = 2.0 \times 10^4$ M^{-1}	105

***Inhibition or association rate constants:

IC_{50}, the half maximal inhibitory concentration, is the amount of inhibitor taken to effectively inhibit half of the initial amount of enzyme. IC_{50} indicates the functional strength of an inhibitor against an enzyme, not the affinity of binding between them.

k_{ass}, the second-order rate constant of an inhibitory reaction, indicates how fast an effective inhibition occurs.

Ki is the dissociation constant of an inhibitory reaction. It indicates the affinity of the binding between inhibitor and enzyme.

Table 4. Synthetic Kgp inhibitors

Inhibitor	Chemical Property	Inhibition or Association Rate Constants	Reference
Cbz-Phe-Lys-CH_2O-CO-2,4,6-Me_3-Ph	Acyloxymethyl ketone	$k_{ass} = 4.2 \times 10^6$ $M^{-1}s^{-1}$	41
Chlorhexidine	Chlohexidine	$Ki = 1.65 \times 10^{-4}$ M	106
A71561	Amide	$Ki = 9 \times 10^{-10}$ M	107
KYT-36	Ketopeptide	$Ki = 1.3 \times 10^{-10}$ M	108
aza-peptide Michael acceptor (with Lys)	aza-peptide Michael acceptor	$k_{ass} = 3.3 \times 10^6$ $M^{-1}s^{-1}$	109
Rational-designed inhibitors	Chloromethyl ketone	$k_{ass} = 10^6 \sim 10^7$ $M^{-1}s^{-1}$	110

enzyme-substrate-based Kgp inhibitors, which achieve a higher range of k_{ass} values, yet to be developed. In agreement with this hypothesis, some small dipeptide chloromethyl ketone Kgp inhibitors, recently developed by Białas et al by structure-based rational design, for the first time were found to exhibit rapid interaction with Kgp, with k_{ass} values reaching 10^7 $M^{-1}s^{-1}$.[110] Importantly, Białas et al also highlighted the critical role of the P2-residue, which acts as a hydrophobic interactant in the Kgp inhibitors.[110] This finding reflected the necessity of determining the full specificity of the active site of Kgp to allow the development of optimal Kgp inhibitors.

CONCLUSION

It is evident that the lysine-specific gingipain, Kgp, is a major virulence factor of the anaerobe, *P. gingivalis*, which in turn is a major pathogen of adult periodontitis. The development of inhibitors of this enzyme will therefore be of critical importance to combat this disease. Understanding of the catalytic mechanism, transition state intermediates and the full specificity of the enzyme appear to be prerequisites for the development of further selective, potent inhibitors of Kgp.

REFERENCES

1. Brown LJ, John BA, Wall TP. The economics of periodontal diseases. Periodontol 2000. 2002; 29:223-234.
2. Spencer AJ, Wright FAC, Brown DF et al. A socio-dental study of adult periodontal health: Melbourne 1985. Melbourne: The University of Melbourne, 1988.
3. Shultis WA, Weil EJ, Looker HC et al. Effect of periodontitis on overt nephropathy and end-stage renal disease in type 2 diabetes. Diabetes Care 2007; 30:306-311.
4. DeStefano F, Anda RF, Kahn HS et al. Dental disease and risk of coronary heart disease and mortality. Br Med J 1993; 306:688-691.
5. Jeffcoat MK, Hauth JC, Geurs NC et al. Periodontal disease and preterm birth: result of a pilot intervention study. J Periodontol 2003; 74:1214-1218.
6. Scannapieco FA. Role of oral bacteria in respiratory infection. J Periodontol 1999; 70:793-802.
7. Shapira L, Ayalon S, Brenner T. Effects of Porphyromonas gingivalis on the central nervous system: Activation of glial cells and exacerbation of experimental autoimmune encephalomyelitis. J Periodontol 2002; 73:511-516.
8. Teng YT, Taylor GW, Scannapieco F et al. Periodontal health and systemic disorders. J Can Dent Assoc 2002; 68:188-192.
9. Mercado FB, Marshall RI, Bartold PM. Inter-relationship between rheumatoid arthritis and periodontal disease. A review. J Clin Periodontol 2003; 30:761-772.
10. Meyer MS, Joshipura K, Giovannucci E et al. A review of the relationship between tooth loss, periodontal disease and cancer. Cancer Cause Control 2008; 19:895-907.
11. Seymour GJ, Ford PJ, Cullinan MP et al. Relationship between periodontal infections and systemic disease. Clin Microbiol Infect 2007; 13:3-10.
12. Socransky S, Haffajee A. Evidence of bacterial aetiology: a historical perspective. Periodontol 2000 1994; 5:7-25.
13. Armitage GC. Periodontal diagnoses and classification of periodontal diseases. Periodontol 2000 2004; 34:9-12.
14. Kinane DF. Causation and pathogenesis of periodontal diseases. Periodontol 2000 2001; 25:8-20.
15. Armitage GC. Classifying periodontal diseases—a long-standing dilemma. Periodontol 2000 2002; 30:9-23.
16. Haffajee AD, Socransky SS. Microbial etiological agents of destructive periodontal diseases. Periodontol 2000 1994; 5:78-111.
17. Fitzpatrick RE, Wijeyewickrema LC, Pike RN. The gingipains: scissors and glue of the periodontal pathogen, Porphyrmonas gingivalis. Future Microbiol 2009; 4:471-487.
18. Kasuga Y, Ishihara K, Okuda K. Significance of detection of Porphyromonas gingivalis, Bacteroides forsythus and Treponema denticola in periodontal pockets. Bull Tokyo Dent Coll 2000; 41:109-117.

19. Holt SC, Ebersole JL. Porphyromonas gingivalis, Treponema denticola and Tannerella forsythia: the 'red complex', a prototype polybacterial pathogenic consortium in periodontitis. Periodontol 2000 2005; 38:72-122.
20. White D, Mayrand D. Association of oral bacteroides with gingivitis and adult periodontitis. J Periodontal Res 1981; 16:259-265.
21. Holt SC, Ebersole J, Felton J et al. Implantation of Bacteroides gingivalis in nonhuman primates initiates progression of periodontitis. Science 1988; 239:55-57.
22. Gibson FC, Genco CA. The genus Porphyromonas. In: The Prokaryote. New York: Springer, 2006:428-454.
23. Lamont RJ, Jenkinson HF. Life below the gum line: pathogenic mechanisms of Porphyromonas gingivalis. Microbiol Mol Biol Rev 1998; 62:1244-1263.
24. Mayrand D, Holt SC. Biology of asaccharolytic black-pigmented Bacteroides species. Microbiol Rev 1988; 52:134-152.
25. Smalley JW, Silver J, Marsh PJ et al. The periodontopathogen Porphyromonas gingivalis binds iron protoporphyrin IX in the m-oxo dimeric form: an oxidative buffer and possible pathogenic mechanism. Biochem J 1998; 331:681-685.
26. Shah HN, Williams RAD. Utilization of glucose and amino acids by Bacteroides intermedius and Bacteroides gingivalis. Curr Microbiol 1987; 15:241-246.
27. van Steenbergen TJM, Kastelein P, Touw JJA et al. Virulence of black-pigmented Bacteroides strains from periodontal pockets and other sites in experimentally induced skin lesions in mice. J Periodont Res 1982; 17:41-49.
28. Amano A, Shizukuishi S, Horie H et al. Binding of Porphyromonas gingivalis fimbriae to proline-rich glycoproteins in parotid saliva via a domain shared by major salivary components. Infect Immun 1998; 66:2072-2077.
29. Holt SC, Kesavaku L, Walker S et al. Virulence factors of Porphyromonas gingivalis. Periodontol 2000 1999; 20:168-238.
30. Lamont RJ, Yilmaz O. In or out: the invasiveness of oral bacteria. Periodontol 2000 2002; 30:61-69.
31. Lamont RJ, Bevan CA, Gil S et al. Involvement of Porphyromonas gingivalis fimbriae in adherence to Streptococcus gordonii. Oral Microbiol Immunol 1993; 8:272-276.
32. Isogai H, Isogai E, Yoshimura F et al. Specific inhibition of adherence of an oral strain of Bacteroides gingivalis 381 to epithelial cells by monoclonal antibodies against the bacterial fimbriae. Arch Oral Biol 1988; 33:479-485.
33. Birkedal-Hansen H, Taylor RE, Zambon JJ et al. Characterization of collagenolytic activity from strains of Bacteroides gingivalis. J Periodont Res 1988; 23:258-264.
34. Bleeg HS, Polenik P. Sodium dodecyl sulfate potentiates collagen degradation by proteases from Porphyromonas gingivalis. FEMS Microbiol Ecol 1991; 85:125-132.
35. Kesavalu L, Holt SC, Ebersole JL. Trypsin-like protease activity of Porphyromonas gingivalis as a potential virulence factor in the murine lesion model. Microb Pathog 1996; 20:1-10.
36. Chen Z, Potempa J, Polanowski A et al. Purification and characterization of a 50-kDa cysteine proteinase (gingipain) from Porphyromonas gingivalis. J Biol Chem 1992; 267:18896-18901.
37. Pike R, McGraw W, Potempa J et al. Lysine- and arginine-specific proteinases from Porphyromonas gingivalis. Isolation, characterization and evidence for the existence of complexes with hemagglutinins. J Biol Chem 1994; 269:406-411.
38. Kesavalu L, Holt SC, Ebersole JL. Porphyromonas gingivalis virulence in a murine lesion model: effects of immune alterations. Microb Pathog 1997; 23:317-326.
39. McKee AS, McDermid AS, Wait R et al. Isolation of colonial variants of Bacteroides gingivalis W50 with reduced virulence. J Med Microbiol 1988; 27:59-64.
40. Pavloff N, Pemberton PA, Potempa J et al. Molecular cloning and characterization of Porphyromonas gingivalis lysine-specific gingipain. J Biol Chem 1997; 272:1595-1600.
41. Potempa J, Pike R, Travis J. Titration and mapping of the active site of cysteine proteinases from Porphyromonas gingivalis (gingipain) using peptidyl chloromethanes. J Biol Chem 1997; 378:223-230.
42. Curtis MA, Aduse-Opoku J, Rangarajan M. Cysteine Proteases of Porphyromonas Gingivalis. Crit Rev Oral Biol M 2001; 12:192-216.
43. Curtis MA, Kuramitsu HK, Lantz M et al. Molecular genetics and nomenclature of proteases of Porphyromonas gingivalis. J Periodontal Res 1999; 34:464-472.
44. Tokuda M, Duncan M, Cho MI et al. Role of Porphyromonas gingivalis protease activity in colonization of oral surfaces. Infect Immun 1996; 64:4067-4073.
45. Tokuda M, Karunakaran T, Duncan M et al. Role of Arg-gingipain A in virulence of Porphyromonas gingivalis. Infect Immun 1998; 66:1159-1166.
46. Shi Y, Ratnayake DB, Okamoto K et al. Genetic analyses of proteolysis, hemoglobin binding and hemagglutination of Porphyromonas gingivalis: construction of mutants with a combination of rgpA, rgpB, kgp and hagA. J Biol Chem 1999; 274:17955-17960.

47. O'Brien-Simpson NM, Paolini RA, Hoffmann B et al. Role of RgpA, RgpB and Kgp proteinases in virulence of Porphyromonas gingivalis W50 in a murine lesion model. Infect Immun 2001; 69:7527-7534.
48. Pathirana RD, O'Brien-Simpson NM, Brammar GC et al. Kgp and RgpB, but not RgpA, are important for Porphyromonas gingivalis virulence in the murine periodontitis model. Infect Immun 2007; 75:1436-1442.
49. O'Brien-Simpson NM, Veith PD, Dashper SG et al. Porphyromonas gingivalis gingipains: The molecular teeth of a microbial vampire. Curr Protein Pept Sci 2003; 4:409-426.
50. Barua PK, Dyer DW, Neiders ME. Effect of iron limitation on Bacteroides gingivalis. Oral MicroImmunol 1990; 5:263-268.
51. Shizukuishi S, Tazaki K, Inoshita E et al. Effect of concentration of compounds containing iron on the growth of Porphyromonas gingivalis. FEMS Microbiol Lett 1995; 131:313-317.
52. Lewis JP, Dawson JA, Hannis JC et al. Hemoglobinase activity of the lysine gingipain protease (Kgp) of Porphyromonas gingivalis. J Bacteriol 1999; 181:4905-4913.
53. Okamoto K, Nakayama K, Kadowaki T et al. Involvement of a lysine-specific cysteine proteinase in hemoglobin adsorption and heme accumulation by Porphyromonas gingivalis. J Biol Chem 1998; 273:21225-21231.
54. Pike RN, Potempa J, McGraw W et al. Characterization of the binding activities of proteinase-adhesin complexes from Porphyromonas gingivalis. J Bacteriol 1996; 178:2876-2882.
55. Ryan ME, Golub LM. Modulation of matrix metalloproteinase activities in periodontitis as a treatment strategy. Periodontol 2000 2000; 24:226-238.
56. Decarlo Jr AA, Windsor LJ, Bodden MK et al. Activation and novel processing of matrix metalloproteinases by a thiol-protease from the oral anaerobe Porphyromonas gingivalis. J Dent Res 1997; 76:1260-1270.
57. Imamura T, Potempa J, Pike RN et al. Effect of free and vesicle-bound cysteine proteinases of Porphyromonas gingivalis on plasma clot formation: implications for bleeding tendency at periodontitis sites. Infect Immun 1995; 63:4877-4882.
58. Fletcher J, Reddi K, Poole S et al. Interactions between periodontopathogenic bacteria and cytokines. J Periodont Res 1997; 32:200-205.
59. Yun PLW, DeCarlo AA, Hunter N. Modulation of major histocompatibility complex protein expression by human gamma interferon mediated by cysteine proteinase-adhesin polyproteins of Porphyromonas gingivalis. Infect Immun 1999; 67:2986-2995.
60. Oleksy A, Banbula A, Bugno M et al. Proteolysis of interleukin-6 receptor (IL-6R) by Porphyromonas gingivalis cysteine proteinases (gingipains) inhibits interleukin-6-mediated cell activation. Microb Pathog 2002; 32:173-181.
61. Yun PLW, Decarlo AA, Collyer C et al. Hydrolysis of interleukin-12 by Porphyromonas gingivalis major cysteine proteinases may affect local gamma interferon accumulation and the Th1 or Th2 T-cell phenotype in periodontitis. Infect Immun 2001; 69:5650-5660.
62. Liu RK, Cao CF, Meng HX et al. Polymorphonuclear neutrophils and their mediators in gingival tissues from generalized aggressive periodontitis. J Periodontol 2001; 72:1545-1553.
63. Sugawara S, Nemoto E, Tada H et al. Proteolysis of human monocyte CD14 by cysteine proteinases (gingipains) from Porphyromonas gingivalis leading to lipopolysaccharide hyporesponsiveness. J Immunol 2000; 165:411-418.
64. Kitamura Y, Yoneda M, Imamura T et al. Gingipains in the culture supernatant of Porphyromonas gingivalis cleave CD4 and CD8 on human T-cells. J Periodont Res 2002; 37:464-468.
65. Kitamura Y, Hirofuji T, Yoneda M et al. Cleavage of surface proteins on human T-cells by the culture supernatant of Porphyromonas gingivalis 381. Dent J 2000; 36:109-112.
66. Fishburn CS, Slaney JM, Carman RJ et al. Degradation of plasma proteins by the trypsin-like enzyme of Porphyromonas gingivalis and inhibition of protease activity by a serine protease inhibitor of human plasma. Oral Microbiol Immunol 1991; 6:209-215.
67. Uehara A, Imamura T, Potempa J et al. Gingipains from Porphyromonas gingivalis synergistically induce the production of proinflammatory cytokines through protease-activated receptors with Toll-like receptor and NOD1/2 ligands in human monocytic cells. Cell Microbiol 2008; 10:1181-1189.
68. Huang GTJ, Kim D, Lee JKH et al. Interleukin-8 and intercellular adhesion molecule 1 regulation in oral epithelial cells by selected periodontal bacteria: multiple effects of Porphyromonas gingivalis via antagonistic mechanisms. Infect Immun 2001; 69:1364-1372.
69. Pavloff N, Potempa J, Pike RN et al. Molecular cloning and structural characterization of the arg-gingipain proteinase of Porphyromonas gingivalis. J Biol Chem 1995; 270:1007-1010.
70. Mikolajczyk-Pawlinska J, Kordula T, Pavloff N et al. Genetic variation of Porphyromonas gingivalis genes encoding gingipains, cysteine proteinases with arginine or lysine specificity. Biol Chem 1998; 379:205-211.
71. Potempa J, Sroka A, Imamura T et al. Gingipains, the major cysteine proteinases and virulence factors of Porphyromonas gingivalis: Structure, function and assembly of multidomain protein complexes. Curr Protein Pept Sci 2003; 6:443-450.
72. Slakeski N, Cleal SM, Bhogal PS et al. Characterization of a Porphyromonas gingivalis gene prtK that encodes a lysine-specific thiol proteinase and multiple adhesins. Oral Microbiol Immunol 1999; 14:92-97.

73. Slakeski N, Bhogal PS, O'Brien-Simpson NM et al. Characterization of a second cell-associated Arg-specific cysteine proteinase of Porphyromonas gingivalis and identification of an adhesin binding motif involved in association of the prtR and prtK proteinases and adhesins into large complexes. Microbiol 1998; 144:1583-1592.
74. O'Brien-Simpson NM, Paolini RA, Reynolds EC. RgpA-Kgp peptide-based immunogens provide protection against Porphyromonas gingivalis challenge in a murine lesion model. Infect Immun 2000; 68:4055-4063.
75. O'Brien-Simpson NM, Pathirana RD, Paolini RA et al. An immune response directed to proteinase and adhesin functional epitopes protects against Porphyromonas gingivalis-induced periodontal bone loss. J Immunol 2005; 175:3980-3989.
76. Barkocy-Gallagher GA, Han N, Patti JM et al. Analysis of the prtP gene encoding porphypain, a cysteine proteinase of Porphyromonas gingivalis. J Bacteriol 1996; 178:2734-2741.
77. Lewis JP, Macrina FL. IS195, an insertion sequence-like element associated with protease genes in Porphyromonas gingivalis. Infect Immun 1998; 66:3035-3042.
78. Okamoto K, Kadowaki T, Nakayama K et al. Cloning and sequencing of the gene encoding a novel lysine-specific cysteine proteinase (lys-gingipain) in Porphyromonas gingivalis: Structural relationship with the arginine-specific cysteine proteinase (arg-gingipain). J Biochem 1996; 120:398-406.
79. Li N, Yun P, Nadkarni MA et al. Structure determination and analysis of a haemolytic adhesin domain from Porphyromonas gingivalis. Mol Microbiol 2010; 76:861-873.
80. Bhogal PS, Slakeski N, Reynolds E. Characterization of a cell-associated, protein complex of Porphyromonas gingivalis W50 composed of Arg- and Lys-specific cysteine proteinases and adhesins. Microbiol 1997; 143:2485-2495.
81. Curtis MA, Thickett A, Slaney JM et al. Variable carbohydrate modifications to the catalytic chains of the RgpA and RgpB proteases of Porphyromonas gingivalis W50. Infect Immun 1999; 67:3816-3823.
82. Rangarajan M, Smith SJM, U S et al. Biochemical characterization of the arginine-specific proteases of Porphyromonas gingivalis W50 suggests a common precursor. Biochem J 1997; 323:701-709.
83. Seers CA, Slakeski N, Veith PD et al. The RgpB C-terminal domain has a role in attachment of RgpB to the outer membrane and belongs to a novel C-terminal-domain family found in Porphyromonas gingivalis. J Bacteriol 2006; 188:6376-6386.
84. Pathirana RD, O'Brien-Simpson NM, Veith PD et al. Characterization of proteinase-adhesin complexes of Porphyromonas gingivalis. Microbiol 2006; 152:2381-2394.
85. Fujimura S, Nakamura T. Isolation and characterization of a protease from Bacteroides gingivalis. Infect Immun 1987; 55:716-720.
86. Bedi GS. Comparative study of four proteases from spent culture media of Porphyromonas gingivalis (FAY-19M-1). Prep Biochem 1995; 25:133-154.
87. Takii R, Kadowaki T, Baba A et al. A functional virulence complex composed of gingipains, adhesins and lipopolysaccharide shows high affinity to host cells and matrix proteins and escapes recognition by host immune systems. Infect Immun 2005; 73:883-893.
88. Shoji M, Ratnayake DB, Shi Y et al. Construction and characterization of a nonpigmented mutant of Porphyromonas gingivalis: cell surface polysaccharide as an anchorage for gingipains. Microbiol 2002; 148:1183-1191.
89. Aduse-Opoku J, Davies NN, Gallagher A et al. Generation of Lys-gingipain protease activity in Porphyromonas gingivalis W50 is independent of Arg-gingipain protease activities. Microbiol 2000; 146:1933-1940.
90. Potempa J, Banbula A, Travis J. Role of bacterial proteinases in matrix destruction and modulation of host responses. Periodontol 2000 2000; 24:153-192.
91. Okamoto K, Misumi Y, Kadowaki T et al. Structural characterization of argingipain, a novel arginine-specific cysteine proteinase as a major periodontal pathogenic factor from Porphyromonas gingivalis. Arch Biochem Biophys 1995; 316:917-925.
92. Eichinger A, Beisel HG, Jacob U et al. Crystal structure of gingipain R: an Arg-specific bacterial cysteine proteinase with a caspase-like fold. EMBO J 1999; 18:5453-5462.
93. Schechter I, Berger A. On the size of the active site in proteases. I. Papain. Biochem Biophys Res Commun 1967; 27:157-162.
94. Chen JM, Rawlings ND, Stevens RA et al. Identification of the active site of legumain links it to caspases, clostripain and gingipains in a new clan of cysteine endopeptidases. FEBS Lett 1998; 441:361-365.
95. Howard AE, Kollman PA. OH- versus SH- nucleophilic attack on amides: dramatically different gas-phase and solvation energetics. J Am Chem Soc 1988; 110:7195-7200.
96. Arad D, Langridge R, Kollman PA. A simulation of the sulfur attack in the catalytic pathway of papain using molecular mechanics and semiempirical quantum mechanics. J Am Chem Soc 1988; 112:491-502.
97. Shi Y. Caspase activation: revisiting the induced proximity model. Cell 2004; 117:855-858.
98. Albeck A, Kliper S. Mechanism of cysteine protease inactivation by peptidyl epoxides. Biochem J 1997; 322:879-884.

99. Abe N, Kadowaki T, Okamoto K et al. Biochemical and functional properties of Lysine-specific cysteine proteinase (Lys-gingipain) as a virulence factor of Porphyromonas gingivalis in periodontal disease. J Biochem 1998; 123:305-312.
100. Abe N, Baba A, Kadowaki T et al. Design and synthesis of sensitive fluorogenic substrates specific for Lys-gingipain. J Biochem 2000; 128:877-881.
101. Gusman H, Travis J, Helmerhorst EJ et al. Salivary histatin 5 is an inhibitor of both host and bacterial enzymes implicated in periodontal disease. Infect Immun 2001; 69:1402-1408.
102. Snipas SJ, Stennicke HR, Riedl S et al. Inhibition of distant caspase homologues by natural caspase inhibitors. J Biochem 2001; 357:575-580.
103. Bodet C, Piché M, Chandad F et al. Inhibition of periodontopathogen-derived proteolytic enzymes by a high-molecular-weight fraction isolated from cranberry. J Antimicrob Chemother 2006; 57:685-690.
104. Yamanaka A, Kouchi T, Kasai K et al. Inhibitory effect of cranberry polyphenol on biofilm formation and cysteine proteases of Porphyromonas gingivalis. J Periodont Res 2007; 42:589-592.
105. Bania J, Kubiak A, Wojtachnio K et al. Pancreatic secretory trypsin inhibitor acts as an effective inhibitor of cysteine proteinases gingipains from Porphyromonas gingivalis. J Periodont Res 2008; 43:232-236.
106. Cronan CA, Potempa J, Travis J et al. Inhibition of Porphyromonas gingivalis proteinases (gingipains) by chlorhexidine: synergistic effect of Zn(II). Oral Microbiol Immunol 2006; 21:212-217.
107. Curtis MA, Aduse-Opoku J, Rangarajan M et al. Attenuation of the virulence of Porphyromonas gingivalis by using a specific synthetic Kgp protease inhibitor. Infect Immun 2002; 70:6968-6975.
108. Kadowaki T, Baba A, Abe N et al. Suppression of pathogenicity of Porphyromonas gingivalis by newly developed gingipain inhibitors. Mol Pharmacol 2004; 66:1599-1606.
109. Ekici OD, Götz MG, James KE et al. Aza-peptide Michael acceptors: a new class of inhibitors specific for caspases and other clan CD cysteine proteases. J Med Chem 2004; 47:1889-1892.
110. Białas A, Grembecka J, Krowarsch D et al. Exploring the Sn binding pockets in gingipains by newly developed inhibitors: structure-based design, chemistry and activity. J Med Chem 2006; 49:1744-1753.
111. Ally, N. The specificity of proteinase-adhesins from Porphyromonas gingivalis. PhD thesis. Melbourne: Monash University, 2003.
112. Thomas DA, Francis P, Smith C et al. A broad-spectrum fluorescence-based peptide library for the rapid identification of protease substrates. Proteomics 2006; 6:2112-2120.
113. Birkett AJ, Soler DF, Wolz RL et al. Determination of enzyme specificity in a complex mixture of peptide substrates by N-terminal sequence analysis. Anal Biochem 1991; 196:137-143.
114. Petithory JR, Masiarz FR, Kirsch JF et al. A rapid method for determination of endoproteinase substrate specificity: Specificity of the 3C proteinase from hepatitis A virus. Proc Natl Acad Sci USA 1991; 88:11510-11514.
115. Turk BE, Cantley LC. Using peptide libraries to identify optimal cleavage motifs for proteolytic enzymes. Methods 2004; 32:398-405.
116. Meldal M, Svendsen I, Breddam K et al. Portion-mixing peptide libraries of quenched fluorogenic substrates for complete subsite mapping of endoprotease specificity. P Natl Acad Sci USA 1994; 91:3314-3318.
117. Leon S, Quarrell R, Lowe G. Evaluation of resins for on-bead screening: a study of papain and chymotrypsin specificity using PEGA-bound combinatorial peptide libraries. Bioorg Med Chem Lett 1998; 8:2997-3002.
118. Harris JL, Backes BJ, Leonetti F et al. Rapid and general profiling of protease specificity by using combinatorial fluorogenic substrate libraries. P Natl Acad Sci USA 2000; 97:7754-7759.
119. Matthews DJ, Well JA. Substrate phage: selection of protease substrates by monovalent phage display. Science 1993; 260:1113-1117.
120. Deng SJ, Bickett DM, Mitchell JL et al. Substrate specificity of human collagenase 3 assessed using a phage-displayed peptide library. J Biol Chem 2000; 275:31422-31427.
121. Le Bonniec BF, Myles T, Johnson T et al. Characterization of the P2′ and P3′ specificities of thrombin using fluoresence-quenched substrates and mapping of the subsites by mutagenesis. Biochem 1996; 35:7114-7122.
122. Ally N, Whisstick JC, Sieprawska-Lupa M et al. Characterization of the specificity of Arginine-specific gingipain from Porphyromonas gingivalis reveals active site differences between different forms of the enzymes. Biochem 2003; 42:11693-11700.

CHAPTER 3

FALCIPAINS AND OTHER CYSTEINE PROTEASES OF MALARIA PARASITES

Philip J. Rosenthal

Department of Medicine, San Francisco General Hospital, University of California, San Francisco, California, USA
Email: prosenthal@medsfgh.ucsf.edu

Abstract: A number of cysteine proteases of malaria parasites have been described and many more are suggested by analysis of the *Plasmodium falciparum* genome sequence. The best characterized of these proteases are the falcipains, a family of four papain-family enzymes. Falcipain-2 and falcipain-3 act in concert with other proteases to hydrolyze host erythrocyte hemoglobin in the parasite food vacuole. Disruption of the falcipain-2 gene led to a transient block in hemoglobin hydrolysis and parasites with increased sensitivity to protease inhibitors. Disruption of the falcipain-3 gene was not possible, strongly suggesting that this protease is essential for erythrocytic parasites. Disruption of the falcipain-1 gene did not alter development in erythrocytes, but led to decreased production of oocysts in mosquitoes. Other papain-family proteases predicted by the genome sequence include dipeptidyl peptidases, a calpain homolog and serine-repeat antigens (SERAs). Dipeptidyl aminopeptidase 1 appears to be essential and localized to the food vacuole, suggesting a role in hemoglobin hydrolysis. Dipeptidyl aminopeptidase 3 appears to play a role in the rupture of erythrocytes by mature parasites. The *P. falciparum* calpain homolog gene could not be disrupted, suggesting that the protein is essential and a role in the parasite cell cycle has been suggested. Nine *P. falciparum* SERAs have cysteine protease motifs, but in some the active site Cys is replaced by a Ser. Gene disruption studies suggested that SERA-5 and SERA-6 are essential. Activation of SERA-5 by a serine protease seems to be required for merozoite egress from the erythrocyte. New drugs for malaria are greatly needed and cysteine proteases represent potential drug targets. Cysteine protease inhibitors have demonstrated potent antimalarial effects and the optimization and testing of falcipain inhibitor antimalarials is underway.

Cysteine Proteases of Pathogenic Organisms, edited by Mark W. Robinson and John P. Dalton.

INTRODUCTION

Malaria, particularly disease caused by *Plasmodium falciparum*, is one of the most important infections of humans. There is new optimism regarding malaria control, with increasing use of effective control measures, including insecticide-impregnated bednets, indoor residual spraying of insecticides and new efficacious drug combinations to treat and prevent malaria.[1] However, the control of malaria continues to be challenged by resistance to most available drugs. Thus new antimalarial drugs, ideally directed against new targets, are needed. Among potential new targets for chemotherapy are *Plasmodium* proteases. Many proteolytic enzymes appear to play key roles in the life cycles of malaria parasites. This chapter will focus on cysteine proteases. Recent advances, including the sequencing of *Plasmodium* genomes (http://plasmodb.org) and the development of techniques for manipulating *Plasmodium* genes, have improved our understanding of the cysteine protease repertoire of malaria parasites and have begun to illuminate specific functions of these enzymes. In parallel with biochemistry advances, drug discovery projects designed to identify new antimalarial cysteine protease inhibitors are underway. This chapter, which updates a prior review on this topic,[2] will survey available information on *Plasmodium* cysteine proteases and prospects for exploitation of this class of enzymes as drug targets.

CYSTEINE PROTEASE NOMENCLATURE

Cysteine proteases utilize a catalytic cysteine, which mediates protein hydrolysis via nucleophilic attack on the carbonyl carbon of a susceptible peptide bond. Cysteine proteases are subdivided into clans, which do not share sequence or structural identity and probably arose independently.[3] Clan CA proteases utilize catalytic Cys, His and Asn residues that are invariably in this order in the primary sequence of the protease. Clan CA, Family C1 (papain-family) cysteine proteases are well characterized for many eukaryotic organisms,[4] and these are the best characterized cysteine proteases of *Plasmodium*. In *P. falciparum*, analysis of the genome sequence suggests that clan CA cysteine proteases include four falcipains, three dipeptidyl peptidases, nine proteins related to the serine-rich antigen (SERA) and a calpain homolog (Table 1).[5] Also of interest in *Plasmodium* is clan CD, which utilizes a catalytic His-Cys dyad. Clan CD proteases includes caspases in higher organisms and sequence analyses suggest that members of the C13 and C14 families are present in plasmodia. Clan CE, which is characterized by catalytic residues in the order His, Glu (or Asp), Cys, is also represented in the *P. falciparum* genome.

FUNCTIONS OF *PLASMODIUM* CYSTEINE PROTEASES DETERMINED FROM INHIBITOR STUDIES

Studies with protease inhibitors have provided valuable information regarding the functions of cysteine proteases in malaria parasites.[6] Most available inhibitors are directed toward clan CA proteases, but do not offer marked specificity within this large clan. Therefore, they have been most useful to identify cysteine protease functions rather than the roles of specific enzymes.

The timing of effects of cysteine protease inhibitors provides clues regarding protease functions. In a characterization of the effects of leupeptin and E-64 against cultured

Table 1. Summary of predicted *P. falciparum* cysteine protease genes and characterized cysteine proteases

Clan	Family	Protease/ Homolog	Chr[a]	Gene ID	Purif.[b] Nat	Rec	Activity[c]	KO Phenotype[d]	Putative Function (Eryth Parasites)	Stage of Transcription[e] Eryth	Other	Stage of Protein Expression[f] Immunobl	Proteomic
CA	C1	Falcipain-1	14	PF14_0553		+	CP	Nl development	Not essential	R	M/ Sp/G	R/T/S	Sp
		Falcipain-2	11	PF11_0165	+	+	CP	Block Hb hydrolysis	Hb hydrolysis	T	M/G/ Sp	T	T
		Falcipain-2′	11	PF11_0161		+	CP	Nl development	Hb hydrolysis	R/T	M		T
		Falcipain-3	11	PF11_0162		+	CP	Lethal?	Hb hydrolysis	T/S	M/ Sp/G	T/S	T/G
		Dipeptidyl pep 1	11	PF11_0174	+		CP	Lethal?	Hb hydrolysis	R/T	Sp/ M/G		T/M
		Dipeptidyl pep 2	12	PFL2290w						S	G		T/G
		Dipeptidyl pep 3	4	PFD0230c					Eryth rupture	S	M/Sp		M
		SERA-1	2	PFB0325c				Nl development	Not essential	R	Sp/ M/G		Sp
		SERA-2	2	PFB0330c				Nl development	Not essential	T/S	M/ Sp/G		M/T
		SERA-3	2	PFB0335c				Nl development	Not essential	T/S	M	T/S	None
		SERA-4	2	PFB0340c				Nl development	Not essential	S	M/ Sp/G	T/S	M/T
		SERA-5	2	PFB0345c		+	SP	Lethal?	Eryth rupture	S	M	T/S	Sp
		SERA-6	2	PFB0350c				Lethal?	Eryth rupture	S	M/ Sp/G	T/S	G/T

continued on next page

Table 1. Continued

Clan	Family	Protease/ Homolog	Chr[a]	Gene ID	Purif.[b] Nat	Purif.[b] Rec	Activity[c]	KO Phenotype[d]	Putative Function (Eryth Parasites)	Stage of Transcription[e] Eryth	Stage of Transcription[e] Other	Stage of Protein Expression[f] Immunobl	Stage of Protein Expression[f] Proteomic
		SERA-7	2	PFB0355c				Nl development	Not essential	S	Sp/M/G		None
		SERA-8	2	PFB0360c				Nl development	Not essential	S	Sp/G		None
		SERA-9	9	PFI0135c				Nl development	Not essential	T/S	Sp/M/G		Sp
	C2	Calpain	13	MAL13P1.310				Lethal?	Cell cycle	R	M/Sp/G		None
	C12	UCH1	11	PF11_0177						T/S	G/M/Sp		Sp
		UCH1	14	PF14_0577						T/S	Sp/M/G		T
	C19	UCH2	1	PFA0220w						S/R	M/Sp/G		Sp
		UCH2	4	PFD0165w						R/T/S	Sp/M/G		Sp
		UCH2	4	PFD0680c						T/S	G/M		M
		UCH2	5	PFE1355c						T/S	M/G/Sp		M/T/G
		UCH2	5	PFE0835w						S/R	M		M
		UCH2	7	MAL7P1.147						T	Sp/G/M		G/M/Sp/T
		UCH2	9	PFI0225w						R/T/S	Sp/M/G		Sp
		UCH2	13	PF13_0096						R/T/S	Sp/M		None
		UCH2	14	PF14_0145						R	M/Sp/G		Sp

continued on next page

Table 1. Continued

Clan	Family	Protease/ Homolog	Chr[a]	Gene ID	Purif.[b] Nat	Purif.[b] Rec	Activity[c]	KO Phenotype[d]	Putative Function (Eryth Parasites)	Stage of Transcription[e] Eryth	Stage of Transcription[e] Other	Stage of Protein Expression[f] Immunobl	Stage of Protein Expression[f] Proteomic
CD	C13	GPI:Pr Tr	11	PF11_0298						S	G		G
	C14	Metacas-pase-1	13	PF13_0289					Apoptosis	R	Sp/G		None
		Metacas-pase-2	14	PF14_0363						S	G/ Sp/M		G
CE	C48	Sumo 1	12	PFL1635w						R/T/S	M/ Sp/G		G/M
		Ulp 2	8	MAL8P1.157						T	G/M/ Sp		Sp

Data shown are based on information from multiple sources, including primary research papers referenced in this chapter, genomic[53,54] and proteomic[55] screens of *P. falciparum* and summaries on the PlasmoDB web site (http://plasmodb.org). Blanks indicate that data are unavailable or inconclusive. For genomic and proteomic screens, the parasite stages are listed based on the relative quantities of signals identified, but it is cautioned that the biological relevance of these results are uncertain.

Abbreviations: KO, knockout; Eryth, erythrocyte or erythrocytic; Immunobl, immunoblots; Nat, native; Rec, recombinant; R, ring; T, trophozoite; S, schizont; M, merozoite; G, gametocyte; Sp, sporozoite; CP, cysteine protease; SP, serine protease; Nl, normal

[a] *P. falciparum* chromosome number.

[b] As indicated, active protein has been purified from native (Nat) or recombinant (Rec) material.

[c] General properties of proteases that have been purified and characterized.

[d] For characterizations of KO phenotypes, "Lethal?" indicates that surviving transfectants could not be obtained despite repeated attempts with protocols that successfully generated transfectants with other gene disruptions.

[e] Information is summarized based on two large genomics screens.[53,54]

[f] Data are based on published immunoblot results and a proteomic screen.[55]

P. falciparum parasites, the inhibitors were maximally active against mature trophozoites and schizonts, although 8 hour incubations during any portion of the life cycle elicited some inhibition of parasite development.[7] These results support a critical role for cysteine proteases in trophozoite hemoglobin hydrolysis, but also suggest additional cysteine protease functions.

Hemoglobin Hydrolysis

The best characterized function of *Plasmodium* cysteine proteases is the hydrolysis of hemoglobin. Erythrocytic malaria parasites multiply asexually, reaching large numbers in the circulation and causing the clinical manifestations of malaria. During this cycle, parasites take up erythrocyte cytosol through a specialized organelle, the cytostome, transport the cytosol to an acidic food vacuole and degrade hemoglobin.[8] As hemoglobin is processed, its heme component is converted into hemozoin pigment and globin is hydrolyzed to its constituent amino acids. Hemoglobin hydrolysis appears to be necessary to provide amino acids for parasite protein synthesis,[8-9] to maintain the osmotic stability of malaria parasites,[10] and to provide space for the growing intraerythrocytic parasite. Hemoglobin hydrolysis is likely the result of a cooperative process involving proteases of multiple catalytic classes, including cysteine, aspartic and metallo proteases.[8]

The first evidence for a role of cysteine proteases in hemoglobin hydrolysis came from studies with protease inhibitors. Incubating parasites with broadly active cysteine protease inhibitors, such as leupeptin and E-64, caused the food vacuole to swell and fill with undegraded hemoglobin.[11-13] Although aspartic proteases also have a clear role in hemoglobin hydrolysis,[14] only cysteine protease inhibitors cause the food vacuole swelling that is indicative of a block in hemoglobin processing.[15] Analysis of proteins from cysteine protease inhibitor-treated parasites has identified large quantities of intact native hemoglobin, suggesting that cysteine proteases participate in initial cleavages of hemoglobin.[16]

Cysteine proteases also contribute indirectly to hemoglobin hydrolysis via the processing of aspartic proteases. The processing of the aspartic proteases plasmepsin I and plasmepsin II was inhibited by tripeptide aldehyde cysteine protease inhibitors, but not by other cysteine protease inhibitors (including leupeptin and E-64) suggesting that a cysteine protease other than falcipain-2 or falcipain-3 mediates the processing of plasmepsins.[17] However, in more recent studies, plasmepsin processing was inhibited by the cell permeable cysteine protease inhibitor E-64d and more so by a combination of E-64d and the aspartic protease inhibitor pepstatin.[18] Processing at the natural cleavage site of plasmepsin II was blocked by E-64d, but not pepstatin, indicating that falcipains play a primary role in activation of plasmepsins. Autohydrolysis by plasmepsins also occurs, but with slower kinetics. E-64 inhibited plasmepsin processing in a cell-free system, indicating that the limited ability of falcipain inhibitors to block processing in earlier studies was due to lack of access to intracellular enzyme targets. Purified falcipain-2 and falcipain-3 cleaved plasmepsin II at the native cleavage site, strongly suggesting that these proteases are responsible for plasmepsin processing. These results appear to explain why cysteine, but not aspartic inhibitors cause the accumulation of large quantities of undegraded hemoglobin in *P. falciparum* trophozoites[15] and why these two classes of inhibitors demonstrate synergistic antimalarial activities.[19]

The precise pathways of cleavage of hemoglobin have been defined for plasmepsins,[17] but it has been difficult to do this for falcipains, presumably due to rapid action of these proteases against multiple peptide bonds. Indeed, recent detailed studies utilizing either overlapping peptides representing the sequence of hemoglobin or intact hemoglobin demonstrated that falcipain-2 and falcipain-3 rapidly cleave the protein at multiple sites.[20] A higher degree of specificity was seen with peptide substrates and peptidyl inhibitors, with a preference for cleavage after a P_2 Leu.

Available data suggest a remarkable degree of redundancy in proteases of *P. falciparum* that are responsible for hemoglobin hydrolysis.[21] This process appears to involve multiple falcipains and plasmepsins, as described above and also the metalloprotease falcilysin,[22] dipeptidyl aminopeptidase 1 (see below),[23] and metalloaminopeptidases.[24] This redundancy allows parasites to tolerate loss of certain proteins, including disruption of genes encoding all 4 food vacuole plasmepsins,[25,26] falcipain-2, or falcipain-2′. However, some of these mutant parasites have altered phenotypes, including increased sensitivity to cysteine and/or aspartic protease inhibitors.[27,28] Redundancy between protease classes is not complete and so inhibition of falcipains, plasmepsins and likely other protease classes is deleterious to parasites. Not surprisingly, inhibition of both cysteine and aspartic proteases offers synergistic antimalarial activity.[19] Interpretations of inhibitor studies are complex, as inhibitors may act against proteases in addition to hemoglobinases[28] and as differential effects are, at times, explained by differing cell permeability. Nonetheless, synergistic activity against multiple protease classes offers a promising opportunity for potent antimalarial activity.

Erythrocyte Rupture

Cysteine protease inhibitors also block the rupture of erythrocytes at the completion of the erythrocytic cycle. Cysteine protease activity therefore appears to be required for the release of merozoites, which rapidly invade other erythrocytes to reinitiate the asexual cycle. Older studies showed the accumulation of mature schizonts in cultures treated with leupeptin.[29-31] E-64 blocked lysis of the schizont parasitophorous vacuole membrane, suggesting that cysteine protease activity is required for the hydrolysis of membrane-associated proteins to mediate merozoite release.[32] In a study using different techniques, two other cysteine protease inhibitors (leupeptin and antipain) blocked lysis of the erythrocyte membrane.[33] Considering these results together, it appears that the release of merozoites is a two-step process, requiring hydrolysis of proteins associated with the parasitophorous vacuole and erythrocyte membranes, although the sequence of these two steps is in dispute. Differential effects of studied inhibitors may have been due to differences in inhibitor specificities, suggesting that different proteases act upon targets in the two membranes, or to differences in access to intracellular targets. In any event, these results suggest that cysteine proteases hydrolyze proteins associated with the parasitophorous vacuole and erythrocyte membranes to facilitate erythrocyte rupture. Of note, more recent studies have suggested key roles for SERA-5, dipeptidyl aminopeptidase 3 (see below) and subtilisin-like serine proteases[34] in erythrocyte rupture, but the specific proteases responsible for cleavage of erythrocyte proteins are unknown.

Erythrocyte Invasion

Most reports have not identified an effect of cysteine protease inhibitors on the invasion of erythrocytes by merozoites.[6] Rather, this process has generally been blocked by serine protease inhibitors such as chymostatin, although older literature is confused by some reports that did not clearly distinguish between inhibitor effects on erythrocyte invasion and rupture. In *P. knowlesi* and *P. chabaudi*, species from which invasive merozoites can be obtained, thus simplifying analysis, the serine protease inhibitor chymostatin, but not the cysteine protease inhibitors leupeptin or E-64, inhibited erythrocyte invasion.[30,35] However, arguing for a role for cysteine proteases in invasion, a specific inhibitor of falcipain-1 blocked invasion of erythrocytes.[36] Importantly, interpretation of these results has been complicated by consideration of additional studies of falcipain-1 and this specific inhibitor, as will be discussed below. More recently, a *P. falciparum* serine protease, PfSUB2, has been shown to mediate the cleavage of merozoite surface proteins that is required for erythrocyte invasion.[37] Considering available data, it seems clear that serine proteases are required for erythrocyte invasion by *P. falciparum*, but the role of cysteine proteases in this process is uncertain.

Nonerythrocytic Parasite Stages

During gametogenesis, hydrolysis of the gametocyte surface protein Pfs230 generates two polypeptides which remain associated with the newly-formed gamete.[38] The cleavage of Pfs230 is blocked by E-64d, suggesting that a cysteine protease is responsible for this cleavage, but the relevant enzymes have not been identified.[39] Further, treatment with E-64d[40] and the knockout of falcipain-1[41] both led to markedly decreased oocyst production, suggesting a specific role for this protease in sexual-stage parasites. High concentrations of a compound directed against falcipains blocked microgamete formation, but E-64d did not inhibit this process and inhibitors of other classes of proteases were more potent, suggesting principal roles for other classes in microgamete formation.[42] In a genetic approach, disruption of the gene encoding a *P. berghei* cysteine protease expressed in mosquito stages (egress cysteine protease-1, an ortholog of *P. falciparum* SERA-8) blocked egress of sporozoites from oocysts, proving a role for SERA-family proteases in oocyst rupture.[43]

FALCIPAIN CYSTEINE PROTEASES

The falcipains comprise four papain-family (clan CA, family C1) enzymes of *P. falciparum* including falcipain-1, which is encoded on chromosome 14 and falcipain-2, falcipain-2′ (or falcipain 2B; 99% homology with falcipain-2 in the catalytic domain) and falcipain-3, which are encoded within a 12 kb stretch of chromosome 11. The falcipains are fairly typical papain-family cysteine proteases, but they have some unusual features, including unusually large prodomains, predicted membrane spanning sequences within the prodomains and an unusual insertion between highly conserved residues near the carboxy terminus.[44-46] Homology between falcipain-1 and the other falcipains is relatively low (~40% identity between the catalytic domains). Falcipain-2 and falcipain-3 are much more similar in sequence (68% identity), share similarly sized prodomains (that of falcipain-1 is much longer) and include an unusual amino-terminal extension of the catalytic domain that is not found in falcipain-1. Based on these comparisons and similar findings with homologs in other *Plasmodium* species, it seems appropriate to consider two distinct groups of falcipains, the falcipain-1 and falcipain-2/3 sub-families. Other *Plasmodium* species contain a single homolog of falcipain-1 and varied numbers of falcipain-2/3 homologs.[47] In *P. vivax*, the second most important human malaria parasite, a single homolog of falcipain-1 (72% identity between the catalytic domains)[48] and three homologs of falcipain-2/3 (60-70% identity)[49] have been characterized. Analysis of murine malaria parasites, including *P. berghei* and *P. vinckei*, has identified only single homologs of falcipain-1[50] and falcipain-2/3.[51] Identity between human and rodent parasite proteases is lower than that between *P. falciparum* and *P. vivax* proteases, but identities within sub-families are generally over 50% (in catalytic domains), compared to 30-40% identities between falcipain-1 and falcipain-2/3 sub-families.[47]

Expression of Falcipains by Erythrocytic Parasites

Affinity purification of falcipain-2 demonstrated that the enzyme is responsible for over 90% of the cysteine protease activity that is identified in trophozoite lysates with standard peptidyl substrates.[45] Abundant falcipain-3 is also expressed by erythrocytic parasites, but it is relatively inactive against peptidyl substrates.[46] Falcipain-2 and falcipain-3 differ in their timing of expression, with maximal expression in early trophozoites for falcipain-2 and late trophozoites for falcipain-3.[27,46] Immunoblots have also identified falcipain-1 expression across the erythrocytic cycle.[52] This enzyme was localized to merozoites by immunofluorescence microscopy,[36] and transcription of the gene was maximal in ring forms.[53,54] However, a proteomic analysis identified falcipain-1 peptides only in sporozoites.[55]

Biochemical Characterization

Biochemical characterization of falcipain-2 and falcipain-3 was expedited by the development of efficient systems for heterologous expression in *E. coli* followed by refolding of the active enzymes.[45,46,56] The biochemical features of the two enzymes are very similar, but not identical, which offers clues to potential differences in functions. Both falcipain-2 and falcipain-3 have low pH optima, consistent with activity in the acidic food vacuole. As with most other papain-family proteases, the P_2 position of substrates is most important for specificity and both proteases prefer peptidyl substrates with Leu at this position. Falcipain-2 is much more active against peptidyl substrates, is uniquely able to activate and undergo autohydrolysis at neutral pH and is more stable at neutral pH.[46] Considering specificity for peptide substrates and inhibitors, important differences were seen between falcipain-2 and -3 and homologs from the rodent parasites *P. berghei* and *P. vinckei*.[51] Differences in specificity between falcipain-2 and -3 were less pronounced.[46,57]

Both falcipain-2 and falcipain-3 have been localized to the food vacuole by cell fractionation,[45,46] immunofluorescence and immunoelectron microscopy.[58] Both enzymes are synthesized as membrane-bound proforms that are processed to soluble mature forms, but falcipain-2 is processed to the mature protease much more quickly than falcipain-3, likely due to its susceptibility to autohydrolysis at neutral pH. Cysteine protease inhibitors and brefeldin A, but not aspartic or serine protease inhibitors, blocked the processing of both enzymes, suggesting that falcipain-2 and -3 process by autohydrolysis after exiting the endoplasmic reticulum/Golgi network.[58] These results suggest that the proteases are synthesized as integral membrane proteins in a specific cellular compartment before delivery to the food vacuole and auto-hydrolysis to release soluble active proteases.

Falcipain-2′, a near-identical copy of falcipain-2, has also been heterologously expressed and biochemically characterized.[59-61] No biochemical differences between falcipain-2 and falcipain-2′ have been identified. However, expression of the enzymes is not identical and the knockout of falcipain-2, but not falcipain-2′, has a noteworthy phenotype (see below). Thus, falcipain-2′ cannot replace the function of falcipain-2 but the biological role of this enzyme is unknown. Falcipain-1 has also been heterologously expressed by a number of groups, but in all cases expression has been at low levels, limiting biochemical evaluation.[62-64]

Hydrolysis of Natural Substrates

Both falcipain-2 and falcipain-3 hydrolyze native hemoglobin and denatured globin, with maximal activity near the pH (~5.2) of the food vacuole.[65] As is typical for cysteine proteases, maximal activity requires a reducing environment, although relatively low concentrations of glutathione that are likely physiological are adequate to support this activity. Falcipain-2 also hydrolyzes the erythrocyte cytoskeletal proteins band 4.1 and ankyrin at neutral pH, suggesting an additional role for this protease (which undergoes activation at neutral pH) in erythrocyte rupture.[66,67] However, a biological role for falcipain-2 in this process has not been confirmed.

Functions of Different Falcipain Domains

The mature forms of falcipain-2, falcipain-3 and some homologs from other *Plasmodium* species are capable of refolding to active enzymes after expression in *E. coli* by dilution in alkaline buffer.[56] These are the only papain family proteases known to refold without their prodomains.[68] However, correct folding does require the presence of a small amino terminal extension (17 amino acids in falcipain-2) which can mediate correct folding either when it is included upstream of the recombinant catalytic domain or in the refolding buffer as a separate folding domain-prodomain polypeptide.[69] All falcipain-2/3 proteases, but no other known papain-family proteases (and no falcipain-1 sub-family enzymes) contain an ~20 amino acid extension at the amino terminus of the catalytic domain. The 14-15 amino acids immediately N-terminal of the catalytic domain, beginning with a fully conserved Tyr, are required to mediate folding.[70] The amino terminal domains show only moderate homology between members of the falcipain-2/3 sub-family, but there is functional conservation, as chimeras of the falcipain-2 catalytic domain with folding domains from other members of the family folded with similar kinetics.[70] The folding domain was not required for activity once folding had occurred and interaction between catalytic and folding domains were independent of the active site.[70]

As is the case with other papain-family proteases, the prodomain of falcipain-2 is a potent inhibitor of the enzyme.[69] The inhibitory domain was recently defined as encompassing a C-terminal portion of the prodomain (Leu155-Asp243) that includes previously described "ERFNIN" and GNFD" domains that appear to mediate inhibition by prodomains in many papain family proteases.[71]

N-terminal portions of the falcipain-2 prodomain mediate trafficking of the mature enzyme to the food vacuole, its principal site of action. Studies of chimeras with portions of the prodomain fused to green fluorescent protein and containing instructive deletions and point mutations indicated that both a 20-amino acid stretch of the lumenal portion and a 10-amino acid stretch of the cytoplasmic portion of the falcipain-2 prodomain were required for efficient trafficking to the food vacuole.[72] Mutants with altered trafficking were arrested at the plasma membrane, implicating trafficking via this structure. Photobleaching studies indicated that falcipain-2 is trafficked to the food vacuole via cytostomal vesicles.[73] Thus, falcipains utilize a previously undescribed bipartite motif-dependent mechanism for targeting to the food vacuole via the plasma membrane and cytostomal vesicles.

Falcipains also contain an insertion near the C-terminus of the mature protease that is unusual for papain-family proteases.[74] Removal of this 10 amino acid insertion had no effect on the activity of falcipain-2 against a number of peptide and protein substrates. In contrast, the altered protease had a dramatic loss in activity against hemoglobin. This loss in activity was due to an inability of the mutant protease to bind hemoglobin. Thus, cleavage of hemoglobin by falcipain-2 requires an unusual motif for binding to hemoglobin prior to its hydrolysis.[74]

Biological Roles of Falcipains—RNA Interference

RNAi directed against either falcipain-1 or falcipain-2 inhibited the development of erythrocytic parasites, with accumulation of undegraded hemoglobin and the appearance of abnormal vacuoles that resembled, but were probably not identical to those caused by incubation with cysteine protease inhibitors.[75] Treating mice infected with *P. berghei* with short interfering RNAs encoding the *P. berghei* homologs of falcipain-2 and falcipain-3 led to parasites with large vacuoles, although murine malaria progressed similarly in treated and control animals.[76] RNAi directed against falcipain-2 also arrested parasites at the schizont stage, suggesting a role for the protease in erythrocyte rupture.[77] Taken together, available RNAi experiments suggest specific roles for falcipains, for the most part consistent with models based on biochemical results, but it remains unclear if RNAi systems are present in *P. falciparum* and so interpretations of these studies are not straightforward.

Biological Roles of Falcipains- Disruption of Falcipain Genes

Gene disruption studies have provided the most definitive results on falcipain functions. All four falcipain genes were disrupted by transfecting parasites with plasmids encoding truncated genes. Falcipain-2 knockout trophozoites had markedly diminished cysteine protease activity and swollen, dark-staining food vacuoles, consistent with a block in hemoglobin hydrolysis, as caused by cysteine protease inhibitors.[27] However, more mature knockout parasites had normal morphologies. Expression of other falcipains and plasmepsin aspartic proteases was similar in wild type and knockout parasites. Although the multiplication rates of wild type and knockout parasites were similar, the knockout parasites were about 3-times more sensitive to the cysteine protease inhibitors E-64 and leupeptin and over 50-fold more sensitive to the aspartic protease inhibitor pepstatin. These results assign a specific function for falcipain-2, the hydrolysis of hemoglobin in trophozoites. This is the first proven function for a plasmodial protease. In addition, they highlight the cooperative action of cysteine and aspartic proteases in hemoglobin degradation. In contrast to results for falcipain-2, repeated attempts to knock out falcipain-3 were unsuccessful, although replacement of the protease gene with a tagged functional copy was possible.[78] This result argues strongly that falcipain-3 is essential for erythrocytic parasites. Knockout of falcipain-1 and of falcipain-2′ was successful,[41,52,78] and in each case knockouts demonstrated morphologies and growth rates unchanged from those of wild type strains. For falcipain-1, the compound previously described as a specific inhibitor, YA29,[36] also inhibited falcipain-2 at micromolar concentrations and its effects were the same as those of E-64 and leupeptin, without the inhibition of erythrocyte invasion described previously.[52] In summary, falcipain-3 appears to play an essential role in erythrocytic parasites, falcipain-2 participates in hemoglobin hydrolysis, but is not essential and falcipain-1 and falcipain-2′ have unknown and nonessential functions in erythrocytic parasites. In contrast, knockout of falcipain-1 led to markedly decreased production of oocysts in mosquitoes.[41] Falcipain-1 RNA[46] and protein[52] are expressed in erythrocytic parasites and the protein has been localized to merozoites by immunofluroescence.[36,64] However, the apparent lack of effect of the falcipain-1 knockout on erythrocytic parasites, markedly decreased oocyst production by a falcipain-1 knockout clone,[41] and the identification of falcipain-1 peptides in a proteomic screen only in sporozoites (Table 1) suggest that the principal roles of falcipain-1 are in nonerythrocytic stage parasites.

OTHER CLAN CA CYSTEINE PROTEASES OF MALARIA PARASITES

Dipeptidyl Peptidases

The *P. falciparum* genome includes three sequences with homology with dipeptidyl peptidases which remove dipeptides from the amino-termini of polypeptides (Table 1). Overall identity between the three sequences is only ~30%; identity with the falcipains is 20-25%. Purified *P. falciparum* dipeptidyl aminopeptidase 1 cleaved dipeptide substrates and was localized to the food vacuole, suggesting that it contributes to late steps in hemoglobin hydrolysis.[23] A chemical genetic screen identified specific inhibitors for dipeptidyl aminopeptidases 1 and 3.[79] In studies of cultured parasites, inhibition of dipeptidyl aminopeptidase 1 had nonspecific effects, which were difficult to distinguish from inhibition of falcipains, consistent with a role for the enzyme in hemoglobin hydrolysis. Specific inhibition of dipeptidyl aminopeptidase 3 led to the accumulation of erythrocytes containing mature schizonts, suggesting that this protease plays a role in the rupture of erythrocytes at the conclusion of the erythrocytic cycle. Evaluation of inhibitor effects on protease maturation led to a model for erythrocyte rupture in which dipeptidyl aminopeptidase 3 activates the serine protease PfSUB1, which in turn activates SERA 5 (see below).[79]

Calpain Homolog

Calpains of higher organisms are cysteine proteases that are calcium-dependent and contain both catalytic and calcium binding domains.[80] Homologs in other organisms may not have calcium dependence. The *P. falciparum* genome sequence encodes a single calpain homolog. This protein is localized to the nucleolus of erythrocytic parasites.[81] Disruption or truncation of the calpain gene was not possible, but replacement with a functional copy was successful, strongly suggesting that *P. falciparum* calpain is essential for erythrocytic parasites.[82] Knock down of calpain expression impaired parasite growth and delayed progression through the cell cycle.[82] *P. falciparum* parasites also appear to utilize a host calpain; immunodepletion of calpain-1 from erythrocytes blocked egress of mature parasites, suggesting that parasites utilize erythrocyte calpain-1 to facilitate this process.[83]

SERAs

The serine rich antigen (SERA, now named SERA-5),[84] an immunogenic protein and potential vaccine component, was noted many years ago to have similarities in sequence with cysteine proteases. A second member of the family, serine rich protein homolog (SERPH or SERA-6), was described soon thereafter.[85] Subsequent sequencing has identified an array of 8 SERA-family proteins encoded on chromosome 2 and a single additional member of the family encoded on chromosome 9 (Table 1). In all cases, the proteins contain a "protease domain", equivalent in size and in some sequence features to a papain-family catalytic domain, located within a much larger protein without other apparent similarity to cysteine proteases. Within the protease domain, all SERA-family proteins share modest homology with papain-family proteases, but it is noteworthy that canonical clan CA active site residues are either fully conserved (SERA-6, 7 and 8) or substituted (e.g., a replacement of the catalytic Cys with Ser in SERA-5). Homologs of SERAs are also seen in other *Plasmodium* species.[86-88] A number of the SERAs are expressed in erythrocytic parasites, with maximal expression of SERA-5, but also marked expression of SERA-3, 4 and 6, primarily in the trophozoite and schizont stages.[89,90] SERA-5 and SERA-6 are most likely to be essential, as only these SERA genes could not be disrupted in cultured parasites.[89,91] Recombinant SERA-5, which includes a cysteine protease scaffold but replacement of the canonical Cys by Ser, exhibited serine protease (chymotrypsin-like) activity, with autohydrolysis inhibited by serine protease inhibitors and modest activity against peptidyl serine protease substrates.[88] SERA-6, which contains typical papain-family active site residues, is most likely a more typical cysteine protease, but biochemical evidence to support this conclusion is lacking.

Since cysteine protease inhibitors block erythrocyte rupture (see above) and as both SERA-5 and SERA-6 have been localized to the parasitophorous vacuole that surrounds mature schizonts,[85,92] these proteins and perhaps other SERAs, may be responsible for proteolytic cleavages required for the egress of merozoites from the erythrocyte. The processing of SERA-5 was well studied some years ago; a series of processing steps occurs at the time of erythrocyte rupture, with steps both sensitive and insensitive to cysteine protease inhibitors.[34] Recent work has shown that SERA-5 is activated by the serine protease PfSUB1, which is released into the parasitophorous vacuole space immediately prior to parasite egress from the erythrocyte.[93] As noted above, PfSUB1 may require activation by the cysteine protease dipeptidyl aminopeptidase 3. Although some details remain uncertain, it appears that the egress of mature *P. falciparum* from erythrocytes requires a series of events, culminating in the activation of SERA proteases by the serine protease PfSUB1.[34] However, specific functions of SERAs remain unclear.

Clan CD Cysteine Proteases

Based on evaluation of the genome sequence, *P. falciparum* may express a number of clan CD proteases, but none have been well characterized or demonstrated to have enzymatic activity (e.g., metacaspase1 and 2, Table 1).[94] Considering roles in other biological systems, clan CD proteases are likely to have tighter substrate specificity than clan CA enzymes, suggesting roles in fine regulation of parasite metabolism.[94] This high level of specificity suggests promise for clan CD protease inhibitors as highly specific antimalarial drugs. It has been noted that the *P. falciparum* metacaspase-1 gene encodes an N-terminal caspase recruitment domain, suggesting a role in parasite apoptosis.[95] However, a homolog of this gene from *P. berghei* was not expressed in erythrocytic-stage parasites and knockout of the gene was not deleterious to the murine parasites.[96]

An Endogenous Inhibitor of Cysteine Proteases.

P. falciparum expresses an endogenous cysteine protease inhibitor, falstatin, presumably for control of parasite and/or host protease activity.[97] Recombinant falstatin was a potent reversible inhibitor of falcipain-2, falcipain-3 and related host and parasite proteases. It is expressed in schizonts, merozoites and rings, but not in trophozoites, the stage at which the cysteine protease activity of *P. falciparum* is maximal. Falstatin localizes to the periphery of rings and early schizonts, is diffusely expressed in late schizonts and merozoites and is released upon erythrocyte rupture. Treatment of late schizonts with antibodies that blocked the inhibitory activity of falstatin against falcipain-2 and falcipain-3 dose-dependently decreased the subsequent invasion of erythrocytes by merozoites. These results suggest that *P. falciparum* requires expression of falstatin to limit proteolysis by certain host or parasite cysteine proteases during erythrocyte invasion. This mechanism of regulation of proteolysis suggests new strategies for the development of antimalarial agents that specifically disrupt erythrocyte invasion.

POTENTIAL FOR CYSTEINE PROTEASE INHIBITORS AS ANTIMALARIAL DRUGS

As cysteine proteases play essential roles in erythrocytic malaria parasites, an obvious consideration is the inhibition of these enzymes to treat malaria. A number of older studies have supported this concept, with the demonstration that cysteine protease inhibitors have potent in vitro and in vivo antimalarial effects.[6] Specifically, peptidyl fluoromethyl ketone,[50,98,99] vinyl sulfone,[100-102] and aldehyde[103] inhibitors of falcipains blocked the development of cultured parasites at nanomolar concentrations. Some nonpeptide falcipain inhibitors showed more modest antiparasitic activity.[104-106] The inhibition of parasite development was generally accompanied by a specific block in hemoglobin hydrolysis, marked by the appearance of swollen, dark staining food vacuoles and antiparasitic effects correlated with the degree of inhibition of falcipain-2 and falcipain-3.[102] Many potent inhibitors also blocked homologous enzymes from *P. vivax*[49] and *P. vinckei*,[50] although specificities varied, particularly between proteases of human and murine parasites.

Drug discovery directed against falcipains is now facilitated by available structures of falcipain-2 and falcipain-3 complexed with a number of small molecule and protein inhibitors.[107-110] These structures are similar, but not identical to those predicted by modeling exercises,[111,112] and suggest explanations for subtle biochemical differences between falcipain-2 and falcipain-3.

Cysteine protease inhibitors have also exhibited antimalarial effects in vivo, although these results must be interpreted in light of important differences between homologous proteases of human and rodent parasites.[51,113] Treatment of *P. vinckei*-infected mice with fluoromethyl ketone,[50,100] vinyl sulfone,[100] and aldehyde[103] inhibitors led to partial or complete protection against lethal malaria. Also, inhibitors of cysteine and aspartic proteases showed synergistic antimalarial effects both in vitro[11,19] and in vivo,[19] suggesting the possibility of combined protease inhibitor antimalarial therapy. Extensive studies of the antimalarial activity of nonpeptidyl inhibitors of falcipains are now underway.[114-122] Concerning the potential for resistance to antimalarial cysteine protease inhibitors, parasites were selected for resistance to a vinyl sulfone falcipain inhibitor, but the selection was slow and the mechanism of resistance was complex.[123] This result might indicate that resistance to antimalarial cysteine protease inhibitors will be slow to develop, but the best means of avoiding resistance will likely be the use of combination antimalarial therapy.

Differences between the cysteine proteases of human and rodent malaria parasites challenge standard approaches to drug discovery in which rodent parasite models play a critical role in compound screening.[51,113] This problem might be circumvented by development of recombinant rodent parasites that express the human parasite proteases[124] or the use of new models that allow propagation of *P. falciparum* in immunocompromised mice.[125] The *P. falciparum* mouse model was recently used to demonstrate potent in vivo antimalarial activity of falcipain inhibitors.[126]

CONCLUSION

Our understanding of the cysteine protease repertoire of malaria parasites has increased markedly in recent years. Falcipain-2 and falcipain-3 are key hemoglobinases that are appropriate targets for antimalarial chemotherapy. Efforts to optimize falcipain inhibitors as antimalarials are currently underway. Our understanding of the biological roles of parasite dipeptidyl peptidases, calpain and SERAs is improving and all of these proteases also offer potential targets for chemotherapy. Some proteases have key roles in nonerythrocytic parasite stages that might be relevant in chemoprevention or vaccine strategies. Further characterization of these proteases should expedite efforts to improve the control of malaria.

ACKNOWLEDGEMENTS

Work in the author's laboratory has been supported by the National Institutes of Health, the Medicines for Malaria Venture, the Centers for Disease Control and Prevention and the Doris Duke Charitable Foundation, with which the author is a Distinguished Clinical Scientist.

REFERENCES

1. Greenwood BM, Bojang K, Whitty CJ et al. Lancet 2005; 365:1487-1498.
2. Rosenthal PJ. Cysteine proteases of malaria parasites. Int J Parasitol 2004; 34:1489-1499.
3. Barrett AJ, Rawlings ND. Evolutionary lines of cysteine peptidases. Biol Chem 2001; 382:727-733.
4. Turk V, Turk B, Turk D. Lysosomal cysteine proteases: facts and opportunities. Embo J 2001; 20:4629-4633.
5. Wu Y, Wang X, Liu X et al. Data-mining approaches reveal hidden families of proteases in the genome of malaria parasite. Genome Res 2003; 13:601-616.
6. Rosenthal PJ. Protease inhibitors. In: Rosenthal PJ, ed. Antimalarial Chemotherapy: Mechanisms of Action, Resistance and New Directions in Drug Discovery. Totowa: Humana Press, 2001:325-345.
7. Shenai BR, Semenov AV, Rosenthal PJ. Stage-specific antimalarial activity of cysteine protease inhibitors. Biol Chem 2002; 383:843-847.
8. Francis SE, Sullivan DJ Jr. Goldberg DE. Hemoglobin metabolism in the malaria parasite Plasmodium falciparum. Annu Rev Microbiol 1997; 51:97-123.
9. McKerrow JH, Sun E, Rosenthal PJ et al. The proteases and pathogenicity of parasitic protozoa. Annu Rev Microbiol 1993; 47:821-853.
10. Lew VL, Tiffert T, Ginsburg H. Excess hemoglobin digestion and the osmotic stability of Plasmodium falciparum-infected red blood cells. Blood 2003; 101:4189-4194.
11. Bailly E, Jambou R, Savel J et al. Plasmodium falciparum: differential sensitivity in vitro to E-64 (cysteine protease inhibitor) and Pepstatin A (aspartyl protease inhibitor). J Protozool 1992; 39:593-599.
12. Dluzewski AR, Rangachari K, Wilson RJ et al. Plasmodium falciparum: protease inhibitors and inhibition of erythrocyte invasion. Exp Parasitol 1986; 62:416-422.
13. Rosenthal PJ, McKerrow JH, Aikawa M et al. A malarial cysteine proteinase is necessary for hemoglobin degradation by Plasmodium falciparum. J Clin Invest 1988; 82:1560-1566.
14. Gluzman IY, Francis SE, Oksman A et al. Order and specificity of the Plasmodium falciparum hemoglobin degradation pathway. J Clin Invest 1994; 93:1602-1608.

15. Rosenthal PJ. Plasmodium falciparum: effects of proteinase inhibitors on globin hydrolysis by cultured malaria parasites. Exp Parasitol 1995; 80:272-281.
16. Gamboa de Dominguez ND, Rosenthal PJ. Cysteine proteinase inhibitors block early steps in hemoglobin degradation by cultured malaria parasites. Blood 1996; 87:4448-4454.
17. Francis SE, Banerjee R, Goldberg DE. Biosynthesis and maturation of the malaria aspartic hemoglobinases plasmepsins I and II. J Biol Chem 1997; 272:14961-14968.
18. Drew ME, Banerjee R, Uffman EW et al. Plasmodium food vacuole plasmepsins are activated by falcipains. J Biol Chem 2008; 283:12870-12876.
19. Semenov A, Olson JE, Rosenthal PJ. Antimalarial synergy of cysteine and aspartic protease inhibitors. Antimicrob Agents Chemother 1998; 42:2254-2258.
20. Subramanian S, Hardt M, Choe Y et al. Hemoglobin cleavage site-specificity of the Plasmodium falciparum cysteine proteases falcipain-2 and falcipain-3. PLoS ONE 2009; 4:e5156.
21. Liu J, Istvan ES, Gluzman IY et al. Plasmodium falciparum ensures its amino acid supply with multiple acquisition pathways and redundant proteolytic enzyme systems. Proc Natl Acad Sci USA 2006; 103:8840-8845.
22. Eggleson KK, Duffin KL, Goldberg DE. Identification and characterization of falcilysin, a metallopeptidase involved in hemoglobin catabolism within the malaria parasite Plasmodium falciparum. J Biol Chem 1999; 274:32411-32417.
23. Klemba M, Gluzman I, Goldberg DE. A Plasmodium falciparum dipeptidyl aminopeptidase I participates in vacuolar hemoglobin degradation. J Biol Chem 2004; 279:43000-43007.
24. Dalal S, Klemba M. Roles for two aminopeptidases in vacuolar hemoglobin catabolism in Plasmodium falciparum. J Biol Chem 2007; 282:35978-35987.
25. Bonilla JA, Bonilla TD, Yowell CA et al. Critical roles for the digestive vacuole plasmepsins of Plasmodium falciparum in vacuolar function. Mol Microbiol 2007; 65:64-75.
26. Bonilla JA, Moura PA, Bonilla TD et al. Effects on growth, hemoglobin metabolism and paralogous gene expression resulting from disruption of genes encoding the digestive vacuole plasmepsins of Plasmodium falciparum. Int J Parasitol 2007; 37:317-327.
27. Sijwali PS, Rosenthal PJ. Gene disruption confirms a critical role for the cysteine protease falcipain-2 in hemoglobin hydrolysis by Plasmodium falciparum. Proc Natl Acad Sci USA 2004; 101:4384-4389.
28. Moura PA, Dame JB, Fidock DA. Role of Plasmodium falciparum digestive vacuole plasmepsins in the specificity and antimalarial mode of action of cysteine and aspartic protease inhibitors. Antimicrob Agents Chemother 2009; 53:4968-4978.
29. Debrabant A, Delplace P. Leupeptin alters the proteolytic processing of P126, the major parasitophorous vacuole antigen of Plasmodium falciparum. Mol Biochem Parasitol 1989; 33:151-158.
30. Hadley T, Aikawa M, Miller LH. Plasmodium knowlesi: studies on invasion of rhesus erythrocytes by merozoites in the presence of protease inhibitors. Exp Parasitol 1983; 55:306-311.
31. Lyon JA, Haynes JD, Diggs CL et al. Plasmodium falciparum antigens synthesized by schizonts and stabilized at the merozoite surface by antibodies when schizonts mature in the presence of growth inhibitory immune serum. J Immunol 1986; 136:2252-2258.
32. Salmon BL, Oksman A, Goldberg DE. Malaria parasite exit from the host erythrocyte: a two-step process requiring extraerythrocytic proteolysis. Proc Natl Acad Sci USA 2001; 98:271-276.
33. Wickham ME, Culvenor JG, Cowman AF. Selective inhibition of a two-step egress of malaria parasites from the host erythrocyte. J Biol Chem 2003; 278:37658-37663.
34. Blackman MJ. Malarial proteases and host cell egress: an 'emerging' cascade. Cell Microbiol 2008; 10:1925-34.
35. Breton CB, Blisnick T, Jouin H et al. Plasmodium chabaudi p68 serine protease activity required for merozoite entry into mouse erythrocytes. Proc Natl Acad Sci USA 1992; 89:9647-9651.
36. Greenbaum DC, Baruch A, Grainger M et al. A role for the protease falcipain 1 in host cell invasion by the human malaria parasite. Science 2002; 298:2002-2006.
37. Harris PK, Yeoh S, Dluzewski AR et al. Molecular identification of a malaria merozoite surface sheddase. PLoS Pathog 2005; 1:241-251.
38. Williamson KC, Fujioka H, Aikawa M et al. Stage-specific processing of Pfs230, a Plasmodium falciparum transmission-blocking vaccine candidate. Mol Biochem Parasitol 1996; 78:161-169.
39. Brooks SR, Williamson KC. Proteolysis of Plasmodium falciparum surface antigen, Pfs230, during gametogenesis. Mol Biochem Parasitol 2000; 106:77-82.
40. Eksi S, Czesny B, van Gemert GJ et al. Inhibition of Plasmodium falciparum oocyst production by membrane-permeant cysteine protease inhibitor E64d. Antimicrob Agents Chemother 2007; 51:1064-1070.
41. Eksi S, Czesny B, Greenbaum DC et al. Targeted disruption of Plasmodium falciparum cysteine protease, falcipain 1, reduces oocyst production, not erythrocytic stage growth. Mol Microbiol 2004; 53:243-250.
42. Rupp I, Bosse R, Schirmeister T et al. Effect of protease inhibitors on exflagellation in Plasmodium falciparum. Mol Biochem Parasitol 2008; 158:208-212.

43. Aly AS, Matuschewski K. A malarial cysteine protease is necessary for Plasmodium sporozoite egress from oocysts. J Exp Med 2005; 202:225-230.
44. Rosenthal PJ, Nelson RG. Isolation and characterization of a cysteine proteinase gene of Plasmodium falciparum. Mol Biochem Parasitol 1992; 51:143-152.
45. Shenai BR, Sijwali PS, Singh A et al. Characterization of native and recombinant falcipain-2, a principal trophozoite cysteine protease and essential hemoglobinase of Plasmodium falciparum. J Biol Chem 2000; 275:29000-29010.
46. Sijwali PS, Shenai BR, Gut J et al. Expression and characterization of the Plasmodium falciparum haemoglobinase falcipain-3. Biochem J 2001; 360:481-489.
47. Rosenthal PJ, Sijwali PS, Singh A et al. Cysteine proteases of malaria parasites: targets for chemotherapy. Curr Pharm Des 2002; 8:1659-1672.
48. Rosenthal PJ, Ring CS, Chen X et al. Characterization of a Plasmodium vivax cysteine proteinase gene identifies uniquely conserved amino acids that may mediate the substrate specificity of malarial hemoglobinases. J Mol Biol 1994; 241:312-316.
49. Na BK, Shenai BR, Sijwali PS et al. Identification and biochemical characterization of vivapains, cysteine proteases of the malaria parasite Plasmodium vivax. Biochem J 2004; 378:529-538.
50. Rosenthal PJ, Lee GK, Smith RE. Inhibition of a Plasmodium vinckei cysteine proteinase cures murine malaria. J Clin Invest 1993; 91:1052-1056.
51. Singh A, Shenai BR, Choe Y et al. Critical role of amino acid 23 in mediating activity and specificity of vinckepain-2, a papain-family cysteine protease of rodent malaria parasites. Biochem J 2002; 368:273-281.
52. Sijwali PS, Kato K, Seydel KB et al. Plasmodium falciparum cysteine protease falcipain-1 is not essential in erythrocytic stage malaria parasites. Proc Natl Acad Sci USA 2004; 101:8721-8726.
53. Bozdech Z, Llinas M, Pulliam BL et al. The transcriptome of the intraerythrocytic developmental cycle of Plasmodium falciparum. PLoS Biol 2003; 1:E5.
54. Le Roch KG, Zhou Y, Blair PL et al. Discovery of gene function by expression profiling of the malaria parasite life cycle. Science 2003; 301:1503-1508.
55. Florens L, Washburn MP, Raine JD et al. A proteomic view of the Plasmodium falciparum life cycle. Nature 2002; 419:520-526.
56. Sijwali PS, Brinen LS, Rosenthal PJ. Systematic optimization of expression and refolding of the Plasmodium falciparum cysteine protease falcipain-2. Protein Expr Purif 2001; 22:128-134.
57. Ramjee MK, Flinn NS, Pemberton TP et al. Substrate mapping and inhibitor profiling of falcipain-2, falcipain-3 and berghepain-2: implications for peptidase anti-malarial drug discovery. Biochem J 2006; 399:47-57.
58. Dahl EL, Rosenthal PJ. Biosynthesis, localization and processing of falcipain cysteine proteases of Plasmodium falciparum. Mol Biochem Parasitol 2005; 139:205-212.
59. Singh N, Sijwali PS, Pandey KC et al. Plasmodium falciparum: biochemical characterization of the cysteine protease falcipain-2′. Exp Parasitol 2006; 112:187-192.
60. Goh LL, Sim TS. Characterization of amino acid variation at strategic positions in parasite and human proteases for selective inhibition of falcipains in Plasmodium falciparum. Biochem Biophys Res Commun 2005; 335:762-770.
61. Jeong JJ, Kumar A, Hanada T et al. Cloning and characterization of Plasmodium falciparum cysteine protease, falcipain-2B. Blood Cells Mol Dis 2006; 36:429-435.
62. Salas F, Fichmann J, Lee GK et al. Functional expression of falcipain, a Plasmodium falciparum cysteine proteinase, supports its role as a malarial hemoglobinase. Infect Immun 1995; 63:2120-2125.
63. Goh SL, Goh LL, Sim TS. Cysteine protease falcipain 1 in Plasmodium falciparum is biochemically distinct from its isozymes. Parasitol Res 2005; 97:295-301.
64. Kumar A, Kumar K, Korde R et al. Falcipain-1, a Plasmodium falciparum cysteine protease with vaccine potential. Infect Immun 2007; 75:2026-2034.
65. Shenai BR, Rosenthal PJ. Reducing requirements for hemoglobin hydrolysis by Plasmodium falciparum cysteine proteases. Mol Biochem Parasitol 2002; 122:99-104.
66. Dua M, Raphael P, Sijwali PS et al. Recombinant falcipain-2 cleaves erythrocyte membrane ankyrin and protein 4.1. Mol Biochem Parasitol 2001; 116:95-99.
67. Hanspal M, Dua M, Takakuwa Y et al. Plasmodium falciparum cysteine protease falcipain-2 cleaves erythrocyte membrane skeletal proteins at late stages of parasite development. Blood 2002; 100:1048-1054.
68. Bromme D, Nallaseth FS, Turk B. Production and activation of recombinant papain-like cysteine proteases. Methods 2004; 32:199-206.
69. Sijwali PS, Shenai BR, Rosenthal PJ. Folding of the Plasmodium falciparum cysteine protease falcipain-2 is mediated by a chaperone-like peptide and not the prodomain. J Biol Chem 2002; 277:14910-14915.
70. Pandey KC, Sijwali PS, Singh A et al. Independent intramolecular mediators of folding, activity and inhibition for the Plasmodium falciparum cysteine protease falcipain-2. J Biol Chem 2004; 279:3484-3491.
71. Pandey KC, Barkan DT, Sali A et al. Regulatory elements within the prodomain of falcipain-2, a cysteine protease of the malaria parasite Plasmodium falciparum. PLoS ONE 2009; 4:e5694.

72. Subramanian S, Sijwali PS, Rosenthal PJ. Falcipain cysteine proteases require bipartite motifs for trafficking to the Plasmodium falciparum food vacuole. J Biol Chem 2007; 282:24961-24969.
73. Dasaradhi PV, Korde R, Thompson JK et al. Food vacuole targeting and trafficking of falcipain-2, an important cysteine protease of human malaria parasite Plasmodium falciparum. Mol Biochem Parasitol 2007; 156:12-23.
74. Pandey KC, Wang SX, Sijwali PS et al. The Plasmodium falciparum cysteine protease falcipain-2 captures its substrate, hemoglobin, via a unique motif. Proc Natl Acad Sci USA 2005; 102:9138-9143.
75. Malhotra P, Dasaradhi PV, Kumar A et al. Double-stranded RNA-mediated gene silencing of cysteine proteases (falcipain-1 and -2) of Plasmodium falciparum. Mol Microbiol 2002; 45:1245-1254.
76. Mohmmed A, Dasaradhi PV, Bhatnagar RK et al. In vivo gene silencing in Plasmodium berghei—a mouse malaria model. Biochem Biophys Res Commun 2003; 309:506-511.
77. Dasaradhi PV, Mohmmed A, Kumar A et al. A role of falcipain-2, principal cysteine proteases of Plasmodium falciparum in merozoite egression. Biochem Biophys Res Commun 2005; 336:1062-8.
78. Sijwali PS, Koo J, Singh N et al. Gene disruptions demonstrate independent roles for the four falcipain cysteine proteases of Plasmodium falciparum. Mol Biochem Parasitol 2006; 150:96-106.
79. Arastu-Kapur S, Ponder EL, Fonovic UP et al. Identification of proteases that regulate erythrocyte rupture by the malaria parasite Plasmodium falciparum. Nat Chem Biol 2008; 4:203-213.
80. Perrin BJ, Huttenlocher A. Calpain. Int J Biochem Cell Biol 2002; 34:722-725.
81. Russo I, Oksman A, Goldberg DE. Fatty acid acylation regulates trafficking of the unusual Plasmodium falciparum calpain to the nucleolus. Mol Microbiol 2009; 72:229-245.
82. Russo I, Oksman A, Vaupel B et al. A calpain unique to alveolates is essential in Plasmodium falciparum and its knockdown reveals an involvement in preS-phase development. Proc Natl Acad Sci USA 2009; 106:1554-1559.
83. Chandramohanadas R, Davis PH, Beiting DP et al. Apicomplexan parasites co-opt host calpains to facilitate their escape from infected cells. Science 2009; 324:794-797.
84. Bzik DJ, Li WB, Horii T et al. Amino acid sequence of the serine-repeat antigen (SERA) of Plasmodium falciparum determined from cloned cDNA. Mol Biochem Parasitol 1988; 30:279-288.
85. Knapp B, Nau U, Hundt E et al. A new blood stage antigen of Plasmodium falciparum highly homologous to the serine-stretch protein SERP. Mol Biochem Parasitol 1991; 44:1-13.
86. Kiefer MC, Crawford KA, Boley LJ et al. Identification and cloning of a locus of serine repeat antigen (sera)- related genes from Plasmodium vivax. Mol Biochem Parasitol 1996; 78:55-65.
87. Gor DO, Li AC, Wiser MF et al. Plasmodial serine repeat antigen homologues with properties of schizont cysteine proteases. Mol Biochem Parasitol 1998; 95:153-158.
88. Hodder AN, Drew DR, Epa VC et al. Enzymic, phylogenetic and structural characterization of the unusual papain-like protease domain of Plasmodium falciparum SERA5. J Biol Chem 2003; 278:48169-48177.
89. Miller SK, Good RT, Drew DR et al. A subset of Plasmodium falciparum SERA genes are expressed and appear to play an important role in the erythrocytic cycle. J Biol Chem 2002; 277:47524-47532.
90. Aoki S, Li J, Itagaki S et al. Serine repeat antigen (SERA5) is predominantly expressed among the SERA multigene family of Plasmodium falciparum and the acquired antibody titers correlate with serum inhibition of the parasite growth. J Biol Chem 2002; 277:47533-47540.
91. McCoubrie JE, Miller SK, Sargeant T et al. Evidence for a common role for the serine-type Plasmodium falciparum serine repeat antigen proteases: implications for vaccine and drug design. Infect Immun 2007; 75:5565-5574.
92. Delplace P, Fortier B, Tronchin G et al. Localization, biosynthesis, processing and isolation of a major 126 kDa antigen of the parasitophorous vacuole of Plasmodium falciparum. Mol Biochem Parasitol 1987; 23:193-201.
93. Yeoh S, O'Donnell RA, Koussis K et al. Subcellular discharge of a serine protease mediates release of invasive malaria parasites from host erythrocytes. Cell 2007; 131:1072-1083.
94. Mottram JC, Helms MJ, Coombs GH et al. Clan CD cysteine peptidases of parasitic protozoa. Trends Parasitol 2003; 19:182-187.
95. Meslin B, Barnadas C, Boni V et al. Features of apoptosis in Plasmodium falciparum erythrocytic stage through a putative role of PfMCA1 metacaspase-like protein. J Infect Dis 2007; 195:1852-1859.
96. Le Chat L, Sinden RE, Dessens JT. The role of metacaspase 1 in Plasmodium berghei development and apoptosis. Mol Biochem Parasitol 2007; 153:41-47.
97. Pandey KC, Singh N, Arastu-Kapur S et al. Falstatin, a Cysteine Protease Inhibitor of Plasmodium falciparum, Facilitates Erythrocyte Invasion. PLoS Pathog 2006; 2:e117.
98. Rockett KA, Playfair JH, Ashall F et al. Inhibition of intraerythrocytic development of Plasmodium falciparum by proteinase inhibitors. FEBS Lett 1990; 259:257-259.
99. Rosenthal PJ, Wollish WS, Palmer JT et al. Antimalarial effects of peptide inhibitors of a Plasmodium falciparum cysteine proteinase. J Clin Invest 1991; 88:1467-1472.

100. Olson JE, Lee GK, Semenov A et al. Antimalarial effects in mice of orally administered peptidyl cysteine protease inhibitors. Bioorg Med Chem 1999; 7:633-638.
101. Rosenthal PJ, Olson JE, Lee GK et al. Antimalarial effects of vinyl sulfone cysteine proteinase inhibitors. Antimicrob Agents Chemother 1996; 40:1600-1603.
102. Shenai BR, Lee BJ, Alvarez-Hernandez A et al. Structure-activity relationships for inhibition of cysteine protease activity and development of Plasmodium falciparum by peptidyl vinyl sulfones. Antimicrob Agents Chemother 2003; 47:154-160.
103. Lee BJ, Singh A, Chiang P et al. Antimalarial activities of novel synthetic cysteine protease inhibitors. Antimicrob Agents Chemother 2003; 47:3810-3814.
104. Dominguez JN, Lopez S, Charris J et al. Synthesis and antimalarial effects of phenothiazine inhibitors of a Plasmodium falciparum cysteine protease. J Med Chem 1997; 40:2726-2732.
105. Li R, Kenyon GL, Cohen FE et al. In vitro antimalarial activity of chalcones and their derivatives. J Med Chem 1995; 38:5031-5037.
106. Ring CS, Sun E, McKerrow JH et al. Structure-based inhibitor design by using protein models for the development of antiparasitic agents. Proc Natl Acad Sci USA 1993; 90:3583-3587.
107. Wang SX, Pandey KC, Somoza JR et al. Structural basis for unique mechanisms of folding and hemoglobin binding by a malarial protease. Proc Natl Acad Sci USA 2006; 103:11503-11508.
108. Hogg T, Nagarajan K, Herzberg S et al. Structural and functional characterization of Falcipain-2, a hemoglobinase from the malarial parasite Plasmodium falciparum. J Biol Chem 2006; 281:25425-25437.
109. Kerr ID, Lee JH, Farady CJ et al. Vinyl sulfones as antiparasitic agents and a structural basis for drug design. J Biol Chem 2009; 284:25697-25703.
110. Kerr ID, Lee JH, Pandey KC et al. Structures of falcipain-2 and falcipain-3 bound to small molecule inhibitors: implications for substrate specificity. J Med Chem 2009; 52:852-857.
111. Batra S, Sabnis YA, Rosenthal PJ et al. Structure-based approach to falcipain-2 inhibitors: synthesis and biological evaluation of 1,6,7-Trisubstituted dihydroisoquinolines and isoquinolines. Bioorg Med Chem 2003; 11:2293-2299.
112. Sabnis YA, Desai PV, Rosenthal PJ et al. Probing the structure of falcipain-3, a cysteine protease from Plasmodium falciparum: comparative protein modeling and docking studies. Protein Sci 2003; 12:501-509.
113. Chan C, Goh LL, Sim TS. Differences in biochemical properties of the Plasmodial falcipain-2 and berghepain-2 orthologues: implications for in vivo screens of inhibitors. FEMS Microbiol Lett 2005; 249:315-321.
114. Desai PV, Patny A, Gut J et al. Identification of novel parasitic cysteine protease inhibitors by use of virtual screening. 2. The available chemical directory. J Med Chem 2006; 49:1576-1584.
115. Chipeleme A, Gut J, Rosenthal PJ et al. Synthesis and biological evaluation of phenolic Mannich bases of benzaldehyde and (thio)semicarbazone derivatives against the cysteine protease falcipain-2 and a chloroquine resistant strain of Plasmodium falciparum. Bioorg Med Chem 2007; 15:273-282.
116. Verissimo E, Berry N, Gibbons P et al. Design and synthesis of novel 2-pyridone peptidomimetic falcipain 2/3 inhibitors. Bioorg Med Chem Lett 2008; 18:4210-4214.
117. Capela R, Oliveira R, Goncalves LM et al. Artemisinin-dipeptidyl vinyl sulfone hybrid molecules: design, synthesis and preliminary SAR for antiplasmodial activity and falcipain-2 inhibition. Bioorg Med Chem Lett 2009; 19:3229-3232.
118. Ettari R, Micale N, Schirmeister T et al. Novel peptidomimetics containing a vinyl ester moiety as highly potent and selective falcipain-2 inhibitors. J Med Chem 2009; 52:2157-2160.
119. Li H, Huang J, Chen L et al. Identification of novel falcipain-2 inhibitors as potential antimalarial agents through structure-based virtual screening. J Med Chem 2009; 52:4936-4940.
120. Loser R, Gut J, Rosenthal PJ et al. Antimalarial activity of azadipeptide nitriles. Bioorg Med Chem Lett. 2010; 20:252-255.
121. Breuning A, Degel B, Schulz F et al. Michael acceptor based antiplasmodial and antitrypanosomal cysteine protease inhibitors with unusual amino acids. J Med Chem 2010; 53:1951-1963.
122. Ettari R, Bova F, Zappala M et al. Falcipain-2 inhibitors. Med Res Rev 2010; 30:136-167.
123. Singh A, Rosenthal PJ. Selection of cysteine protease inhibitor-resistant malaria parasites is accompanied by amplification of falcipain genes and alteration in inhibitor transport. J Biol Chem 2004; 279:35236-35241.
124. Singh A, Walker KJ, Sijwali PS et al. A chimeric cysteine protease of Plasmodium berghei engineered to resemble the Plasmodium falciparum protease falcipain-2. Protein Eng Des Sel 2007; 20:171-177.
125. Angulo-Barturen I, Jimenez-Diaz MB, Mulet T et al. A murine model of falciparum-malaria by in vivo selection of competent strains in nonmyelodepleted mice engrafted with human erythrocytes. PLoS ONE 2008; 3:e2252.
126. Coterón JM, Catterick D, Castro J et al. Falcipain inhibitors: optimization studies of the 2-pyrimidinecarbonitrile lead series. J Med Chem. 2010; 53:6129-6152.

CHAPTER 4

CATHEPSIN PROTEASES IN *TOXOPLASMA GONDII*

Zhicheng Dou and Vern B. Carruthers*
Department of Microbiology and Immunology, University of Michigan School of Medicine, Ann Arbor, Michigan, USA
**Corresponding Author: Vern B. Carruthers—Email: vcarruth@umich.edu*

Abstract: Cysteine proteases are important for the growth and survival of apicomplexan parasites that infect humans. The apicomplexan *Toxoplasma gondii* expresses five members of the C1 family of cysteine proteases, including one cathepsin L-like (TgCPL), one cathepsin B-like (TgCPB) and three cathepsin C-like (TgCPC1, 2 and 3) proteases. Recent genetic, biochemical and structural studies reveal that cathepsins function in microneme and rhoptry protein maturation, host cell invasion, replication and nutrient acquisition. Here, we review the key features and roles of *T. gondii* cathepsins and discuss the therapeutic potential for specific inhibitor development.

INTRODUCTION

Peptidases play a critical role in protein catabolism by hydrolysis of peptide bonds in the polypeptides.[1] Peptidases are classified into seven categories based on the principal catalytic residue in the active site: Aspartic, Cysteine, Glutamic, Serine, Threonine, Metallo and Mixed, each of which can be further divided into clans and families. Cysteine peptidases, also called thiol peptidases, use the nucleophilic thiol group of cysteine for hydrolysis. Cathepsin peptidases belonging to the C1 family clan CA of "papain-like" cysteine peptidases are widely distributed in eukaryotic organisms. During catalysis, a basic amino acid in the catalytic triad, usually histidine, de-protonates the cysteine thiol group, which attacks the carbonyl carbon group in the substrate for hydrolysis. The hydrolytic cycle is completed when the newly-derived substrate terminal amine accepts the proton from the active site histidine, thus regenerating the prehydrolysis status of the active site.[2]

Cysteine Proteases of Pathogenic Organisms, edited by Mark W. Robinson and John P. Dalton.

GENERAL ROLES OF CATHEPSINS IN BIOLOGICAL SYSTEMS

In eukaryotic cells, cathepsin proteases act classically as lysosomal hydrolases that digest endogenous and exogenous endocytosed polypeptides.[3] However, increasingly it is appreciated that cathepsins can also play more specialized roles in higher eukaryotic organisms including spermatogenesis, antigen presentation, tumor invasion, degradation of matrix proteins and TNFα-induced apoptosis.[4-6] Upregulated cathepsin L enzyme in tumor cells plays an important role in nonmetastatic tumor cell conversion into a highly invasive metastatic state.[7-9] In unicellular eukaryotes such as *Plasmodium falciparum*, cathepsin L-like proteases (falcipains 2a, 2b and 3) are responsible for hemoglobin digestion within the parasite food vacuole during erythrocyte infection.[10-15] Falcipain 1, which resides in intracellular vesicles, plays a yet-to-be-defined, nonessential role in parasite invasion of erythrocytes,[16,17] but it is required for efficient parasite development within infected mosquitoes.[18] In *Trypanosoma b. brucei* and related species, cathepsin L and cathepsin B localize to endolysosomes.[19] Cathepsin L is not essential but it facilitates traversal of the blood-brain barrier by *T. b. rhodesiense* in experimentally infected mice,[20] possibly by activating a protease-activated receptor.[21] *T. b. brucei* cathepsin B appears to be required for parasite survival[20] perhaps because of its role in degrading endocytosed host transferrin for iron acquisition.[22,23] Accordingly, like higher eukaryotes, protozoan parasites appear to use cathepsin endopeptidases for protein degradation and other specialized roles.

PROPERTIES OF *TOXOPLASMA GONDII* AND ITS CATHEPSINS

T. gondii is a ubiquitous apicomplexan parasite that infects a wide range of warm-blooded animals. It is estimated that almost one-third of the human population is infected by this parasite.[24] Infection with *Toxoplasma* can lead to encephalitis, chorioretinitis and congenital birth defects. AIDS and immunocompromised patients are at especially high risk of developing toxoplasmosis. As an obligate intracellular parasite, *T. gondii* must invade host cells to survive and expand the infection. Unlike intracellular bacteria and viruses, *T. gondii* and other related parasites use a unique gliding motility mechanism for invading host cells. *Toxoplasma* parasites lack specialized appendages for motility and instead utilize an intrapellicular actin-myosin system to slide on a substrate or a host-cell surface.[25] During cell invasion, two subcellular organelles, micronemes and rhoptries, sequentially discharge their contents at the apical end of the parasite to mediate entry.[26] As the parasite invades the host cell, a parasitophorous vacuole (PV) membrane is formed and surrounds the parasite.[27] After the parasite finishes entry, other organelles, termed dense granules (DG), secrete proteins into the PV. DG proteins are thought to function in modification of the PV for nutrient acquisition.[26] While most microneme and rhoptry proteins are subjected to limited proteolysis (a process termed proteolytic maturation) as they traffic to their respective secretory organelles, DG proteins are not processed and follow a constitutive secretion pathway to the PV.[28] Protease inhibitor studies revealed that cysteine proteases are involved in the maturation of microneme and rhoptry proteins and the biogenesis of some subcellular organelles.[29-33]

The endosomal system in *T. gondii* is generally similar to that of other eukaryotic cells, but it also displays some unique features (Fig. 1). Recently, a dynamic Vacuolar

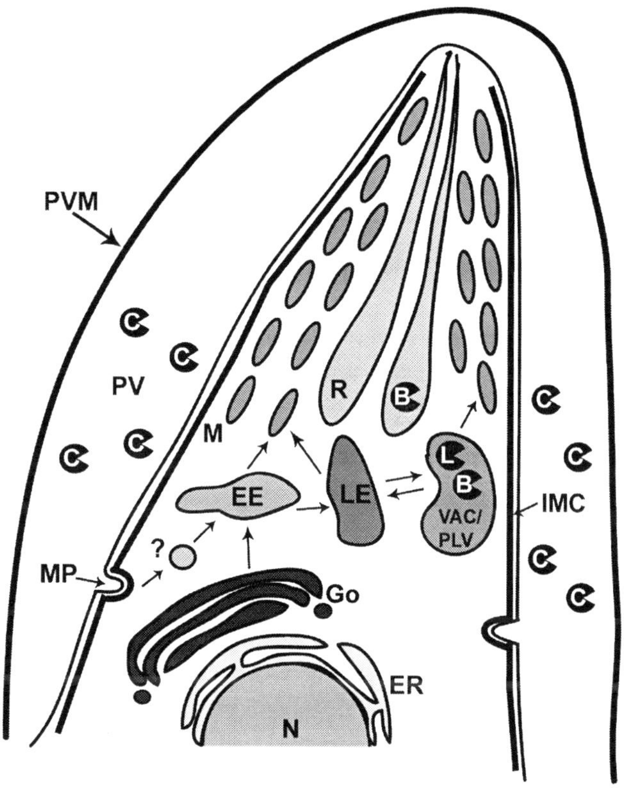

Figure 1. The endosomal system of *T. gondii* and the subcellular locations of cathepsins. TgCPL and TgCPB are predominantly expressed in the VAC and a diminutive amount of TgCPL is seen in the late endosome (LE) where it has been implicated in the maturation of promicroneme proteins.[33,34] TgCPB was also reported to be distributed in the rhoptry to function in the processing of prorhoptry proteins.[30] Immunofluorescence microscopical studies revealed that TgCPC1 protein is secreted into the PV after cell invasion, where it may digest exogenous proteins to meet the parasite's nutrient needs.[39] Exogenous polypeptides may also be endocytosed and trafficked to the VAC for nutrient acquisition. Abbreviations used: B, cathepsin B-like protease; C, cathepsin C-like protease; EE, early endosome; ER, endoplasmic reticulum; Go, Golgi apparatus; IMC, inner membrane complex; L, cathepsin-L like protease, LE, late endosome; M, microneme; MP, micropore; N, nucleus; PLV, plant-like vacuole; PV, parasitophorous vacuole; PVM, parasitophorous vacuole membrane; R, rhoptry; VAC, vacuolar compartment.

Compartment (VAC) (also termed the Plant-Like Vacuole or PLV) was identified in the *Toxoplasma* endosomal system.[33,34] The VAC is often closely associated with the late endosome (LE) in the intermediate apical region of the parasite. The major population of the *T. gondii* cathepsin L (TgCPL) and cathepsin B (TgCPB, also termed toxopain-1 or CP-1) is distributed in the VAC[30,33,35] (and Dou and Carruthers, unpublished data), suggesting this organelle may have a role similar to lysosomes. Although a low proportion of extracellular tachyzoites of *T. gondii* show endocytic ability,[36,37] it remains possible that endocytosis of parasite membrane proteins or host cell polypeptides and oligopeptides is more active in intracellular replicating parasites with a high nutrient demand.

Chromosome Organization

Five cathepsin proteins are encoded in the genome of *T. gondii*: one cathepsin L-like protein (TgCPL), one cathepsin B-like protein (TgCPB) and three cathepsin C-like proteins (TgCPC1, 2 and 3) (Table 1). These five cathepsin genes are distributed among four distinct chromosomes. *Tgcpl* is encoded in chromosome Ib and has four exons while *Tgcpb* consists of seven exons and is encoded on chromosome XII. *Tgcpc1* and *Tgcpc3* are encoded on chromosome IX and therefore could have resulted from intra-chromosomal gene duplication, although they are separated by a considerable distance on the chromosome. *Tgcpc1* and *Tgcpc3* have nine and fourteen exons, respectively. *Tgcpc2* has ten exons on chromosome III. All *T. gondii* cathepsin genes have homologs among the three sequenced strain types of *T. gondii*, except *Tgcpc3*. *Tgcpc3* is not present in the genome sequence of the Type I reference strain GT1 (www.toxodb.org), probably due to incomplete sequence coverage.

Catalytic Residues and Motifs

Toxoplasma cathepsins have almost identical amino acid sequences in their catalytic region as other members of the papain family. The conserved cysteine, histidine and asparagine residues form a triad in the active site that catalyzes peptide cleavage (Fig. 2). TgCPL has several conserved motifs including ERFNIN, which is a signature motif within the prodomain of cathepsin L and H proteases, a KNFD motif (also in the prodomain) and a SPV domain in the mature enzyme.[35] TgCPB has 12 cysteine residues in the mature form, which may participate in the formation of six disulfide bonds. A conserved motif (GCNGG) that exists in the human cathepsin B is also present in TgCPB. Besides endopeptidase activity, TgCPB also shows exopeptidase activity, which is contributed by an intact positively charged occluding loop that can bind to the terminal carboxylic acid of the target substrate.[30]

Like human cathepsin C, also called dipeptidyl peptidase I (DPPI), all mature TgCPCs are composed of an N-terminal residual prodomain (exclusion domain), a heavy chain

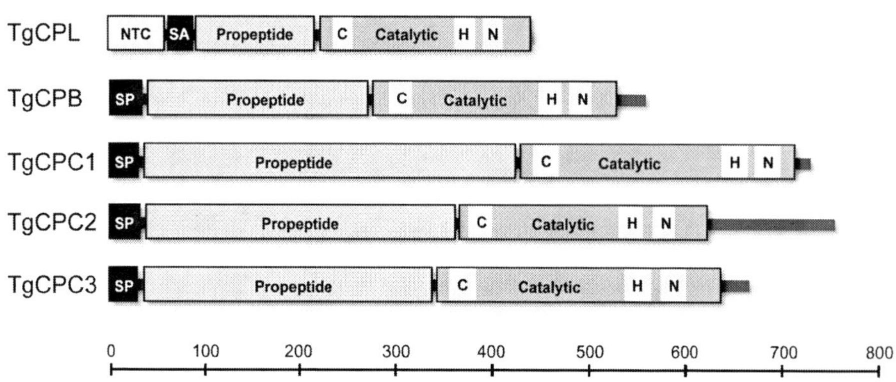

Figure 2. Schematic representations of *T. gondii* cathepsin-like proteases. Domains are represented by rectangles whereas joining regions or extensions are depicted as elongated bars. The positions of catalytic triad residues (C, H and N) are indicated. Abbreviations used: NTC, N-terminal cytosolic domain; SA, signal anchor; SP, signal peptide. Scale is in amino acids.

Table 1. Features of *T. gondii* cathepsin-like proteases

Protease	Length (aa)	Catalytic Domain[1]	Mature MW	Activity[3]	Chromosome	Exons	Optimal pH	Activation	Stage Expression	Substrate Specificity
TgCPL	422	206-420	30 kDa	Endo	Ib	4	5.5-6.0 (polypeptide) 6.5 (synthetic peptides)	Autoactivation in vitro	Tachyzoite and bradyzoite	P2 hydrophobic
TgCPB	569	275-532	28 kDa	Endo/ Exo	XII	7	ND	Autoactivation in vitro	Tachyzoite and bradyzoite (microarray analysis)	P2 hydrophobic
TgCPC1	733	412-709	35 kDa[2]	Exo	IX	9	6.5	No autoactivation in vitro	Tachyzoite	17-fold higher activity for Gly-Arg than human cathepsin-C specific substrate Gly-Phe
TgCPC2	753	357-630	44 kDa[2]	Exo	III	10	ND	No autoactivation in vitro	Tachyzoite	ND
TgCPC3	622	335-647	32 kDa[2]	Exo	IX	14	ND	ND	Sporozoite (expressed sequence tags)	ND

[1]Numbered from the initiator methionine.
[2]MW is based on the size of the catalytic domain without further processing into the heavy and light chains, if occurring.
[3]Endo, endopeptidase; Exo, exopeptidase.

and a light chain. As a dipeptidyl peptidase, the aspartic residue at the N-terminus of the exclusion domain can block formation of the substrate binding cleft with the polypeptide beyond its S2 substrate-binding site.[38,39] Unlike other cathepsin-like enzymes, a halide ion is required for its activity and a conserved tyrosine residue that binds a chloride ion in the crystal structures of the rat and human cathepsin C proteases also exists in *Toxoplasma* CPCs.[40,41] In addition, a tyrosine-based motif, YXXΦ (X is any amino acid and Φ is a bulky hydrophobic amino acid) is present in all *Toxoplasma* CPCs.[39] TgCPC2 and TgCPC3 also display a dileucine-motif, which in higher eukaryotes participates in endosomal/lysosomal protein targeting.[42] TgCPC2 has a 100-residue C-terminal extension, the biological role of which remains unknown.

Three-Dimensional Structure of TgCPL and TgCPB

Recently, Larson et al solved the three dimensional structure of autoactivated TgCPL with its propeptide resident in the active site cleft.[32] The structure of mature TgCPL is very similar to previously determined papain-like cysteine protease structures. TgCPL consists of two domains divided by a deep active site cleft; one domain is primarily α-helical and the other contains a β-barrel-like fold and several α-helices.[32] The Cys_{31}, His_{167} and Asn_{189} residues (numbered according to ref. 32) cluster together and form the active site triad. Three disulfide bonds stabilize the TgCPL structure, a feature that is common in the papain-like cysteine proteases. The propeptide occupies the active cleft in a reverse orientation compared to that of the substrate, thus forming a stable interaction that inhibits its endopeptidase activity until activation and propeptide dissociation occur.[32] An aspartic acid residue occupies the S2 substrate-binding site, which is a major specificity-determining site. Although initial homology modeling predicted that the aspartic acid created an unusually shallow S2 substrate-binding site,[35] the crystal structure of TgCPL revealed that the aspartic acid does not overtly occlude the pocket.[32] Accordingly, TgCPL hydrolyzes synthetic substrates with a variety of hydrophobic amino acids in the P2 position.[33]

Although a crystal structure of TgCPB is not available, homology modeling studies based on human lysosomal and rat cathepsin B show a glutamic residue at the base of the TgCPB S2 pocket,[30] which can interact with positively charged residues, such as arginine at the P2 site. An occluding loop that is close to the substrate-binding cleft blocks the C-terminal end of the active site. Two positively charged histidine residues in this loop can associate with the C-terminal carboxylate group of the P2′ residue to endow this enzyme with exopeptidase activity.[43]

Activation Profile and Optimal pH for Activity

Recombinant TgCPL and TgCPB can be auto-activated at low pH in the presence of a reducing agent.[30,32,33] Recombinant TgCPL is active on both peptide and polypeptide substrates.[33] Recombinant TgCPC1 and TgCPC2 failed to auto-activate in vitro, indicating that they require exogenous cleavage by another protease molecule for activation.[39] The heavy chain and residual prodomain of TgCPC1, which remain associated with the active enzyme, were copurified from a tachyzoite lysate, indicating that the expected activating proteolysis occurs in vivo.[39] Human DPPI protease is trans-activated by cathepsin L cleavage at several sites, but it remains to be determined if TgCPCs are activated by TgCPL.[44]

Almost all cathepsin proteases perform optimally within a low pH range.[6] Recombinant TgCPL efficiently cleaves a recombinant proform of *T. gondii* MIC2-Associated Protein (proTgM2AP) at pH 5.5-6.0,[33] and the optimal pH range for digesting a Lys-Gln-Leu-Arg substrate is 5.5-6.5 with maximum activity at pH 6.5.[35] Recombinant TgCPB expressed in *E. coli* can self-cleave at pH 6.0,[30] but currently the optimal pH for TgCPB proteolytic activity is unknown. TgCPC1 is active within a broad pH range from 4.5 to 8.0 and shows the highest activity near pH 6.5.[39]

PHYLOGENETIC RELATIONSHIPS

A molecular phylogenetic analysis of the catalytic domains of papain-like proteases in apicomplexan parasites is shown in Figure 3. Cathepsin L- and B-like proteins are recognized by the presence of an "ERFNIN" motif in the prodomain or an occluding loop, respectively. Six members of the parvapain family in *Theileria parva* are all cathepsin L-like and are probably the earliest origin of the apicomplexan papain-like proteases. The occluding loop in cathepsin B may have been gained during evolution to grant it exopeptidase activity. Deletion of this loop by mutagenesis completely eliminates this activity.[43] Similarly, the N-terminal residual prodomain in cathepsin Cs (exclusion domain) may also have evolved from the prodomain of cathepsin L to associate with their catalytic heavy and light chains to contribute to their exclusive exopeptidase activity. The activation of human DPPI by cathepsin L and not cathepsin B may also indicate that cathepsin Cs are derived from cathepsin L and co-evolve with cathepsin B.[44] Among the apicomplexan parasites for which genome sequence is available, a cathepsin B-like protease only exists in *T. gondii*. The early ancestor of apicomplexan parasites, *Perkinsus marinus*[45] encodes ten cathepsin B-like proteases, implying that apicomplexans (except *T. gondii*) probably lost cathepsin B, perhaps due to redundancy with cathepsin L- and cathepsin C-like proteases.

LOCALIZATION, PHYSIOLOGICAL FUNCTIONS AND REGULATION

TgCPL is principally expressed within the VAC of tachyzoites and bradyzoites.[33-35] Its function in the VAC is not known, but since the VAC resembles a lysosome or lytic vacuole, TgCPL is proposed to function in protein degradation within this compartment. The VAC contains internal membrane tubules and vesicles typically seen in multivesicular bodies, a major center for membrane protein turnover. TgCPL is also associated with the residual body in the PV after cell division, where it could contribute to the destruction of mother cell organelles that are not partitioned into daughter cells.[33] Additional subpopulations of TgCPL are localised within small cytoplasmic vesicles throughout the cytoplasm and within the late endosomes (LE), which is characterized by the presence of TgRab7, vacuolar pyrophosphatase 1 (TgVP1) and immature microneme proteins also known as proMICs.[46-48] ProMICs display an N-terminal or internal propeptide that is proteolyzed within 15-60 minutes after synthesis.[33,49,50] Parussini et al found that TgCPL contributes to the proteolytic maturation of proTgM2AP and proTgMIC3 based on delayed maturation in a TgCPL-deficient strain and correct processing of recombinant proTgM2AP by recombinant TgCPL in

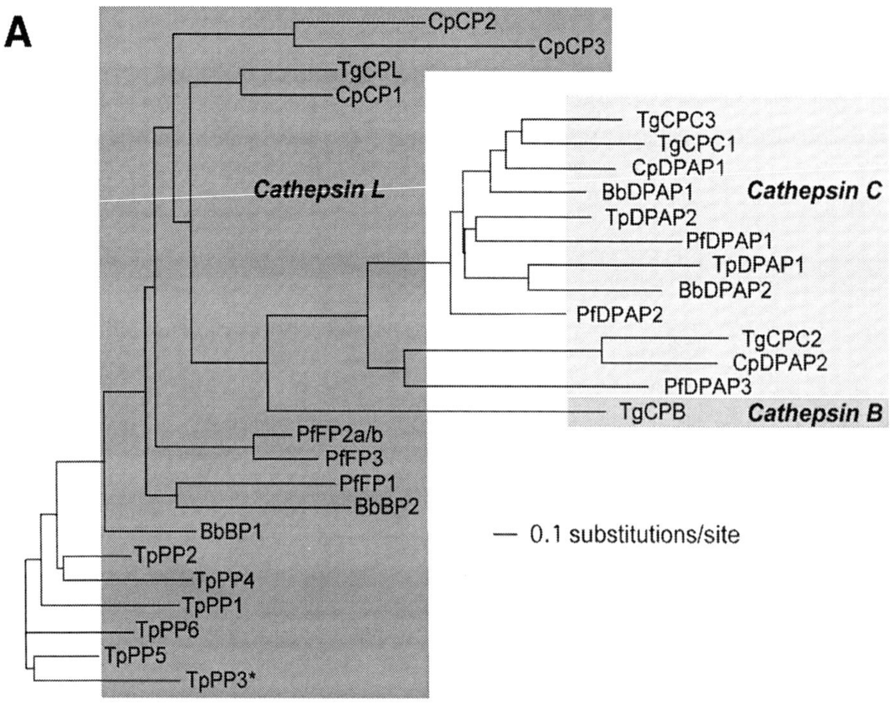

B

```
TgCPC3    TDGPHDTGA--CGSCYAVSSAYILQKRFEIMFS------  FNGWEYTNHAVAVVGWGEDEPDNA--  -TGKPKKFWVVRNTW-
BbDPAP1   NFQTFGQGQ--CGSCYAMAGIYVLTKRIEIMMK------  LNGWEYTNHAIAIVGWGEDEID----  --GIITKYWICKNTW-
CpDPAP1   SDFPFNQGN--CGSCYAVSSVYILAKRAELKLN------  LNGWEYTNHAIAIVGWGE-ENG----  -----IPYWIIRNSW-
TgCPC1    DETVTNQGS--CGSCYAVAATYALQKRFEIAASRMLGQE  LTGWEYTNHAVTIVGWGETDGEN---  --GKPQKYWIVRNTW-
TpDPAP2   DIYNTSQGE--CGSCYIYSSLYVISKRLQILFN------  -HGWEYTNHAIVVVGWGEELVN----  N----VKYWICKNTW-
PfDPAP2   EENVDDQKD--CGSCYSISSVYSLERRFEILF-------  FNGWQQTNHAINIVGWGEQIINQDK-  NNHKLVKYWIIRNTW-
TpDPAP1   NIKVVNQKT--CGGCYTIASLYVLHARFLIRIRELLKKP  LSGMEFTTHAVALIGWGETDEG----  -----FKYWVARNSW-
BbDPAP2   TIPIVNQDS--CGSCHAIASRYVLQSRFLIALERMSNRS  LTGWEYTSHAVAIVGWGQEKVGAR--  ----MIKYWICRNSW-
PfDPAP1   EYEVTNQLL--CGSCYIASQLYAFKRRIEVALTK-----  ITGWDRVNHAIVLLGWGEEEING---  ---KLYKYWIGRNSW-
TgCPC2    IPDAPNQGG--CGSCYAIATGTVLTSRLWIRYA------  NWVWEKVDHAVVISGWGWAKHGDS--  ----WLPYWKVRNSW-
CpDPAP2   IPEVFNQGD--CGDCFAVTAATIITSRLWIKYS------  SIPWHKVDHAMVITGWGWETYGNE--  ----RIPYWIVQNSW-
TgCPL     VTPVKDQRD--CGSCWAFSTTGALEGAHCAKTGK-----  ----TDLDHGVLLVGYG---------  ------DFWIMKNSW-
CpCP1     VNPIRNQKN--CGSCWAFSAVAALEGATCAQTNRG----  ----TKVNHGVVLVEYD---------  ------EYWLVRNSW-
PfFP2a/b  VTPVKDQKN--CGSCWAFSSIGSVESQYAIRKNK-----  ----DELNHAVMLVGFG---------  -----HYYYIIKNSW-
PfFP3     VTPVKDQAL--CGSCWAFSSVGSVESQYAIRKKA-----  ----AAPNHAVILVGYG---------  -----FYYYIIKNSW-
PfFP1     VHEPKDQGL--CGSCWAFASVGNIESVFAKKNKN-----  ----EELNHSVLLVGYGQVEKTKLNY  PDDNIIYYWIIKNSW-
BbBP2     MTPVKDQGN--CGSCWAFSLIGVAEPFFKHKRDI-----  ----DEINHAVVLAGVG---------  ------PFWLIKNSW-
TpPP3     -TYVKDQGT-LAPSDWAVTVIDSVESLFKIYNKER----  ----KSLNHYVLLVGEG---------  -------YFIVKNTLG
TpPP5     VSPVKDQGD-HCGSCWAFSSIGSVESLYRLY-KNK----  ----DALNHAVLLVGEG---------  -------YWIIKNSW-
TpPP2     VSKIKNQGL-ECGSCWAFASVSSVESLYKIY-RNV----  ----SALNHAVLLVGEG---------  -------YWVIKNSW-
TpPP6     VTKVKDQVMDECYSCWAFSTVASVESLYRIY-RDE----  ----HYPNHVVVLVGEG---------  -------YWVLKNSW-
TpPP4     GTRVKDQGLD-CSCCWAFASVAAVESVYQLL-QDD----  ----VNLNHSVLLVGEG---------  -------YWVIKNSW-
BbBP1     VTPVKDQGM--CGSCWAFAAVGSVESLLKR--QKT----  ----SELNHAVLLVGEG---------  -------YWLLKNSW-
TpPP1     -TSVKDQSN--CGGCWAFSTVGSVEGYYMSH-FDK----  ----KSLNHAVVLVGEG---------  -------YWVVQNSWG
CpCP2     VYIPFDQGK--CGGCYAFVVSASISISNCIQKLVL----  --THSELNHAVIMIGYG---------  -------YYVIQNSW-
CpCP3     ITFPDNQGK--CANCYNFSALSAIEGATCRQLGKR----  WCGTRQVDHAVTLIGYG---------  -------YYIMKNSW-
TgCPB     VGHVRDQGD--CGSCWAFASTEAFNDRLCIRSQGKR---  ---LPVGGHAIKIIGWGTENGE----  ------EYWHAVNSW-
PfDPAP3   LDDAIDQKD--CGSCYANSASFIINSRVRIKYNYIKN--  AYIWNKVDHAVVIVGWGEDTLPNFVK  TNNVIIKYWKVLNSW-
```

Figure 3. Sequence-based relationships among *T. gondii* cathepsins. A) Molecular phylogenetic relationships among cathepsin proteases in apicomplexan parasites. The tree was generated by neighbor-joining analysis using POWER (http://power.nhri.org.tw/power/home.htm). Subgroups are shaded according to their similarity to cathepsins based on homology and the presence of an "ERFNIN" motif (cathepsin L-like) or occluding loop (cathepsin B-like). The analysis was restricted to apicomplexan parasites with complete or nearly complete genome sequences. One species from each genera was selected based on the maximal genome sequence coverage. *Plasmodium* SERA proteins were excluded due to their substantial divergence. Figure legend continued on next page.

Figure 3, continued from previous page. The *Plasmodium* falcipains PfFP2a and PfFP2b are identical in sequence and thus are shown in the same dendrite. Note that the *Babesia* and *Theileria* proteases have not been systematically named previously and hence are designated here according to the nomenclature adopted for the *Plasmodium* and *Cryptosporidium* cathepsins. An asterisk indicates a protease missing at least one amino acid involved in catalysis. Abbreviations used: Bb, *Babesia bovis*; BP, Bovipain; Cp, *Cryptosporidium parvum*; CP, Cryptopain; Pf, *Plasmodium falciparum*; FP, falcipain; Tg, *Toxoplasma gondii*; Tp, *Theileria parva*; PP, parvapain. B) Multiple sequence alignment of apicomplexan cathepsin catalytic-proximal sequences. Sequences obtained from the MEROPS database (http://merops.sanger.ac.uk) include only the catalytic domain beginning six amino acids upstream of the catalytic cysteine and ending two amino acids downstream of the active site asparagine. Sequence alignment was compiled using ClustalW (http://www.ebi.ac.uk/Tools/clustalw2/index.html). Fully conserved residues are indicated by asterisks and highly similar and similar residues are indicated by two dots and one dot, respectively. TpPP3 lacks the conserved cysteine residue in the catalytic site, which is replaced with an alanine.

vitro.[33] Moreover, the propeptide cleavage sites of proTgM2AP and proTgMIC3 contain residues favorable for TgCPL recognition and proteolysis based on screening of a peptide substrate library and mapping of the autocatalytic cleavage site of recombinant TgCPL. Maturation of proTgM2AP and proTgMIC3 was not completely abolished in TgCPL-deficient parasites, suggesting the existence of alternative maturase(s) within the *T. gondii* endocytic pathway.

TgCPL proteolytic activity is regulated in several ways. First, like most papain-like enzymes, TgCPL is initially synthesized as an inactive zymogen with the propeptide sterically occluding the active site cleft.[32] This arrangement probably prevents TgCPL activity as it traffics through the proximal secretory system (ER, Golgi), thereby avoiding proteolysis of inappropriate substrates. Second, TgCPL activity is pH regulated, with optimal activity occurring under moderately acidic conditions (pH 5.0-6.5),[33,35] which TgCPL probably encounters as it enters the endocytic system and undergoes maturation to remove the autoinhibitory propeptide. Third, *T. gondii* expresses two inhibitors of cysteine proteases termed toxostatin-1 and toxostatin-2.[35] Overexpression of toxostatin-1 in *T. gondii* tachyzoites inhibited TgCPL and TgCPB activity by 80-90% in parasite extracts. Although it remains unclear if this inhibitor encounters TgCPL or TgCPB within intact cells, toxostatin-1 and toxostatin-2 could serve to prevent unwanted proteolysis by TgCPL or TgCPB that strays to an inappropriate location within the parasite. Finally, TgCPL processing of its substrates is likely regulated spatially by membrane segregation via delivery of substrates to the VAC or shuffling diminutive quantities of TgCPL to the LE for limited proteolysis of proMICs and possibly other substrates.[33]

TgCPB expression and activity have been detected in tachyzoites. It remains uncertain whether TgCPB is expressed in bradyzoite tissue cysts since its transcript was not detected by RT-PCR in cysts recovered from mouse brains,[35] but was shown by microarray analysis to increase upon induction of bradyzoite differentiation in vitro (www.toxodb.org). TgCPB occupies the rhoptries based on immunoelectron and immunofluorescence microscopical studies,[30] although it was not identified in the rhoptry proteome.[51] TgCPB was also detected in the residual body and an electron lucent vacuole that is probably the VAC,[30] raising the possibility that TgCPB works in concert with TgCPL at these locations. Parasites treated with a cathepsin inhibitor showed impaired invasion, altered rhoptry morphology and delayed maturation of TgROP2, implicating TgCPB in ROP protein maturation and parasite invasion.[30] An antisense inhibition approach was used to knockdown the expression of TgCPB, resulting in decreases in parasite cell invasion, replication and parasite tissue burden in a chicken embryo model of infection, indicating that TgCPB contributes to multiple processes during infection.[30]

TgCPC1 and TgCPC2 are expressed in tachyzoites, with TgCPC1 transcripts being ~20-fold more abundant than TgCPC2.[39] None of the TgCPC transcripts including TgCPC3 were detected in bradyzoite tissue cysts. TgCPC3 is expressed in the sporozoite stage based on the presence of expressed sequence tags (www.toxodb.org). TgCPC1 and TgCPC2 have been reported to occupy DGs and the PV during parasite replication.[39] Partial gene disruption of TgCPC1 by replacement of exons 3-5 with a selectable marker resulted in ~4-fold upregulation of TgCPC2 mRNA, suggesting that TgCPC2 compensates for the loss of TgCPC1 and that these proteases might function in the same process or pathway.[39] A selective inhibitor of cathepsin C partially impaired parasite replication and reduced parasite tissue burden in experimentally infected chicken embryos, implicating TgCPCs in parasite replication.[39] The inhibitor also stabilized the expression of *E. coli* β-lactamase within the parasitophorous vacuole of transgenic parasites, suggesting that TgCPCs function in the degradation of exogenous proteins. Proteins derived from the host endoplasmic reticulum were recently shown to be present within the lumen of the PV in infected dendritic cells, indicating that such proteins are available to contribute to the parasites nutritional needs.[52]

THERAPEUTIC POTENTIAL

T. gondii cathepsins are considered potential therapeutic targets based on genetic and inhibitor studies. For example, genetic disruption of TgCPL diminishes parasite cell invasion and growth (ref.[33] and Dou and Carruthers, unpublished data). Also, parasite treatment with the cathepsin inhibitor morpholinurea-leucyl-homophenyl-vinyl sulfone phenyl (LHVS, also known as K11017) impairs cell invasion by blocking secretion of adhesive proteins from parasite micronemes.[31] LHVS principally targets TgCPL based on analysis with a fluorescent derivative of LHVS,[32] but our recent findings suggest that it can also inhibit TgCPB (Dou and Carruthers, unpublished data). Antisense inhibition of TgCPB expression or treatment with cathepsin inhibitors diminished parasite replication, cell invasion and infection in vivo.[30,53] Similarly, targeted deletion or chemical inhibition of TgCPC1 reduced parasite replication and infection.[39] Nevertheless, it should be noted that none of the *T. gondii* cathepsins have been validated as essential enzymes and the cathepsin inhibitors that have been tested to date show relatively low potency with effective concentrations in the low- to mid-micromolar range.[30,39,53,54] It remains unclear whether this is due to poor penetration into *T. gondii* infected cells or partial refractivity to inhibition of the parasite cathepsins within the intracellular environment. Additional genetic ablation studies should provide a clearer picture of the importance of these proteases and their potential for therapeutic development.

CONCLUSION AND FUTURE PERSPECTIVES

The emerging view of *T. gondii* cathepsins is that, like their homologs in other eukaryotes, they function in protein degradation along with playing more specialized roles in the maturation of invasion proteins. However, much remains to be done including the identification of their full range of protein substrates within the parasite, their dependency on one another for activation and their participation in similar or distinct processes within the endocytic system and parasitophorous vacuole. Additional reverse genetic evidence that TgCPB and the TgCPCs are important to parasite survival would further boost their stock as potential targets. The screening, identification and target validation of small molecule inhibitors with greater potency will also be an important avenue of future work. Recent advances in gene tagging for marker identification should illuminate additional features of the *T. gondii* endosomal system and help identify the pathways taken by cathepsins to reach their subcellular locations.[55] This goal will also be facilitated by the use of conditionally expressed dominant negative mutants of membrane trafficking determinants.[56,57] Finally, now that evidence is emerging showing that host ER proteins gain access to the PV, studies of parasite endocytic uptake within infected cells might shed light on whether *T. gondii*, like its kin the malaria parasite, taps host proteins for nutritional gain.

ACKNOWLEDGEMENTS

We thank My-Hang Huynh for critically reading the manuscript. Work in the Carruthers lab is supported by grants R01AI063263 (V.B.C.) from the National Institutes of Health (USA) and grant 04R-796 (V.B.C.) from the Stanley Medical Research Institute (USA).

REFERENCES

1. Lopez-Otin C, Bond JS. Proteases: Multifunctional enzymes in life and disease. J Biol Chem 2008; 283:30433-30437.
2. Rawlings ND, Barrett AJ, Bateman A. MEROPS: The peptidase database. Nucleic Acids Res 2010; 38:D227-D233.
3. Barrett AJ, Kirschke H. Cathepsin B et al. Methods Enzymol 1981; 80 Pt C:535-561.
4. Kirschke H, Barrett AJ, Rawlings ND. Lysosomal Cysteine Proteases. 2nd ed. Oxford ; New York: Oxford University Press; 1998.
5. Guicciardi ME, Deussing J, Miyoshi H et al. Cathepsin B contributes to TNF-alpha-mediated hepatocyte apoptosis by promoting mitochondrial release of cytochrome c. J Clin Invest 2000; 106:1127-1137.
6. Barrett AJ, Rawlings ND, Woessner JF. Handbook of Proteolytic Enzymes. San Diego: Academic Press; 2004.
7. Denhardt DT, Greenberg AH, Egan SE et al. Cysteine proteinase cathepsin L expression correlates closely with the metastatic potential of H-ras-transformed murine fibroblasts. Oncogene 1987; 2:55-59.
8. Frade R, Rodrigues-Lima F, Huang S et al. Procathepsin-L, a proteinase that cleaves human C3 (the third component of complement), confers high tumorigenic and metastatic properties to human melanoma cells. Cancer Res 1998; 58:2733-2736.
9. Amuthan G, Biswas G, Zhang SY et al. Mitochondria-to-nucleus stress signaling induces phenotypic changes, tumor progression and cell invasion. EMBO J 2001; 20:1910-1920.
10. Dluzewski AR, Rangachari K, Wilson RJM et al. Plasmodium falciparum: Protease inhibitors and inhibition of erythrocyte invasion. Exp Parasitol 1986; 62:416-422.
11. Rosenthal PJ, McKerrow JH, Aikawa M et al. A malarial cysteine proteinase is necessary for hemoglobin degradation by Plasmodium falciparum. J Clin Invest 1988; 82:1560-1566.
12. Rosenthal PJ, McKerrow JH, Rasnick D et al. Plasmodium falciparum: Inhibitors of lysosomal cysteine proteinases inhibit a trophozoite proteinase and block parasite development. Mol Biochem Parasitol 1989; 35:177-183.
13. Rosenthal PJ, Wollish WS, Palmer JT et al. Antimalarial effects of peptide inhibitors of a Plasmodium falciparum cysteine proteinase. J Clin Invest 1991; 88:1467-1472.
14. Rosenthal PJ. Plasmodium falciparum: Effects of proteinase inhibitors on globin hydrolysis by cultured malaria parasites. Exp Parasitol 1995; 80:272-281.
15. Rosenthal PJ, Olson JE, Lee GK et al. Antimalarial effects of vinyl sulfone cysteine proteinase inhibitors. Antimicrob Agents Chemother 1996; 40:1600-1603.

16. Greenbaum DC, Baruch A, Grainger M et al. A role for the protease falcipain 1 in host cell invasion by the human malaria parasite. Science 2002; 298:2002-2006.
17. Sijwali PS, Rosenthal PJ. Gene disruption confirms a critical role for the cysteine protease falcipain-2 in hemoglobin hydrolysis by Plasmodium falciparum. Proc Natl Acad Sci USA 2004; 101:4384-4389.
18. Eksi S, Czesny B, Greenbaum DC et al. Targeted disruption of Plasmodium falciparum cysteine protease, falcipain 1, reduces oocyst production, not erythrocytic stage growth. Mol Microbiol 2004; 53:243-250.
19. Caffrey CR, Hansell E, Lucas KD et al. Active site mapping, biochemical properties and subcellular localization of rhodesain, the major cysteine protease of Trypanosoma brucei rhodesiense. Mol Biochem Parasitol 2001; 118:61-73.
20. Abdulla MH, O'Brien T, Mackey ZB et al. RNA interference of Trypanosoma brucei cathepsin B and L affects disease progression in a mouse model. PLoS Negl Trop Dis 2008; 2:e298.
21. Grab DJ, Garcia-Garcia JC, Nikolskaia OV et al. Protease activated receptor signaling is required for african trypanosome traversal of human brain microvascular endothelial cells. PLoS Negl Trop Dis 2009; 3:e479.
22. Mackey ZB, O'Brien TC, Greenbaum DC et al. A cathepsin B-like protease is required for host protein degradation in Trypanosoma brucei. J Biol Chem 2004; 279:48426-48433.
23. O'Brien TC, Mackey ZB, Fetter RD et al. A parasite cysteine protease is key to host protein degradation and iron acquisition. J Biol Chem 2008; 283:28934-28943.
24. Montoya JG, Liesenfeld O. Toxoplasmosis. Lancet 2004; 363:1965-1976.
25. Dobrowolski JM, Sibley LD. Toxoplasma invasion of mammalian cells is powered by the actin cytoskeleton of the parasite. Cell 1996; 84:933-939.
26. Carruthers VB. Proteolysis and Toxoplasma invasion. Int J Parasitol 2006; 36:595-600.
27. Hakansson S, Charron AJ, Sibley LD. Toxoplasma evacuoles: A two-step process of secretion and fusion forms the parasitophorous vacuole. EMBO J 2001; 20:3132-3144.
28. VK, Qi H, Beckers CJ et al. The protozoan parasite Toxoplasma gondii targets proteins to dense granules and the vacuolar space using both conserved and unusual mechanisms. J Cell Biol 1998; 141:1323-1333.
29. Shaw MK, He CY, Roos DS et al. Proteasome inhibitors block intracellular growth and replication of Toxoplasma gondii. Parasitology 2000; 121 (Pt 1):35-47.
30. Que X, Ngo H, Lawton J et al. The cathepsin B of Toxoplasma gondii, toxopain-1, is critical for parasite invasion and rhoptry protein processing. J Biol Chem 2002; 277:25791-25797.
31. Teo CF, Zhou XW, Bogyo M et al. Cysteine protease inhibitors block Toxoplasma gondii microneme secretion and cell invasion. Antimicrob Agents Chemother 2007; 51:679-688.
32. Larson ET, Parussini F, Huynh MH et al. Toxoplasma gondii cathepsin l is the primary target of the invasion inhibitory compound LHVS. J Biol Chem 2009.
33. Parussini F, Coppens I, Shah PP et al. Cathepsin L occupies a vacuolar compartment and is a protein maturase within the endo/exocytic system of Toxoplasma gondii. Mol Microbiol 2010; in press.
34. Miranda K, Pace DA, Cintron R et al. Characterization of a novel organelle in Toxoplasma gondii with similar composition and function to the plant vacuole. Mol Microbiol 2010.
35. Huang R, Que X, Hirata K et al. The cathepsin L of Toxoplasma gondii (TgCPL) and its endogenous macromolecular inhibitor, toxostatin. Mol Biochem Parasitol 2009; 164:86-94.
36. Nichols BA, Chiappino ML, Pavesio CE. Endocytosis at the micropore of Toxoplasma gondii. Parasitol Res 1994; 80:91-98.
37. Botero-Kleiven S, V F, Lindh J et al. Receptor-mediated endocytosis in an apicomplexan parasite (Toxoplasma gondii). Exp Parasitol 2001; 98:134-144.
38. Turk D, Janjic V, Stern I et al. Structure of human dipeptidyl peptidase I (cathepsin C): Exclusion domain added to an endopeptidase framework creates the machine for activation of granular serine proteases. EMBO J 2001; 20:6570-6582.
39. Que X, Engel JC, Ferguson D et al. Cathepsin cs are key for the intracellular survival of the protozoan parasite, Toxoplasma gondii. J Biol Chem 2007; 282:4994-5003.
40. Fruton JS, Mycek MJ. Studies on beef spleen cathepsin C. Arch Biochem Biophys 1956; 65:11-20.
41. McDonald JK, Reilly TJ, Zeitman BB et al. Cathepsin C: A chloride-requiring enzyme. Biochem Biophys Res Commun 1966; 22:771-775.
42. Storch S, Pohl S, Braulke T. A dileucine motif and a cluster of acidic amino acids in the second cytoplasmic domain of the batten disease-related CLN3 protein are required for efficient lysosomal targeting. J Biol Chem 2004; 279:53625-53634.
43. Illy C, Quraishi O, Wang J et al. Role of the occluding loop in cathepsin B activity. J Biol Chem 1997; 272:1197-1202.
44. Dahl SW, Halkier T, Lauritzen C et al. Human recombinant pro-dipeptidyl peptidase I (cathepsin C) can be activated by cathepsins L and S but not by autocatalytic processing. Biochemistry 2001; 40:1671-1678.
45. Leander BS, Clopton RE, Keeling PJ. Phylogeny of gregarines (Apicomplexa) as inferred from small-subunit rDNA and beta-tubulin. Int J Syst Evol Microbiol 2003; 53:345-354.
46. Harper JM, Huynh MH, Coppens I et al. A cleavable propeptide influences Toxoplasma infection by facilitating the trafficking and secretion of the TgMIC2-M2AP invasion complex. Mol Biol Cell 2006; 17:4551-4563.

47. Brydges SD, Harper JM, Parussini F et al. A transient forward-targeting element for microneme-regulated secretion in Toxoplasma gondii. Biol Cell 2008; 100:253-264.
48. El Hajj H, Papoin J, Cerede O et al. Molecular signals in the trafficking of Toxoplasma gondii protein MIC3 to the micronemes. Eukaryot Cell 2008; 7:1019-1028.
49. Rabenau KE, Sohrabi A, Tripathy A et al. TgM2AP participates in Toxoplasma gondii invasion of host cells and is tightly associated with the adhesive protein TgMIC2. Mol Microbiol 2001; 41:537-547.
50. Brydges SD, Sherman GD, Nockemann S et al. Molecular characterization of TgMIC5, a proteolytically processed antigen secreted from the micronemes of Toxoplasma gondii. Mol Biochem Parasitol 2000; 111:51-66.
51. Bradley PJ, Ward C, Cheng SJ et al. Proteomic analysis of rhoptry organelles reveals many novel constituents for host-parasite interactions in Toxoplasma gondii. J Biol Chem 2005; 280:34245-34258.
52. Goldszmid RS, Coppens I, Lev A et al. Host ER-parasitophorous vacuole interaction provides a route of entry for antigen cross-presentation in Toxoplasma gondii-infected dendritic cells. J Exp Med 2009; 206:399-410.
53. Que X, Wunderlich A, Joiner KA et al. Toxopain-1 is critical for infection in a novel chicken embryo model of congenital toxoplasmosis. Infect Immun 2004; 72:2915-2921.
54. Shaw MK, Roos DS, Tilney LG. Cysteine and serine protease inhibitors block intracellular development and disrupt the secretory pathway of Toxoplasma gondii. Microbes Infect 2002; 4:119-132.
55. Huynh MH, Carruthers VB. Tagging of endogenous genes in a Toxoplasma gondii strain lacking Ku80. Eukaryotic Cell 2009; 8:530-539.
56. Agop-Nersesian C, Naissant B, Ben Rached F et al. Rab11A-controlled assembly of the inner membrane complex is required for completion of apicomplexan cytokinesis. PLoS Pathog 2009; 5:e1000270.
57. Breinich MS, Ferguson DJ, Foth BJ et al. A dynamin is required for the biogenesis of secretory organelles in Toxoplasma gondii. Curr Biol 2009; 19:277-286.

CHAPTER 5

ENTAMOEBA HISTOLYTICA CATHEPSIN-LIKE ENZYMES

Interactions with the Host Gut

Vanessa Kissoon-Singh, Leanne Mortimer and Kris Chadee*

Gastrointestinal Research Group, University of Calgary, Calgary, Alberta, Canada

**Corresponding Author: Kris Chadee—Email: kchadee@ucalgary.ca*

Abstract: Cysteine proteases of the protozoan parasite *Entamoeba histolytica* are key virulence factors involved in overcoming host defences. These proteases are cathepsin-like enzymes with a cathepsin-L like structure, but cathepsin-B substrate specificity. In the host intestine, amoeba cysteine proteases cleave colonic mucins and degrade secretory immunoglobulin (Ig) A and IgG rendering them ineffective. They also act on epithelial tight junctions and degrade the extracellular matrix to promote cell death. They are involved in the destruction of red blood cells and the evasion of neutrophils and macrophages and they activate pro-inflammatory cytokines IL-1β and IL-18. In short, amoeba cysteine proteases manipulate and destroy host defences to facilitate nutrient acquisition, parasite colonization and/or invasion. Strategies to inhibit the activity of amoeba cysteine proteases could contribute significantly to host protection against *E. histolytica.*

INTRODUCTION

Cathepsins are globular lysosomal proteases that are involved in protein turnover within the cell.[1] Protein turnover occurs as part of normal cellular function, as is the case with degradation of intracellular proteins or those taken up from other cells. In some disease states such as cancer however, degradation of extracellular matrix proteins facilitates spread of disease.[1,2] In this chapter we examine the cathepsin-like enzymes of *Entamoeba histolytica* and their role in the pathogenesis of amoebiasis.

Cysteine Proteases of Pathogenic Organisms, edited by Mark W. Robinson and John P. Dalton.

The protozoan parasite, *E. histolytica* infects an estimated 500 million people each year, resulting in over 100,000 deaths.[3] Amoebiasis is the second leading cause of mortality due to a protozoan infection—second only to malaria.[4] While the vast majority of cases are asymptomatic,[3,5] 10% result in invasive disease. In the developing world, amoebiasis is one of the leading causes of childhood diarrhoea causing death.[4,6] Amoeba cathepsin-like proteins (cysteine proteases) are prominent virulence factors contributing to the pathogenesis of invasive amoebiasis.[7]

Life Cycle of *E. histolytica*

Humans, the definitive hosts of *E. histolytica*, become infected when they ingest cysts in contaminated food or water.[4] Once ingested, these excyst in the terminal ileum releasing trophozoites, which then migrate to and colonize the colon. In asymptomatic infections, trophozoites colonize the intestinal mucus layer and behave as harmless commensals. The life-cycle is completed when trophozoites encyst and are excreted in stool to perpetuate transmission via fecal-oral spread.[5,8] In invasive amoebiasis, trophozoites invade the mucosal barrier, resulting in amoebic dysentery and/or colitis. In some cases, amoebae enter the portal circulation and disseminate to extra-intestinal sites where they can establish infections such as amoebic liver abscesses (ALA) and/or brain abscesses.

E. histolytica Virulence Factors

In addition to cysteine proteases (CPs), the invasive potential of *E. histolytica* is mediated by a number of virulence factors including the galactose-*N*-acetyl-*D*-galactosamine (Gal/GalNAc) lectin and amoebapores.[4,9-11] The Gal/GalNAc lectin on the parasite surface, binds with high affinity to galactose and *N*-acetyl galactosamine residues on colonic mucins to facilitate colonization of the mucus layer and invasion of the large intestine,[12-14] as well as mediating parasite adherence to host epithelial cells.[10,15,16] Amoebapores are small pore forming molecules with a molecular mass of approximately 8 kDa that are preferentially active at a lower pH.[17] They are nonenzymatic and insert into the membrane of target cells forming pores, that result in cytolysis of both bacterial and eukaryotic cells.[17-19] Other virulence factors include phospholipases and collagenase,[20] EhSTIRP (*E. histolytica* serine-, threonine- and isoleucine-rich protein) and lipophosphopeptidoglycan (LPPS) involved in epithelial cell adhesion and cytotoxicity,[21] the SREHP (serine-rich *E. histolytica* protein) that facilitates binding and phagocytosis of apoptotic cells,[22] Eh arginase which impairs nitric oxide (NO) production by activated macrophages, a superoxide dismutase that detoxifies superoxide to hydrogen peroxide (H_2O_2) and perioxiredoxin which neutralizes H_2O_2.[4,9] While a complete review of *E. histolytica* virulence factors is beyond the scope of this chapter, an excellent review is provided in reference 4.

EhCPs are the major proteases of *E. histolytica* and are expressed both intracellularly and extracellularly.[23] They are key virulence factors in the pathogenesis of invasive amoebiasis[16,24-27] and degrade extracellular matrix proteins,[28] human immunoglobulins IgG and secretory IgA[29] and components of the complement system.[30,31] EhCPs also cleave colonic mucins[32] and cause cell cytolysis,[14] they trigger inflammatory responses, directly destroy tissue and mediate evasion of host defences.[16] CP activity is associated with parasite virulence and the highly virulent HM-1 strain of *E. histolytica* has elevated protease activity and cytotoxicity compared to less virulent strains.[33]

E. HISTOLYTICA CYSTEINE PROTEASES

EhCPs are referred to as cathepsin-like enzymes because their structure is similar to cathepsin-L, however their substrate specificity resembles cathepsin-B.[25,34] All amoebic CPs possess the structural component **Glu**-X_3-**Arg**-X_2-**Ile/Val**-**Phe**-X_2-**Asn**-X_3-**Ile**-X_3-**Asn** (ERFNIN) motif in the pro-sequence.[34-36] In addition, some of these proteases also possess a hydrophobic sequence near the C-terminus which facilitates binding to the extracellular membrane of the trophozoite.[34] The substrate specificity for EhCPs requires two large positively charged amino acids, specifically arginine at the P1 and P2 positions[35] and bind with high affinity to the cathepsin-B synthetic substrate benzyloxycarbonyl–arginine–arginine–4-amino-7-methylcoumarin (Z–Arg–Arg–AMC), but have a low affinity for Z–Arg–AMC or Z–Phe–Arg–AMC, which are substrates for cathepsin H and L respectively.[37]

There is a great deal of ambiguity regarding the nomenclature of EhCPs since they are often referred to as proteases, mucinases or hemoglobinases. For the purposes of this discussion, an EhCP will be defined as a protease that is activated by the CP activators dithiothreitol (DTT) or 2-mercaptoethanol and is inhibited by CP inhibitors and not serine protease inhibitors like phenylmethylsulfonyl fluoride.[37] The most common CP inhibitor is E-64 (L-*trans*-epoxysuccinyl-leucylamido (4-guanidino) butane), but others described in the literature include: *N*-ethylmaleimide (NEM), para-hydroxymercuribenzoate (p-HMB), p-chloromercuribenzoate (p-CMB), α-2 macroglobuline, *N*-tosyl-*L*-lysine chloromethylketone (TLCK), *N*-tosyl-phenylalanine chloromethyl ketone (TPCK), benzyloxycarbonyl-Phe-Phe fluoromethylketone (z-FF.fmk), benzyloxy-carbonyl-Phe-Ala fluoromethylketone (z-FA.fmk) and iodoacetamide.[38] Studies have also identified endogenous *E. histolytica* CP inhibitors, including amoebiasin 1 (EhICP1)[39] and the chagasin-like inhibitor EhICP2.[40]

Recent developments in the *E histolytica* genome project have uncovered over 40 genes encoding CPs, which are moderately expressed, though only 8 of these are expressed by amoeba in culture[34,41-43] and fewer still are expressed at significant levels.[10] Those not expressed in vitro are thought to play a role in infection of the human host, invasion and destruction of host tissue and completion of the parasite life cycle.[25,40] Encystation and excystation, as well as amoebic growth and survival and processes enabling the digestion of red blood cells and bacteria are also thought to involve EhCPs.[34,44-47]

Structure and Function of *E. histolytica* Cysteine Proteases

EhCPs cleave proteins through the hydrolysis of peptide bonds. CPs possess a cysteine residue in the active site whose thiol group binds the substrate to facilitate catalysis,[23] where the thiol acts as a nucleophile donating its electrons to a histidine residue which acts as a general base.[1] They are typically active in most environments over a range of pHs (pH 5.5-7.5) as long as the thiol group is not reduced.[23]

CPs are synthesized as a polypeptide consisting of a hydrophobic presequence (12-20 amino acids), a pro-domain (55-148 amino acids) and a catalytic domain or mature enzyme (109-488 amino acids) (Fig. 1).[25,34] The pro-domain maintains the protease in an inactive state and acts as a template during translation.[25] The enzyme is activated when the pro-domain loses its configuration via cleavage by another protease or exposure to an acidic environment, such as the lysosome.[1]

EhCPs are categorized as EhCP-A or EhCP-B.[34] Proteases of the EhCP-A subfamily typically have shorter pro-domains (72-90 residues) and shorter catalytic domains

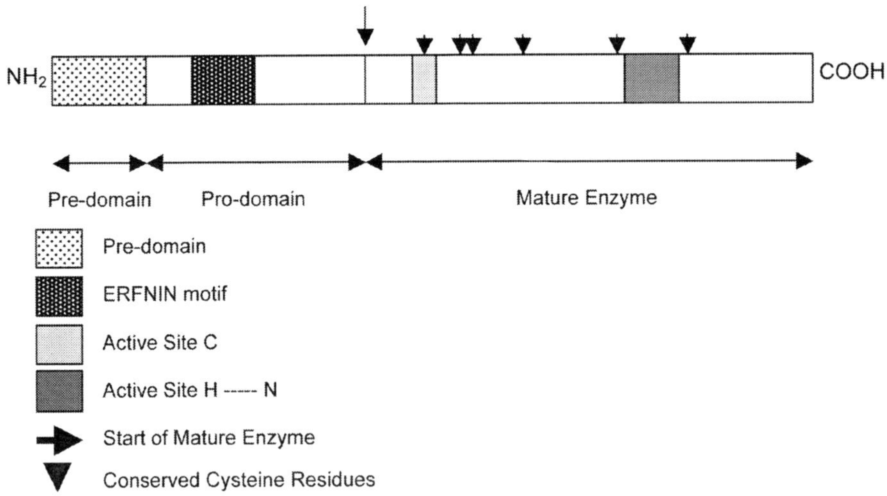

Figure 1. Structure of *Entamoeba histolytica* cysteine proteases. Amoeba cysteine proteinases are synthesized with both a pre- and a pro-domain that are removed to activate the mature enzyme. The predomain is important for translation and trafficking, while the pro-domain maintains the protein in an inactive state. In the lyosome, the pH facilitates dissociation of the pro-domain to leave the mature enzyme. Adapted from references 34, 37 and 124.

(190-254 residues) compared to those of EhCP-B (105-144 residues and 230-353 residues) respectively.[34] There are a few exceptions however. EhCP15 has the shortest pro-domain (55 residues), but it is in the EhCP-B group while EhCP17 with the longest pro-domain (148 residues) is in the EhCP-A group.[34] EhCP-A proteases have a conserved DWR motif (Asp-Trp-Arg) at position 6-8 with an active site cysteine at position 25, while in EhCP-B this is replaced with a cysteine residue at position 8 and active site cysteines at positions 30-38.[34] The catalytic centre of CPs consists of four active site residues: glycine (Gly), cysteine (Cys), histidine (His) and asparagine (Asn).[34]

CPs range in size from 16 to 66kDa.[23] The molecular mass of the twelve mature proteases expressed in vitro range between 24-35 kDa.[34] The three genes *ehcp1*, *ehcp2* and *ehcp5* are responsible for 90% of the total CP activity in *E. histolytica* in vitro.[48] EhCP1 and EhCP2 are membrane bound, while EhCP3 is localized to the cytoplasm.[25] EhCP5 and EhCP112 possess RGD (Arg-Gly-Asp) sequences that anchor them to the membrane,[34] without the need for a transmembrane-spanning domain or a GPI anchor. This allows the proteins to be localized to putative patches on the surface of amoeba.[49,50] The 10 best-known EhCPs are described in greater detail in Table 1.

Role of Cysteine Proteases in Parasite Survival

Studies have suggested that the role of EhCPs is primarily related to survival and not tissue destruction.[9,51] Amoeba CPs appear to aid in protein breakdown prior to ingestion of bacteria or red blood cells (RBC) and upon bacterial phagocytosis, EhCP2, EhCP3 and the 112kDa surface protein are targeted to the phagocytic vesicle, suggesting a role for these in nutrient absorption.[25,49] Furthermore, inhibition of CPs drastically impairs growth of trophozoites,[51] as well as parasite survival and nutrient metabolism.[9] Taken

Table 1. *Entamoeba histolytica* cysteine proteases structure and function

Cysteine Protease	Location	Structure	Function	References
EhCP1 (Amoebapain)	– Found on the surface of amoebic trophozoite –Not found in *E. dispar*	– Very similar to *E. histolytica* CP2 in structure and size	– EhCP1, EhCP2 and EhCP5 contribute 90% of protease activity	48,50
EhCP2 (ACP2) (histolysain)	– On the cell surface as well as internal membranes in the parasite – Membrane-associated – Localized to the surface of *E. dispar*	– Similar to *E. histolytica* CP1 in structure and size	– EhCP1, EhCP2 and EhCP5 contribute 90% of protease activity – After phagocytosis of erythrocytes, colocalized with EhCP3 to phagocytic vesicle – Over expression of EhCP2 in *E. dispar* increased tissue monolayer destruction in vitro – Responsible for approximately 30% of total CP activity	25,48,96
EhCP3 (ACP1)	– Located in cytoplasmic granules –Analog present in *E. dispar*	– Smaller than EhCP1 or EhCP2, but same general structure	– After phagocytosis of erythrocytes, colocalized with EhCP2 to phagocytic vesicle – Not involved in *E. histolytica* virulence	25,48,125
EhCP4	– Analog of EhCP exists in *E. dispar*	– Smaller than EhCP1 or EhCP2, but same general structure – Similar in structure and size to *ehcp3*	– Substrate not known – Functional gene not present in either *E. histolytica* or *E. dispar* – No expression of ehcp4 in vitro	34,48,124
EhCP5	– Punctuate patches on amoeba surface – Also localized to intra–cellular compartments – Not found in *E. dispar*	– RGD domain present in the pro-domain of the peptide – Contains a stretch of hydrophobic amino acids for association with membrane – Similar in structure and size to EhCP1	– EhCP1, EhCP2 and EhCP5 contribute 90% of protease activity – Play a signalling role in development of ALA – Associated with trophozoite membrane and thought to play a role in tissue destruction – Pcp5 induces NF-κB mediated proinflammatory response	48,24,126, 50,98

continued on next page

Table 1. Continued

Cysteine Protease	Location	Structure	Function	References
EhCP6	– Analog of EhCP exists in *E. dispar*	– Similar in structure and size to EhCP5, but lacking the RGD motif – Contains a potential N-glycosylation site	– Substrate not known – Functional gene not present in either *E. histolytica* or *E. dispar* – No expression of ehcp6 detected in vitro	34,48,124
EhCP7		– Very large compared to EhCP1-6 – Contains 2 potential glycosylation sites and a PCN C-domain	– Involved in erythro-phagocytosis	34,102
EhCP8	– Analog of EhCP exists in *E. dispar*	– Similar in structure and size to EhCP1 or EhCP2 – Contains one potential N-glycosylation site	– Substrate not known – Functional gene present in both *E. histolytica* and *E. dispar*	124
EhCP9	– Analog of EhCP exists in *E. dispar*	– Similar in structure and size to EhCP1 or EhCP2	– Substrate not known – Functional gene present in *E. histolytica* but not in *E. dispar*	124
EhCP112	– Localized to the surface of trophozoites – Moves to phagocytic vacuole in RBC phagocytosis, but is recycled to the plasma membrane	– Adds to EhADH112 to form *E. histolytica* adhesin (EhCPADH) involved in trophozoite virulence – Possesses an RGD motif	– Disrupts epithelial monolayers – Cleaves extracellular proteins – Involved in adherence, phagocytosis and tissue destruction – Recombinant CP112 digests in vitro gelatin, Type I collagen, fibronectin and haemoglobin	49,122, 123

together, these data suggest a primary role for CPs in nutrient acquisition, though since they are located on the surface or released extracellularly, they may also be involved in host invasion.[25]

Cysteine Proteases in the Pathogenesis of Invasive Amoebiasis

The role of EhCPs in parasite virulence and invasive disease is underscored by the relative absence of these proteins in the morphologically identical but noninvasive species, *Entamoeba dispar*.[48] Clinical isolates from *E. histolytica* have 10-1,000 times more CP activity in culture than isolates from *E. dispar*.[26] Amoebae expressing the highest levels of EhCP5 were obtained from patients with amoebic liver abscess (ALA).[52] Amoeba virulence has been associated with CP activity in trophozoite extracts[33] and ALA formation was inhibited in the presence of CP inhibitors.[27] Furthermore, CP mRNA levels were significantly higher in more virulent isolates of amoeba[53] and the HM-1 strain was found to have more CP activity than the less virulent HK-9 strain.[54,55] Thus, amoeba CPs are responsible for both parasite survival and nutrient acquisition, as well as causing intestinal inflammation, tissue damage in colitis and extra-intestinal infections.[4,14,56]

HOST DEFENCE AND AMOEBA CYSTEINE PROTEASES

Amoebae CPs are critical in invasion due their ability to dismantle both innate and adaptive host defence mechanisms (Figs. 2 and 3). EhCPs cleave a wide range of host molecules that result in de-polymerization of the mucus layer, destruction of the colonic epithelium with its accompanying tight junctions, inactivation of immunoglobulins and either activation or inactivation of components of the complement cascade.

Cleavage of the Mucus Layer

The intestinal mucus layer forms the first line of host defence against enteric pathogens and is largely composed of the secreted mucin MUC2.[57-59] Suspended in this layer are a number of defence mediators including trefoil peptides (TFF3) and resistin-like molecule beta (RELM-β) secreted by goblet cells, as well as secretory IgA and antimicrobial peptides such as defensins, lysozyme and cathelicidins,[60-63] which collectively form a formidable chemical-physical barrier. In the intestinal lumen, polymeric MUC2 is hydrated to form a gel-like layer that protects intestinal epithelial cells from direct contact with commensal bacteria and pathogenic micro-organisms in the lumen, while acting as a lubricant to allow movement through the intestine.[58,64,65]

We have shown that *E. histolytica* colonizes the large intestine by binding with high affinity to the galactose and *N*-acetyl galactosamine residues on colonic mucins[12,13] (Fig. 3A). In asymptomatic infections, amoebae are proposed to inhabit and graze within the superficial layer of mucus. To invade host tissue however, amoebae must subvert the entire mucus strata to come in contact with underlying cells. In this regard, amoebae CPs play a critical role in penetrating the mucus layer (Fig. 3B). First, *E. histolytica* secretes mucin secretagogues that stimulate goblet cell release of preformed and newly synthesized mucins[66] (Fig. 3F). In vitro, colonic mucins protect against amoeba adherence to epithelial cells.[12,66] However, robust amoeba-induced mucin secretion is proposed to deplete mucin stores and ultimately facilitate parasite invasion via depletion of the mucus layer.[67,68] EhCPs complement this

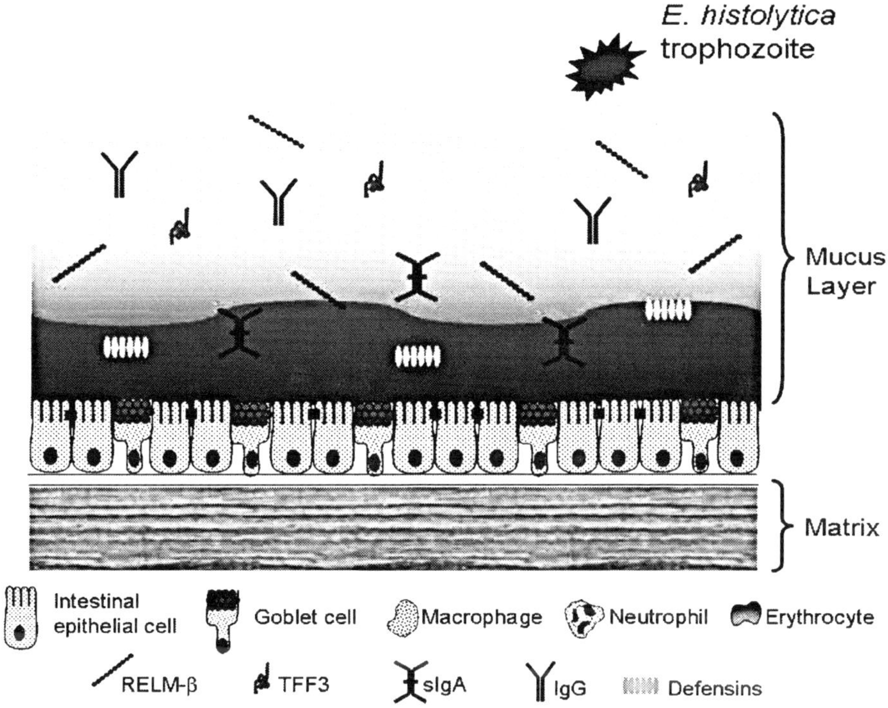

Figure 2. Mucosal host defences in the gut. *Entamoeba histolytica* trophozoites must overcome these defences to successfully cause invasive amoebiasis.

activity by degrading colonic mucins[69] (Fig. 3B). Secreted CPs have been shown to cleave colonic mucins in a dose- and time-dependent manner,[57,69] which can be abrogated with the CP inhibitor E-64. In addition, mucins degraded by amoeba CPs have a lower molecular weight and a lower density than intact mucins[69] and they are markedly less effective than native mucins at inhibiting amoeba adherence to epithelial cells.[57] Subsequently, it was shown that EhCPs cleave MUC2 at the C-terminus at 2 distinct sites,[32] which results in mucus depolymerization. Therefore, CPs degrade colonic mucin, thereby disrupting the mucus layer and facilitating penetration that ultimately leads to invasion.

In addition to the generic pharmacological CP inhibitors, antisense RNAs can be used to inhibit specific CPs. A recent study[70] used antisense RNAs against EhCP5 (to knockdown expression in trophozoites) with a variety of mucin and nonmucin producing cell lines including: chinese hamster ovary (CHO) cells, which is an epithelial cell line that does not produce mucins,[71] LS 174T and HT-29 cells, which are mucin producing cells derived from colorectal adenocarcinomas[71] and the clone HT29Cl.16E, derived from wild type HT-29 cells with increased mucin production. Trophozoites with reduced CP5 activity were ineffective at degrading colonic mucins and could not overcome the mucus barrier to disrupt LS 174T and HT29Cl.16E monolayers, but they could adhere to and disrupt CHO monolayers.[70] In addition, decreased EhCP5 activity resulted in reduced mucin degradation, as illustrated by high levels of mucin in the void volume fractions of the Sepharose 4B column.[70] The mucinase activity of amoeba CPs was further

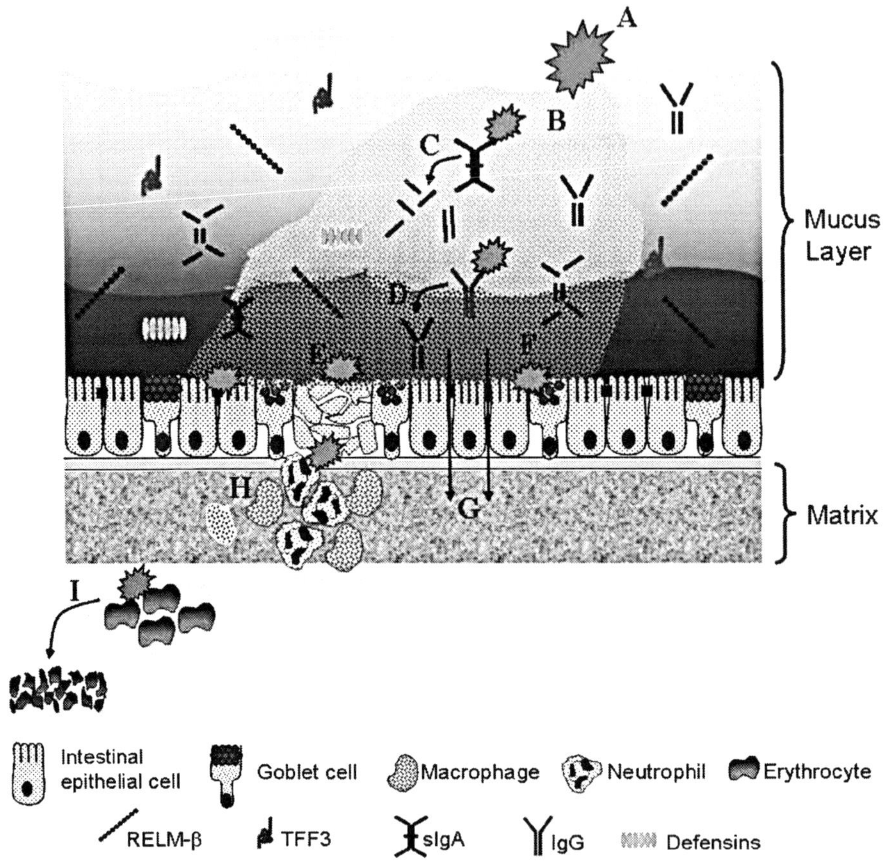

Figure 3. Amoebic cysteine proteases compromise the integrity of mucosal host defences. *E. histolytica* binds to and colonizes the mucus layer (A). EhCPs then degrade colonic mucins (B), cleave secretory IgA (C) and degrade IgG (D). Trophozoites also destroy epithelial cells (E) causes mucin depletion of the goblet cells (F), increase paracellular permeability (G), neutrophil degranulation (H) and erythro-phagoytosis (I).

illustrated by the observation that both wild type and CP5 deficient amoeba were able to bind and destroy CHO cells in the absence of a mucin barrier, but in LS 174T cells only trophozoites with CP5 were able to bind and destroy cells.[70] Interestingly, antisense inhibition of EhCP5 resulted in inhibition of CP5 as well as other CPs, suggesting that they may be co-ordinately regulated.[70]

Degradation of Secretory IgA

IgA is produced by plasma cells in the lamina propria and it is taken up and transported through intestinal epithelial cells with the aid of the polymeric immunoglobulin receptor (pIgR). On the luminal side, the pIgR is cleaved by proteolytic enzymes, leaving intact IgA to diffuse through and bind to the mucus layer. IgA is the most abundant immunoglobulin at mucosal surfaces and is the first line of specific defence against enteric pathogens.[72]

Secretory IgA (sIgA) is released into the intestinal lumen, as a dimer linked by disulphide bonds and surrounded by a carbohydrate rich secretory component,[37] which provides protection against degradation by proteolytic enzymes. sIgA has been shown to reduce mucosal colonization by pathogens and limit their growth, as well as neutralize toxins and enzymes.[73] It may also play an indirect role in antibody-mediated cytotoxicity and increase the action of nonspecific antibacterial factors.[74]

E. histolytica has been shown to elicit a sIgA response and anti-Gal lectin sIgA in the stool is protective against new infections.[75] sIgA antibodies also inhibit amoeba adhesion to target cells,[76] protect against parasite colonization[75] and reduce proteolytic activity in vitro.[77] In addition, sIgA along with apo-lactoferrin and lysozyme, found in both human and bovine milk, showed amoebicidal activity.[78] Two types of sIgA: IgA1 and IgA2 have been identified and they differ in size with IgA1 containing a 12 amino-acid proline-rich sequence in the hinge region that is absent in IgA2[73] that affords IgA2 protection from bacterial proteases. In the colon, approximately 60% of the sIgA is IgA2.[74] Bacterial proteases cleave IgA1 at defined sites in the proline-rich sequence, to produce intact Fab and Fc fragments.[73] Viable trophozoites, parasite sonicates and media conditioned by exposure to live amoebae completely degrade serum IgA in a dose-dependent manner.[79] Amoebae also degrade both IgA1 and IgA2,[72] in conditions that preserved trophozoite viability[79] (Fig. 3C). This degradation required the presence of an activating reducing agent (DTT), but was completely inhibited by the CP inhibitor E-64, indicating that the active enzyme was an amoeba CP.[72]

Much of the function of an antibody depends on its structural integrity,[73] so cleavage by amoeba CPs could compromise the effectiveness of sIgA. The absence of the proline-rich region however rendered IgA2 more resistant to amoebic surface associated CPs since degraded IgA2 was better able to agglutinate amoeba than degraded IgA1.[72] Cleavage of sIgA might also reduce affinity for *E. histolytica* antigens, as well as interfere with antigen disposal or mask the immunogenic determinants by coating them with intact Fab fragments of cleaved IgA.[74] Thus, much of the protective effects of IgA are lost in the presence of cleavage by amoebic CPs.

Interestingly, both serum and secretory IgA are susceptible to degradation by amoeba CPs, but they differ in the cleavage sites.[72] Degradation of serum IgA by trophozoites, led to a loss of the intact heavy chain, but an increase in the immunologically reactive fragments and this could be inhibited with E-64 or IA.[79] Three amoeba strains HM-2, HM-3 and HM-38 all contained a 70 kDa protease capable of cleaving IgA into intact Fab and Fc fragments, analogous to the effect of bacterial proteases.[79] Since the heavy chain is degraded though, the immunologically reactive fragments would coat *E. histolytica* trophozoites as a method of immune evasion without activating complement.

Degradation of IgG

IgG has been described as the most abundant immunoglobulin in blood and tissue. However, it is also present in the intestinal lumen and, together with secretory IgA, is important for humoral immunity across mucosal surfaces.[80] IgG is synthesized and secreted by the plasma B-cells of the lamina propria and transported to the intestinal lumen by the human neonatal Fc receptor.[80] In the intestinal lumen, IgG can bind to its cognate antigen and the IgG-antigen complex can then be transported back (again with the aid of the neonatal Fc receptor) to the lamina propria to be processed by dendritic cells and CD4+ T-cells.[80] IgG has been shown to have a protective role by neutralization of pathogen toxins, complement activation and opsonization.[81]

Over 95% of individuals infected with *E. histolytica* respond by producing IgG regardless of whether the infection is symptomatic or asymptomatic.[29] However, it is difficult to ascribe a protective role to IgG since the level of antibody response correlates with the length of disease, not the clinical response to infection.[82] Furthermore, in the presence of IgG, the parasite is still able to colonize the intestine, leading to invasive amoebiasis and in rare cases, ALA.[29,82] This may be due to the fact that IgG is cleaved by amoeba CPs (Fig. 3D). *E. histolytica* CPs have been likened in their structure and substrate specificity to cruzipain, the major CP of *Trypanosoma cruzi* trypomastigotes.[34] Cruzipain binds nonimmune IgG at the Fab fragments and cleaves the Fc fragments, so the IgG can bind antigenic components, but not activate the immune system.[83] Purified amoeba proteases also cleaved polyclonal human and monoclonal murine IgG in a dose-dependent manner and this was completely inhibited by the CP inhibitor P35017.[29] The major cleavage site for amoeba CPs occurred near the hinge region of IgG, however cleavage also occurred at multiple sites along the heavy chain resulting in almost complete degradation of this portion of the immunoglobulin.[29] Furthermore, the cleaved products were less able to bind to the antigenic portion of amoebae,[29] rendering the IgG nonfunctional and limiting the effectiveness of the host humoral response to the parasite.[29,37] Cleavage of IgG may reduce affinity for and interfere with antigen disposal and is thought to be a key mechanism by which amoebae evade the immune system.[37]

Destruction of Intestinal Microvillar Architecture

Beneath the mucus layer, the next major physical barrier encountered by amoeba is the selectively permeable colonic epithelium, which is sealed by intracellular tight junctions.[84] Intestinal epithelial cells provide an effective barrier against trans-cellular permeation of water soluble molecules, while the tight junctions form an effective barrier against paracellular permeation of luminal contents.[84]

The intestinal epithelial cells are the primary defence that prevent luminal antigens, solutes and micro-organisms from making direct contact with underlying cells. However, this barrier simultaneously allows nonpathogenic bacteria to communicate with the immune system facilitating maturation of the intestinal immune compartment and promotion of immunological tolerance, both of which contribute to intestinal homeostasis.[85-88] In addition, intestinal epithelial cells are required for nutrient absorption, electrolyte transport, mucus glycoprotein production[89] and production of antimicrobial peptides and bacteriolytic enzymes.[62,63,86]

The surface-bound Gal/GalNAc-lectin allows amoeba to directly bind host epithelial cells, bacteria, human colonic mucins and erythrocytes.[12,90,91] This binding is necessary for adherence to and cytolysis of epithelial cells as well as amoeba invasion.[16,90] Indeed, studies have shown that *N*-acetyl-*D*-galactosamine and to a lesser extent galactose, inhibit adherence.[90,91] Trophozoites kill host cells when they come in direct contact[90] and prevention of adherence inhibits destruction of individual cells or cell monolayers, indicating that this is crucial to cytolysis.[90]

Villin is the main actin-bundling protein found in the intestinal microvilli[7,38] and disturbance of the microvilli requires proteolysis of the villin. Proteolysis of villin occurs within one minute of direct contact between live trophozoites and intestinal enterocytes[38] and this activity is dependent on amoebic CPs[7,38] (Fig. 3E). Indeed, pretreatment of trophozoites with the synthetic CP inhibitor z-FA.fmk blocked EhCP activity and prevented villin proteolysis and the associated microvillar defacement.[38] This was further confirmed by the observation that trophozoites with reduced CP activity (pSA8 and SAW760 strains) showed reduced villin proteolysis when cocultured with enterocytes.[38] Similar results were also observed with the inhibitors *N*-tosyl-phenylalanine chloromethyl ketone (TPCK) and *N*-tosyl-*L*-lysine chloromethyl ketone (TLCK).[7] Both TLCK and TPCK were also found to inhibit *E. histolytica* serine protease activity,[33] but subsequent studies revealed that pretreatment of amoebae with these compounds inhibited their ability to cleave the cathepsin-B synthetic substrate z-Arg-Arg-pNA.[7,48] Studies with Caco2 cell monolayers in vitro showed a cauliflower-like microvillar pattern that was completely disrupted when cells were cocultured with live amoebae. This disruption in the microvillar pattern was lost when amoebae were pre-incubated with TLCK or TPCK.[7] In addition, TLCK and TPCK reduced villin proteolysis observed with live trophozoites and prevented trophozoite-induced destruction of cell monolayers in vitro.[7] These results clearly show that in order to invade enteric cell layers, amoeba must overcome the host microvilli barrier, which they effectively accomplish by villin proteolysis.

Destruction of Extracellular Matrix/Architecture

EhCPs have a broad range of substrate specificity and can cleave casein, gelatin and insulin (reviewed in ref. 23). In addition, CPs disrupt monolayers of fibroblasts in culture[33] and they degrade fibronectin, laminin and collagen, all of which are components of the extracellular matrix.[11,17,28,92] This has serious implications since cells die in the absence of an extracellular matrix. Cell apoptosis occurs when cells lose direct contact with each other, or with the basement membrane.[93,94] Amoeba CPs have been shown to disrupt cell monolayers and digest the proteins of the extracellular matrix, as well as induce cells to detach and round up[95] without directly inducing cytolysis[23,96] and this was completely inhibited by CP inhibitors.

A direct relationship was observed between CP activity and the cytopathic effect of amoebae on tissue culture monolayers in vitro.[33,51] Increased cytopathic effects were observed in both *E. histolytica* and *E. dispar* when EhCP2 was over-expressed.[96] The inverse was also true. A protease- and phagocytosis-deficient mutant of *E. histolytica*, L6 was shown to have less CP activity than wild type amoeba, with the level of activity being comparable to that of the noninvasive *E. dispar*.[97] In this mutant, although the *E. histolytica* genes *ehcp1*, *ehcp2*, *ehcp3*, *ehcp5* and *ehcp112* were expressed, there was an 8-10 fold decrease in the levels of EhCP1, EhCP2 and EhCP5 produced,[97] which are the CPs typically responsible for 90% of the protease activity of amoeba.[34] Additionally, this protease-deficient mutant was less virulent and demonstrated reduced cytopathic effects compared to the wild type HM-1 strain.[33] Furthermore, both *E. histolytica* and *E. dispar*, engineered to over-express EhCP5 in vitro, showed no apparent differences in their viability or growth rates, but CP activity was increased 2.8-3.0 fold in engineered *E. dispar* compared to wild type, leading to increased monolayer destruction.[96] In addition, a recent study to examine the effect of various virulence factors on amoebic invasion revealed that trophozoites deficient in EhCP5 were unable to penetrate the submucosa, while amoebae deficient in either the heavy chain of the Gal/GalNAc lectin or amoebapores were still capable of penetrating this layer.[41]

Activity on Tight Junction Proteins

The apical junctional complex (AJC) is composed of tight junctions and adherens junctions and is a dynamic structure that regulates the movement of molecules between epithelial cells.[85] The AJC is particularly sensitive to changes in physiologic conditions and disturbance frequently results in increased epithelial permeability and exposure of intestinal tissue to luminal antigens and/or pathogens.[85] Recent studies in our laboratory have shown that the pro-mature cysteine protease pcp5 of *E. histolytica* is able to induce an NF-κB-mediated pro-inflammatory response through the RGD motif binding to αvβ3 integrin on colonic cells.[98] In addition, this induced expression of the pro-inflammatory cytokines TNF-α, IL-1β, COX-2 and IL-6 in vivo.[98] Since tight junctions are sensitive to inflammation,[85,99] expression of these pro-inflammatory cytokines could induce disassembly of the apical junctional complex and endocytosis of component proteins.

The tight junctions between intestinal epithelial cells have three main functions: they hold adjacent cells together, regulate the passive movement of luminal fluid and solutes through the paracellular spaces and maintain a polarized gradient across the membrane.[85] The tight junctions consist of a series of transmembrane proteins including the claudins, occludin, zonula occludens, junctional adhesion molecules (JAMs), E-cadherin, cingulin and actin.[85] The barrier formed by these tight junctions between cells, is very sensitive to regulatory changes in response to inflammation, complement factors and toxins.[99]

Amoebae weaken cellular tight junctions by disturbing them and cleaving tight junction proteins[7] (Fig. 3G). The integrity of the tight junction barrier can be measured by the trans-epithelial resistance (TER),[100] where the TER is inversely proportional to the paracellular permeability and a low TER is reflective of increased permeability and compromised barrier integrity. Cells cocultured (on filters) with amoeba have a reduced TER. However, this is prevented by pre-incubation of amoeba with CP inhibitors TLCK or TPCK,[7] which delayed (TLCK) or inhibited (TPCK) functional disturbance of the tight junctions,[7] indicating that *E. histolytica* CPs are involved in tight junction disturbance (Fig. 3G). In addition, both TPCK and TLCK prevent proteolysis of the tight junction protein ZO-1, but only TPCK prevents proteolysis of ZO-2 in cocultures with enteric cells.[7]

Erythro-Phagocytosis: Destruction of RBCs

Amoebae CPs are involved in parasite survival as well as virulence and they play a role in the phagocytosis of erythrocytes. Erythro-phagocytosis was reduced by 63% in amoeba deficient in CPs as compared to wild type amoebae.[101] Over-expression of EhCP2 was associated with a slight decrease in erythro-phagocytosis,[96] while another study showed that EhCP5, though not involved in growth rate, cytopathology or haemolysis, plays a role in the phagocytosis of RBCs.[101] More recently, amoebae deficient in EhCP1 and EhCP7 were shown to have impaired digestion and phagocytosis of RBCs.[102] *E. histolytica* possesses enzymes that are able to lyse the membrane of erythrocytes and release their haemoglobin;[103] these amoeba haemoglobinases are neutral proteases that are activated by DTT and inhibited by E-64, NEM and pHMB, suggesting that they are in fact CPs.[103] Strains of the parasite isolated from patients with ALA have more protease activity and virulent strains of the parasite have the same pattern of haemoglobinases.[103] The ability to phagocytose RBCs may be physiologically important for nutrient acquisition, as amoeba continue to grow in culture if ferric ammonium citrate is replaced with human haemoglobin.[103]

EhCP5 appears particularly important for ingestion of RBCs, as gene silencing using antisense mRNA reduced amoeba erythro-phagocytosis.[101] A stable transfectant of the highly virulent strain HM-1:IMSS was generated using the gene encoding EhCP5. The transfectant had the *ehcp5* gene inserted in the opposite orientation, between the 5′ and 3′ un-translated regions and CP activity was decreased by 90%.[101] Though the transfected amoebae maintained their ability to destroy tissue culture monolayers and their haemolytic activity in vitro,[101] they had reduced erythro-phagocytic ability and in other studies this has been linked to nutrient acquisition.[103]

Immune Evasion: Neutrophils and Macrophages

Neutrophils and macrophages are the first cells to respond to the site of infection and tissue invasion.[19] Neutrophils in particular are the first cells of the host immune system to interact with invading amoeba and they play a protective role in the early host response in the liver.[104] CBA mice are partially resistant to amoebic challenge, but when pretreated with the neutrophil depleting antibody (anti-Gr1), this resistance was lost and severe intestinal pathology was observed.[105] In addition, intestinal epithelial cells produce a number of pro-inflammatory cytokines including IL-1β and IL-8 which further attract neutrophils and macrophages to the site of invasion.[19,106,107] Upon arrival, these cells are activated by tumour necrosis factor (TNF-α) and/or interferon gamma (IFN-γ) and activated cells are able to kill amoebae in vitro.[108-110] Interestingly, studies have also shown reduced intestinal inflammation in SCID mice depleted of neutrophils in response to amoebic infection, suggesting a role for these cells in the propagation of an inflammatory response.[106] Amoebae have been shown to lyse neutrophils to release cell mediators that can lead to hepatocyte death.[19] In this section we examine the role of amoebic CPs on various pro-inflammatory cell mediators, including IL-1β and IL-18.

Activity on Pre IL-1β

Interleukin 1 beta (IL-1β) is a pro-inflammatory cytokine produced abundantly by activated macrophages as well as many other cell types.[81] Increased levels of IL-1β are associated with intestinal inflammation including that found in patients with inflammatory bowel disease. In addition, at physiologically relevant concentrations, IL-1β causes an increase in intestinal epithelial tight junction permeability, leading to inflammation.[84] IL-1β is synthesized as an inactive precursor of 31 kDa, which is proteolytically cleaved by IL-1β converting enzyme (ICE; also known as caspase-1) to form the active 17 kDa protein.[111] *E. histolytica* lysates or purified amoebic CPs are capable of mimicking the activity of ICE and cleaving pro-IL-1β to form the active mature cytokine.[56] In this regard, intestinal epithelial cells that are damaged/lysed by amoeba release pro-IL-1β that is cleaved and activated by extracellular EhCPs, to amplify the inflammatory process.[8,56,112] EhCPs cleave pro-IL-1β at a site consistent with a basic amino acid (Arg), thereby generating a product that is 5 amino acids shorter than that produced by ICE.[56]

IL-1α and IL-1β stimulate production of pro-inflammatory cytokines including IL-8, IL-6, growth related oncogene α (GRO-α), COX-2 and granulocyte macrophage colony stimulating factor (GM-CSF) by adjacent intestinal cells through activation of the NF-κB pathway.[106,113,114] Blockade of intestinal cell IL-1β and IL-8 production reduced *E. histolytica* induced gut inflammation and tissue damage, as measured by neutrophil influx, histology and changes in intestinal barrier permeability.[112] In the gut, IL-1β signalling increases intestinal permeability to luminal contents including pathogenic and nonpathogenic micro-organisms, which will in turn cause further inflammation and prolong a heightened inflammatory state. In addition, the trans-epithelial resistance (TER) is compromised in a prolonged inflammatory state, resulting in increased damage to host cells.

IL-18 Cleavage

The IL-1 superfamily of cytokines includes IL-1α, IL-1β and IL-18, all of which are involved in promoting inflammation and enhancing immune responses.[115] IL-18 is produced by antigen presenting and intestinal epithelial cells among others and the IL-18 receptor is found in a variety of organs in the body, including the small and large intestine. IL-18 induces IL-12 receptor expression on naïve T-cells and the reverse is also true. In the presence of both IL-18 and IL-12, there is a synergistic effect, leading to increased T-cell proliferation and IFN-γ production (reviewed in ref. 115). IFN-γ in turn acts synergistically with IL-12 to promote Th1 cell development. IL-18 is also a co-inducer of Th1 type responses, stimulating IFN-γ and activating macrophages capable of lysing trophozoites.[110] IL-18, like IL-1β is a pro-inflammatory cytokine, which is secreted as a precursor protein that must be cleaved by caspase-1 to form the mature cytokine.[111,115] Unlike IL-1β, EhCPs cleave both pro- and mature IL-18, to form a series of biologically inactive fragments.[111] This is analogous to the effect of caspase 3 on IL-18,[115] but the cleavage sites are different.[111]

This dichotomy where EhCPs have opposite effects on two very similar pro-inflammatory cytokines is particularly interesting. While no studies have specifically addressed this issue, it has been suggested that IL-18 enhances host defence in a number of infections including those against parasitic, fungal and bacterial pathogens.[111,115] In particular, IL-18 mRNA is induced in mice pericutaneously-injected with the helminth parasite *Schistosoma mansoni*. Mice injected with a plasmid for both IL-18 and *S. mansoni* glutathione S-transferase exhibited increased antigen specific IFN-γ in spleen cells and this correlated with protection against schistosomiasis.[115] However, excess IL-18 is detrimental and may cause injury to the host, so degradation of IL-18 by EhCPs may be associated with promoting parasite or host survival, or both.

Cleavage of C3 to Activate Complement

The complement protein C3 is a 115 kDa glycoprotein composed of an α chain and a β chain, linked by disulphide bonds and noncovalent forces.[116] Upon activation of complement, C3 is cleaved in the α chain by C3 convertases, to generate C3a a 9 kDa protein with anaphylatoxin activity and C3b composed of the remainder of the α chain, the α′ chain and the entire β chain.[31,116] EhCPs mimic this cleavage and cleave between amino acid residues 78 and 79 as compared to the C3 convertases, which cleave between residues 77 and 78. The CP degradation of C3 is both time- and dose-dependent.[31] As with the fragments generated by the C3 convertases, the C3b fragment generated by amoebic CPs, is susceptible to Factor I inactivation through the binding of Factor H.[31]

While studies have shown that *E. histolytica* is not susceptible to attack by the complement membrane attack complex due to its sequence homology to CD59[117] others indicate that this is not always the case.[30] The majority of the C3 is present in the fluid phase but some is bound through intact α chains, to the surface of both complement sensitive and resistant amoebae. Interestingly, this binding is not mediated by either the thiol ester, or a specific receptor on the surface of the amoebae.[30] Additionally, this bound C3 is not internalized by the parasite, nor is it susceptible to the action of C3 convertases, or other proteins in the fluid phase.[30] On activation of complement, EhCPs cleave the C3 present in the fluid phase, but the α chains of the C3 bound to the amoeba surface remains intact.[30] C3 is very sensitive to cleavage by EhCPs since low levels of CP as in those present in noninvasive strains, is sufficient to cleave C3 in the fluid phase.[30,31] Cleavage of C3 in the fluid phase generates C3b, instrumental to the formation of the membrane attack complex (C5b-C9).[30,118] These terminal complement components kill susceptible, nonpathogenic strains, but not resistant, pathogenic amoebae.[30] Moreover, amoeba resistance to complement is inhibited by preincubating trophozoites with the irreversible CP inhibitor E-64.[30]

Degradation of Anaphylatoxins C3a and C5a

The complement anaphylatoxins, C3a, C4a and C5a are fragments produced by the activation of the complement system. They produce a localized inflammatory response by causing degranulation of mast cells, basophils and phagocytes.[118] In addition, anaphylatoxins also stimulate smooth muscle cell contraction and increased permeability of blood capillaries.[118] This results in an influx of antibodies and phagocytic cells to the site of antigen entry, through leukocyte chemotaxis along a concentration gradient of anaphylatoxins. Two of the main components of the complement system are the glycoproteins C3 and C5, both of which are composed of an α and a β chain.[118] As mentioned above, C3 is cleaved to form C3a and C3b. Similarly, C5 is cleaved to produce C5a, a protein of 11 kDa and C5b.[119] Both C3a and C5a bind to receptors on mast cells and basophils to stimulate degranulation and histamine release along with smooth muscle contraction and increased vascular permeability.[118] C5a also induces chemotaxis of neutrophils, macrophage activation and the release of proinflammatory cytokines including IL-1, IL-6 and IL-8 (reviewed in see ref. 119).

EhCPs activate the alternative pathway of the complement system by cleaving both C3 and C5 to generate C3a- and C5a-like proteins.[119] In addition, these CPs circumvent normal host immunity and inactivate the pro-inflammatory factors C3a and C5a.[119] Amoebic CPs cleave C3 between amino acids 78 and 79, to generate C3a- and C3b-like molecules.[31,119] The α chain of the C3 was cleaved at one site, while the C3b-like molecule did not undergo further digestion and remained haemolytically active.[31] The C3a-like cleavage fragment was degraded in as little as 5 minutes of incubation with the protease, though this was concentration-dependent.[119] In addition, the C3a-like molecules did not maintain their reactivity, since they were unable to induce platelet aggregation in a bioassay for the activity of complement.[119]

Interestingly the concentration of EhCPs required to cleave C5 was five to ten times less than that which was necessary to cleave C3, but this level was easily attainable under physiological conditions.[119] Proteolysis in this case generated a lower molecular weight C5a protein versus the C5 convertase and increased protease concentration or time of incubation resulted in complete degradation of the α chain.[119] Thus, C5a was cleaved in a dose-dependent manner and it was unable to induce human neutrophil chemotaxis in a bioassay for complement activity.[119]

Thus EhCPs activated the alternative pathway of complement and generated nonimmunoreactive C3a- and C5a-like molecules, as well as a haemolytic C3b-like protein. Interestingly, this C3b-like protein caused passive lysis of nonpathogenic but not pathogenic strains of *E. histolytica*.[30] This may help to explain the observations that clinical isolates of patients with invasive amoebiasis are resistant to complement-mediated lysis,[120] as well as the fact that pathogenic, but not nonpathogenic strains of *E. histolytica* participate in the haemolysis of host RBCs.[103] They may also help to explain the relative paucity of acute inflammation seen in patients with prolonged infection, or the lack of neutrophils in patients with ALA.

CYSTEINE PROTEASES AS TARGETS FOR DISEASE INTERVENTION

EhCPs forms a particularly good target for disease prevention and intervention for a number of reasons. Firstly, they are essential for the parasite life cycle in terms of encystation, excystation and nutrient acquisition, so any disruption in the CPs involved in these processes would affect parasite viability. Furthermore, CPs are one of the key virulence factors involved in intestinal colonization and invasion, disruption of host tissue and modulation of cell-mediated immune responses,[53,121] so prevention of these would have a significant impact on disease pathogenesis. Finally, EhCPs are easily inhibited by chemical means, or silenced with the use of antisense RNA and the presence of conserved sites provides potential targets for vaccine development. A number of natural and synthetic inhibitors against EhCPs have been identified; a summary of which can be found in reference 7. Thus *E. histolytica* CPs provide attractive targets for novel chemotherapeutic intervention[122] and CP inhibition is a viable target for antiparasitic therapies against parasite invasion, as has been shown with *Trypanosoma cruzi*, *Plasmodium falciparum* and *Cryptosporidium parvum*.[37]

Vaccines and Antibodies to Amoeba Cysteine Proteases

Specific IgG responses have been observed in over 95% of patients with invasive amoebiasis, or asymptomatic colonization,[82] but the level of antibody correlates to the length of the disease as opposed to the severity or clinical response to infection.[29,82] In addition, secretory IgA is protective against adhesion and colonization[75] and EhCP112 was found to be immunogenic in patients with amoebiasis.[123] In fact, recent studies have shown that animals inoculated with a plasmid mixture containing *Ehadh112* and *Ehcp112* DNA had a higher survival rate and significant prevention against liver abscess formation.[123] Interestingly though, a poor humoral response but a strong cell-mediated response was observed in these animals.[123] Taken together with the various strategies to inhibit EhCP expression, vaccine control using CP targets, is an exciting area worth further investigation.[122]

CONCLUSION

Although these findings indicate the potential for pharmacological CP inhibitors and antisense RNA use in disease prevention and intervention, a number of unanswered questions remain. Previous studies have found that 90% of expressed CP protein is derived from EhCP1, EhCP2 and EhCP5 combined.[24,48,101] However, inhibition of a single gene, *ehcp5* reduced CP activity by 90%, suggesting that inhibition of EhCP5, resulted in reduced expression of EhCP1 and EhCP2.[96] An understanding of transcriptional/

translational regulation of EhCP expression would facilitate development of strategies for disease prevention.

In addition, the majority of EhCPs present within the parasite's genome are not expressed in vitro and consequently little is known about their function. A better understanding of the specific CPs involved in parasite encystation and excystation presents the opportunity to develop strategies that prevent either of these processes thereby disrupting the parasite life cycle. In addition, a number of other eukaryotes produce cystatins, which are thought to play a role in protecting cells from their own CPs.[39] Further studies to delineate the structures of these with a view to determining the binding requirements of the parasite enzyme[37] and candidates for anti-amoebiasis drugs are needed.[39] In addition, the pro-regions of the CPs or the amino acid residues at the N-terminal end, inhibit the enzyme and provide valuable information on protease folding and trafficking,[37] all of which may be exploited in amoebiasis control. While some studies have investigated the use of synthetic protease inhibitors that are able to enter the cell more readily than E-64,[53,111] more work needs to be done in the area. Ultimately, a better understanding of EhCPs and their role in the parasite life cycle and virulence would facilitate their use as treatment or prevention mechanisms.

ACKNOWLEDGEMENTS

Dr. Chadee is a Canada Research Chair (Tier 1) in Gastrointestinal Inflammation and his research is supported by grants from the Natural Sciences and Engineering Research Council of Canada, the Canadian Institute for Health Research, the Canadian Foundation for Innovation, the Crohn's and Colitis Foundation of Canada and the Canadian Association of Gastroenterology-Astra Zeneca-CIHR Research and Fellowship Awards.

REFERENCES

1. Turk B, Turk D, Turk V. Lysosomal cysteine proteases: more than scavengers. Biochim Biophys Acta 2000; 1477(1-2):98-111.
2. Berdowska I. Cysteine proteases as disease markers. Clin Chim Acta 2004; 342(1-2):41-69.
3. WHO. Amoebiasis. Weekly Epidemiological Record. Vol 72. Geneva; 1997:97-100.
4. Lejeune M, Rybicka JM, Chadee K. Recent discoveries in the pathogenesis and immune response toward Entamoeba histolytica. Future Microbiol 2009; 4:105-118.
5. Haque R, Huston CD, Hughes M et al. Amebiasis. N Engl J Med 2003; 348(16):1565-1573.
6. Haque R, Mondal D, Kirkpatrick BD et al. Epidemiologic and clinical characteristics of acute diarrhea with emphasis on Entamoeba histolytica infections in preschool children in an urban slum of Dhaka, Bangladesh. Am J Trop Med Hyg 2003; 69(4):398-405.
7. Lauwaet T, Oliveira MJ, Callewaert B et al. Proteinase inhibitors TPCK and TLCK prevent Entamoeba histolytica induced disturbance of tight junctions and microvilli in enteric cell layers in vitro. Int J Parasitol 2004; 34(7):785-794.
8. Dey I, Chadee K. Prostaglandin E2 produced by Entamoeba histolytica binds to EP4 receptors and stimulates interleukin-8 production in human colonic cells. Infect Immun 2008; 76(11):5158-5163.
9. Perez-Tamayo R, Montfort I, Garcia AO et al. Pathogenesis of acute experimental liver amebiasis. Arch Med Res 2006; 37(2):203-209.
10. Stanley SL, Jr. The Entamoeba histolytica genome: something old, something new, something borrowed and sex too? Trends Parasitol 2005; 21(10):451-453.
11. Tannich E. Royal society of tropical medicine and hygiene meeting at manson house, London, 1998. Amoebic disease. Entamoeba histolytica and E. dispar: comparison of molecules considered important for host tissue destruction. Trans R Soc Trop Med Hyg 1998; 92(6):593-596.

12. Chadee K, Petri WA, Jr., Innes DJ et al. Rat and human colonic mucins bind to and inhibit adherence lectin of Entamoeba histolytica. J Clin Invest 1987; 80(5):1245-1254.
13. Chadee K, Johnson ML, Orozco E et al. Binding and internalization of rat colonic mucins by the galactose/N-acetyl-D-galactosamine adherence lectin of Entamoeba histolytica. J Infect Dis 1988; 158(2):398-406.
14. Gilchrist CA, Petri WA. Virulence factors of Entamoeba histolytica. Curr Opin Microbiol 1999; 2(4):433-437.
15. Petri WA Jr, Snodgrass TL, Jackson TF et al. Monoclonal antibodies directed against the galactose-binding lectin of Entamoeba histolytica enhance adherence. J Immunol 1990; 144(12):4803-4809.
16. Solaymani-Mohammadi S, Petri WA, Jr. Intestinal invasion by Entamoeba histolytica. Subcell Biochem 2008; 47:221-232.
17. Horstmann RD, Leippe M, Tannich E. Host tissue destruction by Entamoeba histolytica: molecules mediating adhesion, cytolysis and proteolysis. Mem Inst Oswaldo Cruz 1992; 87 Suppl 5:57-60.
18. Andra J, Herbst R, Leippe M. Amoebapores, archaic effector peptides of protozoan origin, are discharged into phagosomes and kill bacteria by permeabilizing their membranes. Dev Comp Immunol 2003; 27(4):291-304.
19. Stanley SL, Jr. Amoebiasis. Lancet 2003; 361(9362):1025-1034.
20. Munoz ML, Calderon J, Rojkind M. The collagenase of Entamoeba histolytica. J Exp Med 1982; 155(1):42-51.
21. MacFarlane RC, Singh U. Identification of an Entamoeba histolytica serine-, threonine- and isoleucine-rich protein with roles in adhesion and cytotoxicity. Eukaryot Cell 2007; 6(11):2139-2146.
22. Teixeira JE, Huston CD. Participation of the serine-rich Entamoeba histolytica protein in amebic phagocytosis of apoptotic host cells. Infect Immun 2008; 76(3):959-966.
23. McKerrow JH, Sun E, Rosenthal PJ et al. The proteases and pathogenicity of parasitic protozoa. Annu Rev Microbiol 1993; 47:821-853.
24. Ankri S, Stolarsky T, Bracha R et al. Antisense inhibition of expression of cysteine proteinases affects Entamoeba histolytica-induced formation of liver abscess in hamsters. Infect Immun 1999; 67(1):421-422.
25. Que X, Brinen LS, Perkins P et al. Cysteine proteinases from distinct cellular compartments are recruited to phagocytic vesicles by Entamoeba histolytica. Mol Biochem Parasitol 2002;119(1):23-32.
26. Reed SL, Keene WE, McKerrow JH. Thiol proteinase expression and pathogenicity of Entamoeba histolytica. J Clin Microbiol 1989; 27(12):2772-2777.
27. Stanley SL Jr, Zhang T, Rubin D et al. Role of the Entamoeba histolytica cysteine proteinase in amebic liver abscess formation in severe combined immunodeficient mice. Infect Immun 1995; 63(4):1587-1590.
28. Li E, Yang WG, Zhang T et al. Interaction of laminin with Entamoeba histolytica cysteine proteinases and its effect on amebic pathogenesis. Infect Immun 1995; 63(10):4150-4153.
29. Tran VQ, Herdman DS, Torian BE et al. The neutral cysteine proteinase of Entamoeba histolytica degrades IgG and prevents its binding. J Infect Dis 1998; 177(2):508-511.
30. Reed SL, Gigli I. Lysis of complement-sensitive Entamoeba histolytica by activated terminal complement components. Initiation of complement activation by an extracellular neutral cysteine proteinase. J Clin Invest 1990; 86(6):1815-1822.
31. Reed SL, Keene WE, McKerrow JH et al. Cleavage of C3 by a neutral cysteine proteinase of Entamoeba histolytica. J Immunol 1989; 143(1):189-195.
32. Lidell ME, Moncada DM, Chadee K et al. Entamoeba histolytica cysteine proteases cleave the MUC2 mucin in its C-terminal domain and dissolve the protective colonic mucus gel. Proc Natl Acad Sci USA 2006; 103(24):9298-9303.
33. Keene WE, Hidalgo ME, Orozco E et al. Entamoeba histolytica: correlation of the cytopathic effect of virulent trophozoites with secretion of a cysteine proteinase. Exp Parasitol 1990; 71(2):199-206.
34. Bruchhaus I, Loftus BJ, Hall N et al. The intestinal protozoan parasite Entamoeba histolytica contains 20 cysteine protease genes, of which only a small subset is expressed during in vitro cultivation. Eukaryot Cell 2003; 2(3):501-509.
35. Brinen LS, Que X, McKerrow JH et al. Homology modeling of Entamoeba histolytica cysteine proteinases reveals the basis for cathepsin L-like structure with cathepsin B-like specificity. Arch Med Res 2000; 31(4 Suppl):S63-64.
36. Karrer KM, Peiffer SL, DiTomas ME. Two distinct gene subfamilies within the family of cysteine protease genes. Proc Natl Acad Sci USA 1993; 90(7):3063-3067.
37. Que X, Reed SL. Cysteine proteinases and the pathogenesis of amebiasis. Clin Microbiol Rev 2000; 13(2):196-206.
38. Lauwaet T, Oliveira MJ, Callewaert B et al. Proteolysis of enteric cell villin by Entamoeba histolytica cysteine proteinases. J Biol Chem 2003; 278(25):22650-22656.
39. Riekenberg S, Witjes B, Saric M et al. Identification of EhICP1, a chagasin-like cysteine protease inhibitor of Entamoeba histolytica. FEBS Lett 2005; 579(7):1573-1578.
40. Saric M, Vahrmann A, Bruchhaus I et al. The second cysteine protease inhibitor, EhICP2, has a different localization in trophozoites of Entamoeba histolytica than EhICP1. Parasitol Res 2006; 100(1):171-174.

41. Bansal D, Ave P, Kerneis S et al. An ex-vivo human intestinal model to study Entamoeba histolytica pathogenesis. PLoS Negl Trop Dis 2009; 3(11):e551.
42. Loftus B, Anderson I, Davies R et al. The genome of the protist parasite Entamoeba histolytica. Nature 2005; 433(7028):865-868.
43. MacFarlane RC, Shah PH, Singh U. Transcriptional profiling of Entamoeba histolytica trophozoites. Int J Parasitol 2005; 35(5):533-542.
44. Das S, Gillin FD. Chitin synthase in encysting Entamoeba invadens. Biochem J 1991; 280 (Pt 3):641-647.
45. Ebert F, Bachmann A, Nakada-Tsukui K et al. An Entamoeba cysteine peptidase specifically expressed during encystation. Parasitol Int 2008; 57(4):521-524.
46. Makioka A, Kumagai M, Kobayashi S et al. Entamoeba invadens: cysteine protease inhibitors block excystation and metacystic development. Exp Parasitol 2005; 109(1):27-32.
47. Sharma M, Hirata K, Herdman S et al. Entamoeba invadens: characterization of cysteine proteinases. Exp Parasitol 1996; 84(1):84-91.
48. Bruchhaus I, Jacobs T, Leippe M et al. Entamoeba histolytica and Entamoeba dispar: differences in numbers and expression of cysteine proteinase genes. Mol Microbiol 1996; 22(2):255-263.
49. Garcia-Rivera G, Rodriguez MA, Ocadiz R et al. Entamoeba histolytica: a novel cysteine protease and an adhesin form the 112 kDa surface protein. Mol Microbiol 1999; 33(3):556-568.
50. Jacobs T, Bruchhaus I, Dandekar T et al. Isolation and molecular characterization of a surface-bound proteinase of Entamoeba histolytica. Mol Microbiol 1998; 27(2):269-276.
51. de Meester F, Shaw E, Scholze H et al. Specific labeling of cysteine proteinases in pathogenic and nonpathogenic Entamoeba histolytica. Infect Immun 1990; 58(5):1396-1401.
52. Freitas MA, Fernandes HC, Calixto VC et al. Entamoeba histolytica: cysteine proteinase activity and virulence. Focus on cysteine proteinase 5 expression levels. Exp Parasitol 2009; 122(4):306-309.
53. Reed S, Bouvier J, Pollack AS et al. Cloning of a virulence factor of Entamoeba histolytica. Pathogenic strains possess a unique cysteine proteinase gene. J Clin Invest 1993; 91(4):1532-1540.
54. Gadasi H, Kessler E. Correlation of virulence and collagenolytic activity in Entamoeba histolytica. Infect Immun 1983; 39(2):528-531.
55. Gadasi H, Kobiler D. Entamoeba histolytica: correlation between virulence and content of proteolytic enzymes. Exp Parasitol 1983; 55(1):105-110.
56. Zhang Z, Wang L, Seydel KB et al. Entamoeba histolytica cysteine proteinases with interleukin-1 beta converting enzyme (ICE) activity cause intestinal inflammation and tissue damage in amoebiasis. Mol Microbiol 2000; 37(3):542-548.
57. Moncada D, Keller K, Chadee K. Entamoeba histolytica cysteine proteinases disrupt the polymeric structure of colonic mucin and alter its protective function. Infect Immun 2003; 71(2):838-844.
58. Moncada DM, Chadee, K. Production, Structure and function of gastrointestinal mucins. In: Blaser MJ, Smith, P.J., Ravdin, I.J., Greenberg, H.R. and Guerrant, R.L., ed. Infections of the Gastroinestinal Tract Philadelphia: Williams and Wilkins; 2002:57-79.
59. Dharmani P, Srivastva, V, Kissoon Singh et al. Role of intestinal mucins in innate host defense mehanisms against Pathogens. Journal of Innate Immunity 2009; 1:123-135.
60. Moncada DM. Production, Structure and function of gastrointestinal mucins. Montreal, McGill University; 2006.
61. Tse SK, Chadee K. The interaction between intestinal mucus glycoproteins and enteric infections. Parasitol Today 1991; 7(7):163-172.
62. Bevins CL, Martin-Porter E, Ganz T. Defensins and innate host defence of the gastrointestinal tract. Gut 1999; 45(6):911-915.
63. Zanetti M. Cathelicidins, multifunctional peptides of the innate immunity. J Leukoc Biol 2004; 75(1):39-48.
64. Bergstrom KS, Guttman JA, Rumi M et al. Modulation of intestinal goblet cell function during infection by an attaching and effacing bacterial pathogen. Infect Immun 2008; 76(2):796-811.
65. Johansson ME, Thomsson KA, Hansson GC. Proteomic analyses of the two mucus layers of the colon barrier reveal that their main component, the Muc2 mucin, is strongly bound to the Fcgbp protein. J Proteome Res 2009; 8(7):3549-3557.
66. Belley A, Keller K, Grove J et al. Interaction of LS174T human colon cancer cell mucins with Entamoeba histolytica: an in vitro model for colonic disease. Gastroenterology 1996; 111(6):1484-1492.
67. Chadee K, Meerovitch E. Entamoeba histolytica: early progressive pathology in the cecum of the gerbil (Meriones unguiculatus). Am J Trop Med Hyg 1985; 34(2):283-291.
68. Tse SK, Chadee K. Biochemical characterization of rat colonic mucins secreted in response to Entamoeba histolytica. Infect Immun 1992; 60(4):1603-1612.
69. Moncada D, Yu Y, Keller K et al. Entamoeba histolytica cysteine proteinases degrade human colonic mucin and alter its function. Arch Med Res 2000; 31(4 Suppl):S224-225.
70. Moncada D, Keller K, Ankri S et al. Antisense inhibition of Entamoeba histolytica cysteine proteases inhibits colonic mucus degradation. Gastroenterology 2006; 130(3):721-730.

71. ATTC. American Type Culture Colletion.
72. Garcia-Nieto RM, Rico-Mata R, Arias-Negrete S et al. Degradation of human secretory IgA1 and IgA2 by Entamoeba histolytica surface-associated proteolytic activity. Parasitol Int 2008; 57(4):417-423.
73. Woof JM, Kerr MA. The function of immunoglobulin A in immunity. J Pathol 2006; 208(2):270-282.
74. Kilian M, Mestecky J, Russell MW. Defense mechanisms involving Fc-dependent functions of immunoglobulin A and their subversion by bacterial immunoglobulin A proteases. Microbiol Rev 1988; 52(2):296-303.
75. Haque R, Ali IM, Sack RB et al. Amebiasis and mucosal IgA antibody against the Entamoeba histolytica adherence lectin in Bangladeshi children. J Infect Dis 2001; 183(12):1787-1793.
76. Carrero JC, Diaz MY, Viveros M et al. Human secretory immunoglobulin A anti-Entamoeba histolytica antibodies inhibit adherence of amebae to MDCK cells. Infect Immun 1994; 62(2):764-767.
77. Guerrero-Manriquez GG, Sanchez-Ibarra F, Avila EE. Inhibition of Entamoeba histolytica proteolytic activity by human salivary IgA antibodies. Apmis 1998; 106(11):1088-1094.
78. Leon-Sicairos N, Lopez-Soto F, Reyes-Lopez M et al. Amoebicidal activity of milk, apo-lactoferrin, sIgA and lysozyme. Clin Med Res 2006; 4(2):106-113.
79. Kelsall BL, Ravdin JI. Degradation of human IgA by Entamoeba histolytica. J Infect Dis 1993; 168(5):1319-1322.
80. Yoshida M, Claypool SM, Wagner JS et al. Human neonatal Fc receptor mediates transport of IgG into luminal secretions for delivery of antigens to mucosal dendritic cells. Immunity 2004; 20(6):769-783.
81. Murphy K, Travers P, Walport M. Janeway's Immunobiology. 7th ed. New York, New York: Garland Science, Taylor and Francis Group; 2008.
82. Jackson TF, Gathiram V, Simjee AE. Seroepidemiological study of antibody responses to the zymodemes of Entamoeba histolytica. Lancet 1985; 1(8431):716-719.
83. Bontempi E, Cazzulo JJ. Digestion of human immunoglobulin G by the major cysteine proteinase (cruzipain) from Trypanosoma cruzi. FEMS Microbiol Lett 1990; 58(3):337-341.
84. Al-Sadi R, Boivin M, Ma T. Mechanism of cytokine modulation of epithelial tight junction barrier. Front Biosci 2009; 14:2765-2778.
85. Laukoetter MG, Bruewer M, Nusrat A. Regulation of the intestinal epithelial barrier by the apical junctional complex. Curr Opin Gastroenterol 2006; 22(2):85-89.
86. Muller CA, Autenrieth IB, Peschel A. Innate defenses of the intestinal epithelial barrier. Cell Mol Life Sci 2005; 62(12):1297-1307.
87. Artis D. Epithelial-cell recognition of commensal bacteria and maintenance of immune homeostasis in the gut. Nat Rev Immunol 2008; 8(6):411-420.
88. Sansonetti PJ. War and peace at mucosal surfaces. Nat Rev Immunol 2004; 4(12):953-964.
89. Goke M, Podolsky DK. Regulation of the mucosal epithelial barrier. Baillieres Clin Gastroenterol 1996; 10(3):393-405.
90. Ravdin JI, Guerrant RL. Role of adherence in cytopathogenic mechanisms of Entamoeba histolytica. Study with mammalian tissue culture cells and human erythrocytes. J Clin Invest 1981; 68(5):1305-1313.
91. Petri WA Jr, Smith RD, Schlesinger PH et al. Isolation of the galactose-binding lectin that mediates the in vitro adherence of Entamoeba histolytica. J Clin Invest 1987; 80(5):1238-1244.
92. Schulte W, Scholze H. Action of the major protease from Entamoeba histolytica on proteins of the extracellular matrix. J Protozool 1989; 36(6):538-543.
93. Frisch SM, Francis H. Disruption of epithelial cell-matrix interactions induces apoptosis. J Cell Biol 1994; 124(4):619-626.
94. Taupin DR, Kinoshita K, Podolsky DK. Intestinal trefoil factor confers colonic epithelial resistance to apoptosis. Proc Natl Acad Sci USA 2000; 97(2):799-804.
95. Keene WE, Petitt MG, Allen S et al. The major neutral proteinase of Entamoeba histolytica. J Exp Med 1986; 163(3):536-549.
96. Hellberg A, Nickel R, Lotter H et al. Overexpression of cysteine proteinase 2 in Entamoeba histolytica or Entamoeba dispar increases amoeba-induced monolayer destruction in vitro but does not augment amoebic liver abscess formation in gerbils. Cell Microbiol 2001; 3(1):13-20.
97. Carpeniseanu S, Hirata K, Que X et al. L6: a proteinase- and phagocytosis-deficient mutant of Entamoeba histolytica. Arch Med Res 2000; 31(4 Suppl):S237-238.
98. Hou Y, Kissoon-Singh V, Fodor I et al. Entamoeba histolytica CP5 targets integrin to induce NF-kB proinflammatory signaling. In press. 2009.
99. Fasano A, Baudry B, Pumplin DW et al. Vibrio cholerae produces a second enterotoxin, which affects intestinal tight junctions. Proc Natl Acad Sci USA 1991; 88(12):5242-5246.
100. Leroy A, Lauwaet T, De Bruyne G et al. Entamoeba histolytica disturbs the tight junction complex in human enteric T84 cell layers. Faseb J 2000; 14(9):1139-1146.

101. Ankri S, Stolarsky T, Mirelman D. Antisense inhibition of expression of cysteine proteinases does not affect Entamoeba histolytica cytopathic or haemolytic activity but inhibits phagocytosis. Mol Microbiol 1998; 28(4):777-785.
102. Irmer H, Tillack M, Biller L et al. Major cysteine peptidases of Entamoeba histolytica are required for aggregation and digestion of erythrocytes but are dispensable for phagocytosis and cytopathogenicity. Mol Microbiol 2009; 72(3):658-667.
103. Serrano-Luna JJ, Negrete E, Reyes M et al. Entamoeba histolytica HM1:IMSS: hemoglobin-degrading neutral cysteine proteases. Exp Parasitol 1998; 89(1):71-77.
104. Seydel KB, Zhang T, Stanley SL, Jr. Neutrophils play a critical role in early resistance to amebic liver abscesses in severe combined immunodeficient mice. Infect Immun 1997; 65(9):3951-3953.
105. Asgharpour A, Gilchrist C, Baba D et al. Resistance to intestinal Entamoeba histolytica infection is conferred by innate immunity and Gr-1+ cells. Infect Immun 2005; 73(8):4522-4529.
106. Seydel KB, Li E, Zhang Z et al. Epithelial cell-initiated inflammation plays a crucial role in early tissue damage in amebic infection of human intestine. Gastroenterology 1998; 115(6):1446-1453.
107. Chadee K, Moreau F, Meerovitch E. Entamoeba histolytica: chemoattractant activity for gerbil neutrophils in vivo and in vitro. Exp Parasitol 1987; 64(1):12-23.
108. Denis M, Chadee K. Human neutrophils activated by interferon-gamma and tumour necrosis factor-alpha kill Entamoeba histolytica trophozoites in vitro. J Leukoc Biol 1989; 46(3):270-274.
109. Lin JY, Keller K, Chadee K. Entamoeba histolytica proteins modulate the respiratory burst potential by murine macrophages. Immunology 1993; 78(2):291-297.
110. Seguin R, Mann BJ, Keller K et al. The tumor necrosis factor alpha-stimulating region of galactose-inhibitable lectin of Entamoeba histolytica activates gamma interferon-primed macrophages for amebicidal activity mediated by nitric oxide. Infect Immun 1997; 65(7):2522-2527.
111. Que X, Kim SH, Sajid M et al. A surface amebic cysteine proteinase inactivates interleukin-18. Infect Immun 2003; 71(3):1274-1280.
112. Stanley SL Jr, Reed SL. Microbes and microbial toxins: paradigms for microbial-mucosal interactions. VI. Entamoeba histolytica: parasite-host interactions. Am J Physiol Gastrointest Liver Physiol 2001; 280(6):G1049-1054.
113. Eckmann L, Reed SL, Smith JR et al. Entamoeba histolytica trophozoites induce an inflammatory cytokine response by cultured human cells through the paracrine action of cytolytically released interleukin-1 alpha. J Clin Invest 1995; 96(3):1269-1279.
114. Huston CD. Parasite and host contributions to the pathogenesis of amebic colitis. Trends Parasitol 2004; 20(1):23-26.
115. Biet F, Locht C, Kremer L. Immunoregulatory functions of interleukin 18 and its role in defense against bacterial pathogens. J Mol Med 2002; 80(3):147-162.
116. Lundwall A, Eggertsen G. Isolation of human complement factors C3, C5 and H. J Immunol Methods 1985; 81(1):147-160.
117. Ventura-Juarez J, Campos-Rodriguez R, Jarillo-Luna RA et al. Trophozoites of Entamoeba histolytica express a CD59-like molecule in human colon. Parasitol Res 2009; 104(4):821-826.
118. Kindt TJ, Goldsby RA, Osborne BA. Immunology. 6th ed. New York: W.H. Freeman and Company; 2007.
119. Reed SL, Ember JA, Herdman DS et al. The extracellular neutral cysteine proteinase of Entamoeba histolytica degrades anaphylatoxins C3a and C5a. J Immunol 1995; 155(1):266-274.
120. Reed SL, Curd JG, Gigli I et al. Activation of complement by pathogenic and nonpathogenic Entamoeba histolytica. J Immunol 1986; 136(6):2265-2270.
121. Campbell D, Chadee K. Survival strategies of Entamoeba histolytica: Modulation of cell-mediated immune responses. Parasitol Today 1997; 13(5):184-190.
122. Ocadiz R, Orozco E, Carrillo E et al. EhCP112 is an Entamoeba histolytica secreted cysteine protease that may be involved in the parasite-virulence. Cell Microbiol 2005; 7(2):221-232.
123. Martinez MB, Rodriguez MA, Garcia-Rivera G et al. A pcDNA-Ehcpadh vaccine against Entamoeba histolytica elicits a protective Th1-like response in hamster liver. Vaccine 2009; 27(31):4176-4186.
124. Moncada DM. E. histolytica cysteine proteinases. Montreal, McGill University; 2006.
125. Bruchhaus I, Tannich E. A gene highly homologous to ACP1 encoding cysteine proteinase 3 in Entamoeba histolytica is present and expressed in E. dispar. Parasitol Res 1996; 82(2):189-192.
126. Tillack M, Nowak N, Lotter H et al. Increased expression of the major cysteine proteinases by stable episomal transfection underlines the important role of EhCP5 for the pathogenicity of Entamoeba histolytica. Mol Biochem Parasitol 2006; 149(1):58-64.

CHAPTER 6

CYSTEINE PEPTIDASES OF KINETOPLASTID PARASITES

Conor R. Caffrey,[1,*] Ana-Paula Lima[2] and Dietmar Steverding[3]

[1]Sandler Center for Drug Discovery, California Institute for Quantitative Biosciences, Byers Hall, University of California San Francisco, San Francisco, USA; [2]Instituto de Biofísica Carlos Chagas Filho, Universidade Federal do Rio de Janeiro, Bloco G, C.C.S., Cidade Universitária, Ilha do Fundão, Rio de Janeiro, Brazil; [3]BioMedical Research Centre, School of Medicine, Health Policy and Practice, University of East Anglia, Norwich, UK

**Corresponding Author: Conor R. Caffrey—Email: caffrey@cgl.ucsf.edu*

Abstract: We review Clan CA Family C1 peptidases of kinetoplastid parasites (*Trypanosoma* and *Leishmania*) with respect to biochemical and genetic diversity, genomic organization and stage-specificity and control of expression. We discuss their contributions to parasite metabolism, virulence and pathogenesis and modulation of the host's immune response. Their applications as vaccine candidates and diagnostic markers as well as their chemical and genetic validation as drug targets are also summarized.

INTRODUCTION

Kinetoplastids are a group of flagellated protozoa that include free-living and parasitic organisms. Of health and economic importance are the trypanosomatids, a collection of obligatory parasites within the kinetoplastids. In particular, *Trypanosoma* and *Leishmania* species afflict millions of people and animals throughout the world. In sub-Saharan Africa, *Trypanosoma brucei gambiense* and *Trypanosoma brucei rhodesiense* are the causative agents of human African trypanosomiasis (HAT; sleeping sickness) whereas *T. b. brucei*, *Trypanosoma congolense* and *Trypanosoma vivax* are responsible for 'Nagana' disease in cattle. In Central and South America, *Trypanosoma cruzi* is the pathogen for Chagas' disease in humans and the leading cause of heart disease. *Leishmania* parasites produce a spectrum of disease depending on the species and the immunological status of the human host (see ref 1. for review on kinetoplastids and disease).

Cysteine Proteases of Pathogenic Organisms, edited by Mark W. Robinson and John P. Dalton.

This chapter reflects that most of the research on kinetoplastid peptidases has involved Clan CA family C1 papain-like cysteine peptidases (CPs; http://merops.sanger.ac.uk/) and cathepsin L-like peptidases more so than cathepsins B. The salient features of these peptidases are summarized in Table 1. Other CPs not covered here, but that have been described in these organisms, include members of the Clan CA family C2 (calpains) and the Clan CD families C13 (GPI:protein transamidase), C14 (metacaspase) and C50 (separase). The reader is referred to comprehensive reviews of protozoan Clan CD peptidases[2] and CPs in parasites in general.[3] Finally, the topics of kinetoplastid Clan C1 peptidases and their endogenous protein inhibitors have been very recently covered in detail.[4-5]

HISTORY OF KINETOPLASTID CP DISCOVERY

The early characterizations of peptidolytic activity in kinetoplastids that appeared from the late 1970s though the 1980s all pointed to the presence of significant CP activity as biochemically dissected through the use of synthetic peptidyl substrates, reducing agents (such as dithiothreitol) and protein- and peptidyl-based inhibitors.[6-13] Essentially, biochemical activity similar to that of mammalian cathepsin L and localized to the endo-lysosomal system[6,14-16] was responsible for the majority of the CP activity measured. Also, for both *T. congolense*[14] and *Leishmania,*[16] low levels of enzyme have been localized to the parasite surface and flagellar pocket, the latter a major site for endocytosis as well as membrane and protein recycling in kinetoplastids.

Expression of the cathepsin L-like enzymes is developmentally-regulated among the different parasitic stages in the intermediate and definitive host. Thus, for *T. brucei,* greater expression is found in the bloodstream 'short stumpy' (insect-infective stage) forms relative to either the bloodstream 'long-slender' or insect procyclic forms.[17,18] For *T. congolense*, activity is likewise greater in the bloodstream forms compared to the insect-stage epimastigotes.[19] In various *Leishmania* species, the cathepsin L-like enzymes are mainly expressed in intracellular amastigotes,[20,21] yet in *Leishmania mexicana*, whether expression is greater in the human-infective metacyclics or amastigotes depends on the gene isoform in question.[22] Expression of *T. cruzi* cathepsin L (aka: cruzain or cruzipain) is greater in epimastigotes and amastigotes than in the extracellular human-dwelling trypomastigotes, whereas cruzipain 2 activity is greatest in the latter stage. For both enzymes RNA levels are similar among the three stages.[23] There is also evidence that individual *T. cruzi* cathepsin L isoforms are preferentially expressed at different life stages,[24] suggesting that functional redundancy is not absolute.

With the application of DNA sequencing to kinetoplastids in the late 1980s and early 1990s, the genes for cathepsin L-like peptidases were identified in the *T. brucei* subspecies,[25,26] *T. congolense,*[27] *T. cruzi*[28] and *Leishmania.*[29-31] Most of the enzymes are unusual in possessing an 11-13 kDa C-terminal extension downstream of the peptidase's catalytic domain and for which a function(s) is still unknown 20 years after the original identification.[25] Only recently for *Leishmania pifanoi* has there been evidence to suggest that the extension may assist penetration of the host macrophage.[32]

Soon after the sequencing of cathepsin L-like genes, cathepsin B activities and sequences were identified in *Leishmania.*[33-35] In 1998, a cathepsin B from *T. cruzi* was characterized and sequenced.[36,37] It was not until 2004 that a similar cathepsin B was sequenced and characterized in *T. brucei* as a direct result of the availability of improved genome sequence information.[38] The same was true in 2008 when 13 cathepsin B genes

Table 1. Summary of kinetoplastid clan CA family C1 cysteine peptidases

Species	Peptidase Name (and synonyms)	Activity	Stage(s) Expressed	Distribution	Synthetic/Natural Inhibitors	Assigned Functions	Selected References
Trypanosoma cruzi	*Tcr*CATL1 (cruzain, cruzipain)	CATL-like and CATB-like, Acidic to alkaline pH	Epimastigote (E), Trypomastigote (T), Amastigote (A)	Golgi (E), Cell Surface(E), Reservosomes (E), Flagellar pocket (T), Vesicles (A) Surface (A)	E-64, Peptidylfluoromethyl ketones, Diazomethyl ketones, Vinyl sulfones, Triterpenoids, *Croton* oil, Diketopiperazines	Metacyclogenesis, Host cell invasion, Host cell signaling, Amastigote survival/nutrition, Generation of kinins, Inflammation, Ag Presentation	23,24,28,45,53, 74,78,97,103
	*Tcr*CATL2 (cruzipain 2)	CATS-like, Acidic to alkaline pH	T	n.d.	n.d.	Host cell invasion	
	*Tcr*CATB	n.d.	E	Reservosomes (E)	E-64, CA-074	n.d.	36,37
Trypanosoma congolense	*Tco*CATL1 (CP1)	CATL-like, Acidic pH	Bloodstream form (BSF)	Flagellar pocket, Cell surface Lysosome(?)	E-64	Trypanotolerance in cattle	42,69,70,83
	*Tco*CATL2 (trypanopain-Tc, congopain, CP2)	CATL-like, Acidic to alkaline pH					

continued on next page

Table 1. Continued

Species	Peptidase Name (and synonyms)	Activity	Stage(s) Expressed	Distribution	Synthetic/Natural Inhibitors	Assigned Functions	Selected References
Trypanosoma congolense (continued)	*Tco*CATB1-13	CATB-like, Acidic pH (CATB7-13 are likely inactive due to active site mutation)	RNA level: CATB1-5—BSF CATB6—Procyclic (P), E, M CATB7-11—P, E, M CATB12—all	Lysosome	E-64, CA-074	Degradation of endocytosed proteins	39
Trypanosoma brucei brucei	*Tbb*CATL (trypanopain-Tb, brucipain)	CATL-like, Acidic to alkaline pH	P but greater in BSF	Lysosome	Peptidyl fluoromethyl ketones, Diazomethyl ketones, Vinyl sulfones, Non-peptidyl chalcones,	Degradation of anti-VSG IgG, Traversal of the blood-brain barrier	18,44,54,95, 104,114-116
Trypanosoma brucei rhodesiense	*Tbr*CATL (trypanopain-Tb, rhodesain)				Acyl hydrazides Aziridine-2, 3-dicarboxylates, Thiosemicarbazones, Purine-derived nitriles, Diketopiperazines	Infection in mice, Interaction with brain endothelium (gene expression)	

continued on next page

Table 1. Continued

Species	Peptidase Name (and synonyms)	Activity	Stage(s) Expressed	Distribution	Synthetic/Natural Inhibitors	Assigned Functions	Selected References
Trypanosoma brucei rhodesiense (continued)	*Tbr*CATB (tbcatB)	CATB-like, Acidic pH	All life stages	Lysosome	E-64, Thiosemicarbazones, purine-derived nitriles, diketopiperazines	Essential gene protein turnover in lysosomes, Transferrin degradation, Cytokenesis Infection in mice	38,112,115,117
Leishmania mexicana	*Lme*CATL-B1-2 (CPB1-2)	CATL-like, Acidic pH	Metacyclic (M)	Micro-vesicular tubule (M), Flagellar pocket (M)	E-64, Vinyl sulfones	Th-2 immune response, Parasite differentiation, Autophagy, NF-kB degradation,	22,32,46,49,52, 79,82,120,124, 125,128
	*Lme*CATL-B3-18 (CPB3-18)		A	Lysosome (A)		Macrophage infection, Lesion progression	
	*Lme*CATL-A (CPA)	CATL-like	All life stages (greatest in A)	Same as above	E-64	Macrophage infection (*L. infantum*), In vivo infection (*L. infantum*)	
	*Lme*CATB (CPC)	CATB-like, Acidic pH	All life stages	?	n.d.	Infection of macrophages, Infection in mice, Lesion progression	33,40,126

n.d., not determined; CATL, cathepsin L; CATB, cathepsin B; CATS, cathepsin S; Ag, antigen were characterized in *T. congolense,*[39] one of which may have been described earlier.[11] The cathepsins B of *T. congolense* are mainly expressed in bloodstream forms and, at least in those isoforms studied, are localised in the lysosomal compartment.[39] The same seems to be the case for *T. brucei.*[38] The cathepsins B from *T. cruzi, Leishmania chagasi* and *Leishmania donovani* are expressed in all life-cycle stages.[36,37,40]

NOMENCLATURE

As the research summarized above progressed, different groups assigned various names to the cathepsin L-like activities under study. Often the suffix '-ain' or '-ipain' was used to indicate relatedness to the archetypal plant enzyme, papain. For example with the African trypanosomes, the term 'Trypanopain-Tb'[41] has appeared to distinguish this enzyme from that found in *T. congolense* called 'Trypanopain-Tc.'[42] Further, for *T. brucei* the terms, 'rhodesain' and 'brucipain' have been employed to distinguish the respective enzymes found in *T. b. brucei* and *T. b. rhodesiense.*[18] The *ad hoc* nature of these names, while understandable, is neither consistent nor helpful to those in the field, let alone to those new to the area. Thus, the present authors recently proposed an unambiguous nomenclature system for kinetoplastid C1 peptidases (both cathepsin B and L)[4] based on that described earlier (Table 1).[43] The enzyme in question is designated with three to six capital letters, i.e., CATB and CATL for kinetoplastid cathepsin B- and L-like peptidases, respectively. Then, the first letter of the genus followed by the first two letters of the species (all in italics) is placed in front to indicate the organism, e.g., *Tco*CATL for the CATL of *T. congolense*. For sub-species, the first letter each of the genus, species and subspecies is taken, e.g., *Tbb*CATL and *Tbr*CATL, for the CATL of *T. b. brucei* (brucipain) and *T. b. rhodesiense* (rhodesain), respectively. Isoforms can be identified using numbers, e.g., *Tco*CATB1, *Tco*CATB2, etc. for the CATB isoforms of *T congolense.* In the case of the *L. mexicana* CPA and CPB 1-18 isoforms, the isoenzymes can be distinguished by combinations of capital letters and numbers, e.g., *Lme*CATL-A and *Lme*CATL-B1, *Lme*CATL-B2 etc.

GENOMIC ORGANIZATION AND CONTROL OF GENE EXPRESSION

In trypanosomatids, key differences from other eukaryotes include the organization of the genome into polycistronic gene clusters,[44-46] a simplified transcriptional machinery and mRNA trans-splicing coupled with polyadenylation.[47] In *T. cruzi*, 12% of protein-encoding genes are found in clusters (of more than two genes), 46 of which (3836 genes) contain 20 or more paralogues. For example, many housekeeping genes occur in highly-conserved tandem clusters throughout the genome.[45] In *Leishmania major,* 36% of the protein-encoding genes are found as clusters of gene families, mostly in repeat units of two or more.[46] In *T. brucei*, the chromosomes contain long, nonoverlapping gene clusters that are probably transcribed as polycistrons.[44] In like fashion, kinetoplastid CATL genes occur as tandem arrays containing multiple gene copies that are regarded as one polycistronic unit, however, stage-regulated expression of their products is often observed (see below).

In *T. brucei*, CATL is encoded by more than 20 genes arranged in tandem.[25,44] The genome of *T. cruzi* contains a number of clusters each containing 3-5 gene copies that are spread across different chromosomes.[45] Also, *T. cruzi* contains a second CATL gene isoform (cruzipain-2) that shares 88% identity with the other CATL genes that are 93-99% homologous to one another.[48] The CATL gene cluster in *L. chagasi* and *L. donovani* comprises five tandemly organised gene copies, but the genome of the latter species also contains an additional CATL gene that is not part of the cluster.[49] In *L. major*, CATL genes occur in multiple copies on a single chromosome of about 0.6 Mb but the exact genomic organization and the number of copies are unknown.[35] In *L. mexicana* and *L. pifanoi,* CATL genes are arranged as single- and multi-copy genes. The multi-copy genes are organised as tandem arrays of 18 and 8-20 copies in *L. mexicana* and *L. pifanoi*, respectively.[29,50,51] In addition, single- and multi-copy CATL genes in *L. pifanoi* have distinct chromosomal locations.[29]

*Lme*CATL-B1-18 (also termed cpb1-18[31]) differ from the single copy *Lme*CATL-A (termed cpa)[30] in that the latter possesses a 3-amino acid insertion at the amino terminal of the predicted mature peptidase. The tandemly-arranged CATL genes are polycistronically transcribed but the expression of the products of individual gene copies of the array is stage-regulated. Specifically, *Lme*CATL-B1 and B2 (cpb1 and cpb2) are expressed predominantly in metacyclics, whereas copies *Lme*CATL-B3-18 are expressed mainly in amastigotes.[22] The regulation of expression is achieved by posttranscriptional mechanisms that were investigated with the re-integration of individual *Lme*CATL genes in *L. mexicana* mutant lines lacking the whole CATL-B gene array. It was found that the 3′-untranslated region of *Lme*CATL-B3-18 and the intercistronic sequence downstream of *Lme*CATL-B1 and B2 act as *cis*-regulatory elements in post-transcriptional stage-specific gene expression.[22] It was proposed that an insertion at the 3′-untranslated region of *Lme*CATL-B1 and B2 mediates metacyclic-specific stage-regulated expression by affecting the maturation of polycistronic pre-mRNA. The expression of the single *Lme*CATL-A gene is stage-regulated at the RNA and protein levels. The gene is expressed in all life-cycle stages but most so in amastigotes and stationary phase promastigote cultures which contain the infective metacyclic form of the parasite.[30]

The organization of kinetoplastid CATB genes is usually much simpler. In *T. cruzi* and *Leishmania* spp. they occur as single- or double-copy genes.[33,35,37,40] *Tbr*CATB occurs as a single-copy gene and its expression is greatest in bloodstream forms[18,38] being regulated at the RNA level.[38] In contrast, for *T. congolense*, three evolutionary clusters of 13 CATB genes have been identified.[39] These genes are not tandemly linked but scattered across different chromosomes. Also, at the RNA level, stage-specificity of expression depends on the gene in question.[39]

PEPTIDASE TRAFFICKING, PROCESSING AND BIOCHEMISTRY

Similar to mammalian Clan CA peptidases, those from kinetoplastids are synthesized as zymogens that are peptidolytically processed to maturity by removal of an N-terminal pro-domain. Maturation of CATL is thought to occur in different compartments of the secretory pathway, either by self (auto)-activation or *trans*-activation.[52,53] In *T. cruzi*, pro-*Tcr*CATL seems to be mainly activated in the Golgi and then sorted to the reservosomes or flagellar pocket.[53] In *L. mexicana*, the CATL-B zymogen is first secreted to the flagellar pocket where it undergoes *trans*-activation by CATL isoforms (including CATL-A), before being endocytosed and targeted to the multi-vesicular tubular structure.[52] Although it is not known exactly where pro-*Tbr*CATL is converted to the mature form, disruption of lysosomal trafficking by RNA interference (RNAi) directs a considerable amount of the zymogen to the flagellar pocket[54] suggesting that, similar to *T. cruzi*, conversion might occur intracellularly between the Golgi and lysosomes. In every case, processing occurs in nonacidic sub-cellular compartments which is different to the situation for mammalian cathepsins.[55,56] Indeed, although auto-processing of recombinant trypanosomatid pro-CATL is most efficient at acid pH, spontaneous processing is also observed at neutral-alkaline pH.[24,57,58]

The pro-domain of *Tcr*CATL and *Lme*CATL is responsible for accurate lysosomal trafficking of the peptidases[59] and for *Tbr*CATL, also involves a trans-membrane receptor.[54] Pro-domain released by peptidolysis re-associates tightly to inhibit mature *Tcr*CATL and *Tbr*CATL, however, the inhibition can be overcome at acidic pH.[60] This finding may explain the paradox that despite peptidolytic excision of the pro-domain early in the secretory pathway, the domain can mediate subsequent targeting to lysosomes. The activity of mature CPs can be further regulated by the chagasin family (Clan JL, Family I42) of protein inhibitors.[61,62] Produced by a few protozoa, these are tight-binding inhibitors of Clan CA peptidases and have been shown to interact with CPs in *T. cruzi* and *T. brucei*.[63,64]

Unlike the mammalian lysosomal C1 peptidases, most kinetoplastid CATLs are stable and active at neutral through alkaline pH values and possess broader, if individually varying, subsite specificities against small molecule substrates and inhibitors.[18,58,65-69] The stability at neutral pH may reflect the action of those peptidases in diverse sub-cellular locations, particularly at the parasite-host interface, including the flagellar pocket. *Tcr*CATL is unique among kinetoplastid CATLs in possessing both CATL- and CATB-like specificities by accepting either hydrophobic or positively charged substrate residues in the P2 position—a phenomenon that is pH dependent.[66,70,71] *Tcr*CATL and *Lme*CATL also act as carboxypeptidases over a broad pH range in vitro[72] and their activities are similar to the carboxydipeptidase activities of mammalian CATB and mammalian CATL, respectively.[73] In the presence of glycosaminoglycans, the activity of *Tcr*CATL toward peptidyl substrates is increased and its sensitivity to protein CP inhibitors reduced,[74] possibly a physiologically-relevant mechanism to ensure that *Tcr*CATL secreted at the host-parasite interface is maximally active. A similar characteristic has been observed for recombinant *Tbr*CATL that shows increased activity towards peptidyl substrates in the presence of glycosaminoglycans (Costa TFR and Lima AP, unpublished data).

For kinetoplastid CATBs, there is little information available regarding processing and trafficking. Biochemically, recombinant *Tco*CATB1 and *Tco*CATB6 are optimally active at acidic (but different) pH values and retain activity at neutral pH.[39] These and other data demonstrating considerable differences in the substrate and inhibitor specificities of kinetoplastid CPs suggest unique physiological functions, including outside of the acidic lysosome. Most interestingly, one group of these genes (*Tco*CATB7-13) encodes a serine in place of the predicted catalytic residue cysteine, likely rendering the translation products peptidolytically inactive,[39] or, possibly, imparting an entirely different catalytic activity, as has been demonstrated for the *Plasmodium faliciparum* SERA5 protein.[75]

VACCINES AND DIAGNOSIS

Kinetoplastid CATB and CATL enzymes are potential candidates for vaccine development, either by preventing infection or decreasing pathology.[4] Immunization of cattle with recombinant *Tco*CATL did not prevent infection with *T. congolense* but the vaccinated cattle maintained or gained weight and had less-pronounced anaemia during the chronic phase of the disease.[76] Injection of mice with *Tcr*CATL DNA stimulated cytotoxic T-lymphocytes that recognized and killed *T. cruzi*-infected cells.[77] In other experiments, immunization of mice with recombinant *Tcr*CATL conferred protection against experimental infection.[78] Likewise, immunization with DNA plasmids encoding *Leishmania* CATLs resulted in protective immunity in murine models of cutaneous and visceral leishmaniasis.[79-81] Finally, vaccination of BALB/c mice with CATB was partially protective against *Leishmania infantum* infections.[82]

As *Tco*CATB elicits a strong immune response in cattle infected with *T. congolense*, it was suggested that this peptidase represents a candidate antigen for diagnosis of Nagana disease.[39] High levels of anti-*Tco*CATL antibodies are found in trypano-tolerant but not trypano-susceptible cattle indicating that *Tco*CATL is a marker of trypano-tolerance in cattle.[83] In patients with Chagas' disease, *Tcr*CATL is immunogenic and could be employed as a specific antigen for routine screening of *T. cruzi* infections in blood donors.[84-86] *Tcr*CATL's immuno-dominant domain corresponds to the C-terminal extension.[87]

DRUG TARGETS

The few drugs that are available to treat kinetoplastid diseases suffer from a variety of problems, including limited efficacy, toxicity, the need for parenteral administration and the establishment of drug resistance.[1] New options for safe, effective and orally-administered treatments are urgently needed.

Since the first proof-of-principle study in 1993,[88] Family C1 peptidases have been validated as drug targets in many parasitic organisms, including kinetoplastids.[89-91] At first, the idea that these peptidases were attractive drug targets at all seemed counter-intuitive in respect of the fears for off-target toxicity through inhibition of orthologous host C1 peptidases. However, over time and experimentation, an understanding has arisen that these concerns can be offset by (i) the propensity of many parasitic organisms to rapidly accumulate molecules (including drugs) from their environment, (ii) the lack of functional redundancy, both qualitative and quantitative, in parasite C1 peptidases relative to the mammalian host (for example the lysosomal concentration of cathepsins B, L and D is as high as 1 mM[92]) and (iii) the projection that short course therapies would be sufficient (under one month in the case of *T. cruzi*) to effect cure such that any off-target host toxicity might be managed.

Initial demonstrations of small molecule inhibition of parasite CPs that limited kinetoplastid survival in vitro came in the early 1990s using peptidyl fluoromethyl ketone inhibitors against *T. cruzi* trypomastigotes and *T. brucei* bloodstream forms,[93] and diazomethyl ketones against *T. brucei* procyclics.[12] These studies were followed with critical in vivo experiments for *T. brucei* demonstrating that inhibition of CP activity (with diazomethyl ketones) decreased parasitaemia to below detectable levels, at least for a few days upon cessation of treatment and that infected mice exposed to the inhibitor lived longer (68 days versus 39 days without inhibitor).[94] Importantly, the inhibition of CP activity correlated with antiparasite activity including a phenotype of an enlarged lysosome suggesting that peptidolysis in the lysosome was being inhibited.[94,95]

The medicinal chemistry focus to design more specific and less toxic inhibitors to *Tcr*CATL (cruzain) over the last 15 years, has identified a variety of peptidomimetic and nonpeptidyl inhibitors[96-102] of which to date, the vinyl sulfone, K11777, is making progress to clinical candidacy (see also the chapter entitled 'Cruzain: The Path from Target

Validation to the Clinic').[103] K11777 also had potent in vitro and in vivo bioactivity against *T. brucei,*[104] presumably by targeting *Tbr*CATL, although it is possible that *Tbr*CATB may also be a target given that the inhibitor is nonselective. In addition, for *T. brucei*, the search for potent bioactive chemical entities has identified nonpeptidyl chalcones, acyl hydrazides and amides that inhibit *Tbb*CATL up to 50-100-fold more efficiently than mammalian CATL, some of which protect mice from otherwise lethal infections.[105] Also characterized as *Tbr*CATL inhibitors was a series of aziridine-2,3-dicarboxylates,[106] and of both *Tbr*CATL and *Tbr*CATB, thiosemicarbazones[107] and purine-derived nitriles.[108] Natural products have been studied as inhibitors of *Tbr*CATL and *Tbr*CATB; diketopiperazines encountered in deep water sediment-derived fungi,[109] and triterpenoids[110] and *Croton* oil from Costa Rican flora.[111] Finally, given that *Tbr*CATB is an essential gene (discussed below) there has been a recent effort to develop peptidyl nitrile inhibitors discriminating this peptidase from *Tbr*CATL.[112] However, in the longer view of drug development, including the time and resources required, it may not be necessary to derive absolute specificity for either *Tb*CATL or *Tb*CATB[4] considering the biological advantages offered by the parasite (discussed above) and the excellent selectivity indices already achieved for *T. brucei* in vitro using nonspecific, 'off-the-shelf' CP inhibitors.[94,113]

The vinyl sulfone, K11777, has also been employed as a tool to study trypanosome pathogenesis. Specifically, the inhibitor limited the ability of *T.b. rhodesiense* to cross an in vitro model of the human blood brain barrier, thereby implicating *Tbr*CATL in parasite entry into the brain.[114] This chemically-defined potential function was supported by RNAi of *Tbr*CATL in *T. brucei.*[115] Also, CATL secreted from *T.b. rhodesiense* induced calcium signaling in brain endothelial cells in vitro.[114] The gene expression profiles in these cells exposed to *T. b rhodesiense* in vitro revealed alterations in several gene groups, particularly those associated with inflammation[116] indicating additional contributions by CPs to pathogenesis.

Though it is clear that CP inhibitors kill kinetoplastids, to date they have not provided the chemical selectivity that is required to determine peptidase essentiality or accurately interrogate the biological function of individual CPs. The alternative genetic approaches of targeted gene disruption and RNAi have improved our understanding in both these regards, particularly for *Leishmania* and *T. brucei*. For the latter, RNAi has shown that *Tbr*CATB is essential to parasite survival both in vitro[38] and in vivo (mice were cured of infection).[115] Subsequent studies of a *T. brucei* cell line defective in one *Tbr*CATB allele resulted in slower growth and swelling of the flagellar pocket and endocytotic compartment.[117] Whether *Tbr*CATL is essential was unclear given that gene suppression by RNAi in vitro was not total.[38] In vivo, partial suppression of CATL expression by RNAi did not cure mice, but prolonged their survival,[115] perhaps indicating that CATL may be more a virulence factor than an essential gene.

VIRULENCE FACTORS AND IMMUNOMODULATORS

Over a decade of comprehensive gene deletion studies in *L. mexicana*, targeting CATL-A, CATL-B, CATB or various combinations thereof, has shown that they are not essential but are important as virulence factors that modulate both pathology and the host immune response.[33,51,118-121] Deletion of the entire *Lme*CATL-B array reduced promastigote infectivity to macrophages in vitro, whereas amastigotes infected macrophages with the same kinetics as wild type parasites.[51] Significant restoration of virulence and a return to a

Th2-type immune response in BALB/c mice required the reinsertion of the entire CATL-B array; re-introduction of single components of the array, namely the amastigote-specific CATL-B2.8 gene or the metacyclic stage-specific CATL-B2 gene, failed in both respects.[122] The findings suggest that *Lme*CATL-B genes are not entirely redundant. Similarly, deletion of *Lme*CATL-A did not generate morphological or behavioral phenotypes that differed from controls (growth rate or infectivity to macrophages).[123] Thus, this particular *Lme*CATL was not considered important for parasite virulence. In contrast, *L. infantum* lacking CATL-A were significantly less virulent in vitro and in vivo and the re-introduction of just one allele was sufficient to improve infectivity to human macrophages,[124] thus suggesting that this CATL isoform is important in visceral leishmaniasis.

L. mexicana mutants lacking both CATL-A and CATL-B genes did not induce lesion growth in BALB/c mice, indicating that a combination of enzymes with CATL-like activities was required for establishment of infection.[118] Also, in mice infected with such mutants there was a shift from a Th2-type to a Th1-type immune response. When these CATL-deficient parasites were tested as vaccines in different strains of mice they induced various degrees of protection from challenges with wild type *L. mexicana*.[118] In hamsters the CATL-deficient mutants delayed onset of disease as recorded by smaller lesions and lower parasite burden.[125] Finally, for *Lme*CATB mutants, infectivity to macrophages was decreased in vitro and in vivo and the resulting lesions were smaller.[126]

More recent studies have demonstrated direct immuno-modulatory effects of *Leishmania* CPs. *Lme*CATL-B2.8 increased IL-4 production, polarized splenocyte anti-CD3-stimulated responses toward a Th2-type bias, induced IgE production, cleaved CD23 and CD25 from murine lymphocytes and enhanced lesion size in mice.[121] *Lme*CATL-B suppressed the antileishmanial Th-1 type protective immune response[119] and was central to the parasite's ability to modulate signaling via NF-κB and consequently inhibit IL-12 production in macrophages.[127] Finally, *Lme*CATL-A and *Lme*CATL-B participate in the autophagy pathway and their genetic deletion prevents cell differentiation from the metacyclic to the amastigote form providing an explanation as to why *Lme*CATL-A and *Lme*CATL-B mutants lack virulence in vitro and in vivo.[128]

CONCLUSION

Over the last 30 years, we have witnessed tremendous advances in our understanding of the importance and diverse functionality of Clan CA Family C1 peptidases in the biology and pathogenesis of parasitic kinetoplastids. Fundamental research with *Leishmania*, specifically the genetic ablation of CP genes, has elucidated the contributions of CPs to cell differentiation, virulence, modulation of the hosts' immune response and not least, the species-specific idiosyncracies that arise. The applicability of CPs as vaccine candidates is also an active area of research. Likewise, the continuing progress in the design of small-molecule CP inhibitors offers hope for improved therapies of human African trypanosomiasis and Chagas' disease.

ACKNOWLEDGEMENTS

CRC is supported by the Sandler Foundation and APL by CNPq, FAPERJ and The Wellcome Trust.

REFERENCES

1. Stuart K, Brun R, Croft S et al. Kinetoplastids: related protozoan pathogens, different diseases. J Clin Invest 2008; 118(4):1301-1310.
2. Mottram JC, Helms MJ, Coombs GH et al. Clan CD cysteine peptidases of parasitic protozoa. Trends Parasitol 2003; 19(4):182-187.
3. Atkinson HJ, Babbitt PC, Sajid M. The global cysteine peptidase landscape in parasites. Trends Parasitol 2009; 25(12):573-581.
4. Caffrey CR, Steverding D. Kinetoplastid papain-like cysteine peptidases. Mol Biochem Parasitol 2009; 167(1):12-19.
5. Lima AP, Mottram JC. Trypanosome-encoded inhibitors of peptidases: unique structural features and possible roles as virulence factors. The Open Parasitology Journal 2010.
6. Lonsdale-Eccles JD, Grab DJ. Lysosomal and nonlysosomal peptidyl hydrolases of the bloodstream forms of Trypanosoma brucei brucei. Eur J Biochem 1987; 169(3):467-475.
7. Itow S, Camargo EP. Proteolytic activites in cell extracts of Trypanosoma cruzi. J Protozool 1977; 24(4):591-595.
8. North MJ, Coombs GH. Proteinases of Leishmania mexicana amastigotes and promastigotes: analysis by gel electrophoresis. Mol Biochem Parasitol 1981; 3(5):293-300.
9. North MJ, Coombs GH, Barry JD. A comparative study of the proteolytic enzymes of Trypanosoma brucei, T. equiperdum, T. evansi, T. vivax, Leishmania tarentolae and Crithidia fasciculata. Mol Biochem Parasitol 1983; 9(2):161-180.
10. North MJ, Mottram JC, Coombs GH. Cysteine proteinases of parasitic protozoa. Parasitol Today 1990; 6(8):270-275.
11. Rautenberg P, Schadler R, Reinwald E et al. Study on a proteolytic enzyme from Trypanosoma congolense. Purification and some biochemical properties. Mol Cell Biochem 1982; 47(3):151-159.
12. Robertson CD, North MJ, Lockwood BC et al. Analysis of the proteinases of Trypanosoma brucei. J Gen Microbiol 1990; 136(5):921-925.
13. Steiger RF, Van Hoof F, Bontemps J et al. Acid hydrolases of trypanosomatid flagellates. Acta Trop 1979; 36(4):335-341.
14. Mbawa ZR, Webster P, Lonsdale-Eccles JD. Immunolocalization of a cysteine protease within the lysosomal system of Trypanosoma congolense. Eur J Cell Biol 1991; 56(2):243-250.
15. Pupkis MF, Tetley L, Coombs GH. Leishmania mexicana: amastigote hydrolases in unusual lysosomes. Exp Parasitol 1986; 62(1):29-39.
16. Duboise SM, Vannier-Santos MA, Costa-Pinto D et al. The biosynthesis, processing and immunolocalization of Leishmania pifanoi amastigote cysteine proteinases. Mol Biochem Parasitol 1994; 68(1):119-132.
17. Pamer EG, So M, Davis CE. Identification of a developmentally regulated cysteine protease of Trypanosoma brucei. Mol Biochem Parasitol 1989; 33(1):27-32.
18. Caffrey CR, Hansell E, Lucas KD et al. Active site mapping, biochemical properties and subcellular localization of rhodesain, the major cysteine protease of Trypanosoma brucei rhodesiense. Mol Biochem Parasitol 2001; 118(1):61-73.
19. Mbawa ZR, Gumm ID, Fish WR et al. Endopeptidase variations among different life-cycle stages of African trypanosomes. Eur J Biochem 1991; 195(1):183-190.
20. Leon LL, Temporal RM, Soares MJ et al. Proteinase activities during temperature-induced stage differentiation of species complexes of Leishmania. Acta Trop 1994; 56(4):289-298.
21. Pral EM, Bijovsky AT, Balanco JM et al. Leishmania mexicana: proteinase activities and megasomes in axenically cultivated amastigote-like forms. Exp Parasitol 1993; 77(1):62-73.
22. Brooks DR, Denise H, Westrop GD et al. The stage-regulated expression of Leishmania mexicana CPB cysteine proteases is mediated by an intercistronic sequence element. J Biol Chem 2001; 276(50):47061-47069.
23. Tomas AM, Kelly JM. Stage-regulated expression of cruzipain, the major cysteine protease of Trypanosoma cruzi is independent of the level of RNA1. Mol Biochem Parasitol 1996; 76(1-2):91-103.
24. Lima AP, dos Reis FC, Serveau C et al. Cysteine protease isoforms from Trypanosoma cruzi, cruzipain 2 and cruzain, present different substrate preference and susceptibility to inhibitors. Mol Biochem Parasitol 2001; 114(1):41-52.
25. Mottram JC, North MJ, Barry JD et al. A cysteine proteinase cDNA from Trypanosoma brucei predicts an enzyme with an unusual C-terminal extension. FEBS Lett 1989; 258(2):211-215.
26. Pamer EG, Davis CE, Eakin A et al. Cloning and sequencing of the cysteine protease cDNA from Trypanosoma brucei rhodesiense. Nucleic Acids Res 1990; 18(20):6141.
27. Fish WR, Nkhungulu ZM, Muriuki CW et al. Primary structure and partial characterization of a life-cycle-regulated cysteine protease from Trypanosoma (Nannomonas) congolense. Gene 1995; 161(1):125-128.
28. Campetella O, Henriksson J, Aslund L et al. The major cysteine proteinase (cruzipain) from Trypanosoma cruzi is encoded by multiple polymorphic tandemly organized genes located on different chromosomes. Mol Biochem Parasitol 1992; 50(2):225-234.

29. Traub-Cseko YM, Duboise M, Boukai LK et al. Identification of two distinct cysteine proteinase genes of Leishmania pifanoi axenic amastigotes using the polymerase chain reaction. Mol Biochem Parasitol 1993; 57(1):101-115.
30. Mottram JC, Robertson CD, Coombs GH et al. A developmentally regulated cysteine proteinase gene of Leishmania mexicana. Mol Microbiol 1992; 6(14):1925-1932.
31. Souza AE, Waugh S, Coombs GH et al. Characterization of a multi-copy gene for a major stage-specific cysteine proteinase of Leishmania mexicana. FEBS Lett 1992; 311(2):124-127.
32. Marin-Villa M, Vargas-Inchaustegui DA, Chaves SP et al. The C-terminal extension of Leishmania pifanoi amastigote-specific cysteine proteinase Lpcys2: a putative function in macrophage infection. Mol Biochem Parasitol 2008; 162(1):52-59.
33. Bart G, Coombs GH, Mottram JC. Isolation of lmcpc, a gene encoding a Leishmania mexicana cathepsin-B-like cysteine proteinase. Mol Biochem Parasitol 1995; 73(1-2):271-274.
34. Robertson CD, Coombs GH. Cathepsin B-like cysteine proteases of Leishmania mexicana. Mol Biochem Parasitol 1993; 62(2):271-279.
35. Sakanari JA, Nadler SA, Chan VJ et al. Leishmania major: comparison of the cathepsin L- and B-like cysteine protease genes with those of other trypanosomatids. Exp Parasitol 1997; 85(1):63-76.
36. Garcia MP, Nobrega OT, Teixeira AR et al. Characterisation of a Trypanosoma cruzi acidic 30 kDa cysteine protease. Mol Biochem Parasitol 1998; 91(2):263-272.
37. Nobrega OT, Santos Silva MA, Teixeira AR et al. Cloning and sequencing of tccb, a gene encoding a Trypanosoma cruzi cathepsin B-like protease. Mol Biochem Parasitol 1998; 97(1-2):235-240.
38. Mackey ZB, O'Brien TC, Greenbaum DC et al. A cathepsin B-like protease is required for host protein degradation in Trypanosoma brucei. J Biol Chem 2004; 279(46):48426-48433.
39. Mendoza-Palomares C, Biteau N, Giroud C et al. Molecular and biochemical characterization of a cathepsin B-like protease family unique to Trypanosoma congolense. Eukaryot Cell 2008; 7(4):684-697.
40. Somanna A, Mundodi V, Gedamu L. Functional analysis of cathepsin B-like cysteine proteases from Leishmania donovani complex. Evidence for the activation of latent transforming growth factor beta. J Biol Chem 2002; 277(28):25305-25312.
41. Troeberg L, Pike RN, Morty RE et al. Proteases from Trypanosoma brucei brucei. Purification, characterisation and interactions with host regulatory molecules. Eur J Biochem 1996; 238(3):728-736.
42. Mbawa ZR, Gumm ID, Shaw E et al. Characterisation of a cysteine protease from bloodstream forms of Trypanosoma congolense. Eur J Biochem 1992; 204(1):371-379.
43. Clayton C, Adams M, Almeida R et al. Genetic nomenclature for Trypanosoma and Leishmania. Mol Biochem Parasitol 1998; 97(1-2):221-224.
44. Berriman M, Ghedin E, Hertz-Fowler C et al. The genome of the African trypanosome Trypanosoma brucei. Science 2005; 309(5733):416-422.
45. El-Sayed NM, Myler PJ, Bartholomeu DC et al. The genome sequence of Trypanosoma cruzi, etiologic agent of Chagas disease. Science 2005; 309(5733):409-415.
46. Ivens AC, Peacock CS, Worthey EA et al. The genome of the kinetoplastid parasite, Leishmania major. Science 2005; 309(5733):436-442.
47. Matthews KR, Tschudi C, Ullu E. A common pyrimidine-rich motif governs trans-splicing and polyadenylation of tubulin polycistronic premRNA in trypanosomes. Genes Dev 1994; 8(4):491-501.
48. Lima AP, Tessier DC, Thomas DY et al. Identification of new cysteine protease gene isoforms in Trypanosoma cruzi. Mol Biochem Parasitol 1994; 67(2):333-338.
49. Mundodi V, Somanna A, Farrell PJ et al. Genomic organization and functional expression of differentially regulated cysteine protease genes of Leishmania donovani complex. Gene 2002; 282(1-2):257-265.
50. Mottram JC, Brooks DR, Coombs GH. Roles of cysteine proteinases of trypanosomes and Leishmania in host-parasite interactions. Curr Opin Microbiol 1998; 1(4):455-460.
51. Mottram JC, Souza AE, Hutchison JE et al. Evidence from disruption of the lmcpb gene array of Leishmania mexicana that cysteine proteinases are virulence factors. Proc Natl Acad Sci USA 1996; 93(12):6008-6013.
52. Brooks DR, Tetley L, Coombs GH et al. Processing and trafficking of cysteine proteases in Leishmania mexicana. J Cell Sci 2000; 113(22):4035-4041.
53. Engel JC, Doyle PS, Hsieh I et al. Cysteine protease inhibitors cure an experimental Trypanosoma cruzi infection. J Exp Med 1998; 188(4):725-734.
54. Tazeh NN, Silverman JS, Schwartz KJ et al. Role of AP-1 in developmentally regulated lysosomal trafficking in Trypanosoma brucei. Eukaryot Cell 2009; 8(9):1352-1361.
55. Mach L, Mort JS, Glossl J. Maturation of human procathepsin B. Proenzyme activation and proteolytic processing of the precursor to the mature proteinase, in vitro, are primarily unimolecular processes. J Biol Chem 1994; 269(17):13030-13035.
56. Menard R, Carmona E, Takebe S et al. Autocatalytic processing of recombinant human procathepsin L. Contribution of both intermolecular and unimolecular events in the processing of procathepsin L in vitro. J Biol Chem 1998; 273(8):4478-4484.

57. Eakin AE, Mills AA, Harth G et al. The sequence, organization and expression of the major cysteine protease (cruzain) from Trypanosoma cruzi. J Biol Chem 1992; 267(11):7411-7420.
58. Sanderson SJ, Pollock KG, Hilley JD et al. Expression and characterization of a recombinant cysteine proteinase of Leishmania mexicana. Biochem J 2000; 347(2):383-388.
59. Huete-Perez JA, Engel JC, Brinen LS et al. Protease trafficking in two primitive eukaryotes is mediated by a prodomain protein motif. J Biol Chem 1999; 274(23):16249-16256.
60. Reis FC, Costa TF, Sulea T et al. The propeptide of cruzipain—a potent selective inhibitor of the trypanosomal enzymes cruzipain and brucipain and of the human enzyme cathepsin F. FEBS J 2007; 274(5):1224-1234.
61. Monteiro AC, Abrahamson M, Lima AP et al. Identification, characterization and localization of chagasin, a tight-binding cysteine protease inhibitor in Trypanosoma cruzi. J Cell Sci 2001; 114(21):3933-3942.
62. Sanderson SJ, Westrop GD, Scharfstein J et al. Functional conservation of a natural cysteine peptidase inhibitor in protozoan and bacterial pathogens. FEBS Lett 2003; 542(1-3):12-16.
63. Santos CC, Coombs GH, Lima AP et al. Role of the Trypanosoma brucei natural cysteine peptidase inhibitor ICP in differentiation and virulence. Mol Microbiol 2007; 66(4):991-1002.
64. Santos CC, Sant'anna C, Terres A et al. Chagasin, the endogenous cysteine-protease inhibitor of Trypanosoma cruzi, modulates parasite differentiation and invasion of mammalian cells. J Cell Sci 2005; 118(5):901-915.
65. Authie E, Muteti DK, Mbawa ZR et al. Identification of a 33-kilodalton immunodominant antigen of Trypanosoma congolense as a cysteine protease. Mol Biochem Parasitol 1992; 56(1):103-116.
66. Lima AP, Scharfstein J, Storer AC et al. Temperature-dependent substrate inhibition of the cysteine proteinase (GP57/51) from Trypanosoma cruzi. Mol Biochem Parasitol 1992; 56(2):335-338.
67. Murta AC, Persechini PM, Padron Tde S et al. Structural and functional identification of GP57/51 antigen of Trypanosoma cruzi as a cysteine proteinase. Mol Biochem Parasitol 1990; 43(1):27-38.
68. Nery ED, Juliano MA, Meldal M et al. Characterization of the substrate specificity of the major cysteine protease (cruzipain) from Trypanosoma cruzi using a portion-mixing combinatorial library and fluorogenic peptides. Biochem J 1997; 323 (2):427-433.
69. Boulange A, Serveau C, Brillard M et al. Functional expression of the catalytic domains of two cysteine proteinases from Trypanosoma congolense. Int J Parasitol 2001; 31(13):1435-1440.
70. Chagas JR, Authie E, Serveau C et al. A comparison of the enzymatic properties of the major cysteine proteinases from Trypanosoma congolense and Trypanosoma cruzi. Mol Biochem Parasitol 1997; 88(1-2):85-94.
71. Gillmor SA, Craik CS, Fletterick RJ. Structural determinants of specificity in the cysteine protease cruzain. Protein Sci 1997; 6(8):1603-1611.
72. Judice WA, Puzer L, Cotrin SS et al. Carboxydipeptidase activities of recombinant cysteine peptidases. Cruzain of Trypanosoma cruzi and CPB of Leishmania mexicana. Eur J Biochem 2004; 271(5):1046-1053.
73. Nagler DK, Tam W, Storer AC et al. Interdependency of sequence and positional specificities for cysteine proteases of the papain family. Biochemistry 1999; 38(15):4868-4874.
74. Lima AP, Almeida PC, Tersariol IL et al. Heparan sulfate modulates kinin release by Trypanosoma cruzi through the activity of cruzipain. J Biol Chem 2002; 277(8):5875-5881.
75. Hodder AN, Drew DR, Epa VC et al. Enzymic, phylogenetic and structural characterization of the unusual papain-like protease domain of Plasmodium falciparum SERA5. J Biol Chem 2003; 278(48):48169-48177.
76. Authie E, Boulange A, Muteti D et al. Immunisation of cattle with cysteine proteinases of Trypanosoma congolense: targetting the disease rather than the parasite. Int J Parasitol 2001; 31(13):1429-1433.
77. Schnapp AR, Eickhoff CS, Scharfstein J et al. Induction of B- and T-cell responses to cruzipain in the murine model of Trypanosoma cruzi infection. Microbes Infect 2002; 4(8):805-813.
78. Cazorla SI, Frank FM, Becker PD et al. Prime-boost immunization with cruzipain co-administered with MALP-2 triggers a protective immune response able to decrease parasite burden and tissue injury in an experimental Trypanosoma cruzi infection model. Vaccine 2008; 26(16):1999-2009.
79. Ferreira JH, Gentil LG, Dias SS et al. Immunization with the cysteine proteinase Ldccys1 gene from Leishmania (Leishmania) chagasi and the recombinant Ldccys1 protein elicits protective immune responses in a murine model of visceral leishmaniasis. Vaccine 2008; 26(5):677-685.
80. Rafati S, Salmanian AH, Taheri T et al. A protective cocktail vaccine against murine cutaneous leishmaniasis with DNA encoding cysteine proteinases of Leishmania major. Vaccine 2001; 19(25-26):3369-3375.
81. Rafati S, Zahedifard F, Nazgouee F. Prime-boost vaccination using cysteine proteinases type I and II of Leishmania infantum confers protective immunity in murine visceral leishmaniasis. Vaccine 2006; 24(12):2169-2175.
82. Khoshgoo N, Zahedifard F, Azizi H et al. Cysteine proteinase type III is protective against Leishmania infantum infection in BALB/c mice and highly antigenic in visceral leishmaniasis individuals. Vaccine 2008; 26(46):5822-5829.
83. Authie E. Trypanosomiasis and trypanotolerance in cattle: a role for congopain? Parasitol Today 1994; 10(9):360-364.

84. Scharfstein J, Rodrigues MM, Alves CA et al. Trypanosoma cruzi: description of a highly purified surface antigen defined by human antibodies. J Immunol 1983; 131(2):972-976.
85. Martinez J, Campetella O, Frasch AC et al. The major cysteine proteinase (cruzipain) from Trypanosoma cruzi is antigenic in human infections. Infect Immun 1991; 59(11):4275-4277.
86. Scharfstein J, Luquetti A, Murta AC et al. Chagas' disease: serodiagnosis with purified Gp25 antigen. Am J Trop Med Hyg 1985; 34(6):1153-1160.
87. Martinez J, Campetella O, Frasch AC et al. The reactivity of sera from chagasic patients against different fragments of cruzipain, the major cysteine proteinase from Trypanosoma cruzi, suggests the presence of defined antigenic and catalytic domains. Immunol Lett 1993; 35(2):191-196.
88. Rosenthal PJ, Lee GK, Smith RE. Inhibition of a Plasmodium vinckei cysteine proteinase cures murine malaria. J Clin Invest 1993; 91(3):1052-1056.
89. Croft SL. Kinetoplastida: new therapeutic strategies. Parasite 2008; 15(3):522-527.
90. McKerrow JH, Rosenthal PJ, Swenerton R et al. Development of protease inhibitors for protozoan infections. Curr Opin Infect Dis 2008; 21(6):668-672.
91. Sajid M, McKerrow JH. Cysteine proteases of parasitic organisms. Mol Biochem Parasitol 2002; 120(1):1-21.
92. Turk B, Turk D, Turk V. Lysosomal cysteine proteases: more than scavengers. Biochim Biophys Acta 2000; 1477(1-2):98-111.
93. Ashall F, Angliker H, Shaw E. Lysis of trypanosomes by peptidyl fluoromethyl ketones. Biochem Biophys Res Commun 1990; 170(2):923-929.
94. Scory S, Caffrey CR, Stierhof YD et al. Trypanosoma brucei: killing of bloodstream forms in vitro and in vivo by the cysteine proteinase inhibitor Z-Phe-Ala-CHN2. Exp Parasitol 1999; 91(4):327-333.
95. Scory S, Stierhof YD, Caffrey CR et al. The cysteine proteinase inhibitor Z-Phe-Ala-CHN2 alters cell morphology and cell division activity of Trypanosoma brucei bloodstream forms in vivo. Kinetoplastid Biol Dis 2007; 6:2.
96. Du X, Hansell E, Engel JC et al. Aryl ureas represent a new class of anti-trypanosomal agents. Chem Biol 2000; 7(9):733-742.
97. Bryant C, Kerr ID, Debnath M et al. Novel nonpeptidic vinylsulfones targeting the S2 and S3 subsites of parasite cysteine proteases. Bioorg Med Chem Lett 2009; 19(21):6218-6221.
98. Chen YT, Lira R, Hansell E et al. Synthesis of macrocyclic trypanosomal cysteine protease inhibitors. Bioorg Med Chem Lett 2008; 18(22):5860-5863.
99. Jaishankar P, Hansell E, Zhao DM et al. Potency and selectivity of P2/P3-modified inhibitors of cysteine proteases from trypanosomes. Bioorg Med Chem Lett 2008; 18(2):624-628.
100. Roush WR, Cheng J, Knapp-Reed B et al. Potent second generation vinyl sulfonamide inhibitors of the trypanosomal cysteine protease cruzain. Bioorg Med Chem Lett 2001; 11(20):2759-2762.
101. Roush WR, Gonzalez FV, McKerrow JH et al. Design and synthesis of dipeptidyl 'alpha,beta'-epoxy ketones, potent irreversible inhibitors of the cysteine protease cruzain. Bioorg Med Chem Lett 1998; 8(19):2809-2812.
102. Ferreira RS, Bryant C, Ang KK et al. Divergent Modes of Enzyme Inhibition in a Homologous Structure-Activity Series. J Med Chem 2009; 52(16):5005-5008.
103. McKerrow JH, Doyle PS, Engel JC et al. Two approaches to discovering and developing new drugs for Chagas disease. Mem Inst Oswaldo Cruz 2009; 104:263-269.
104. Caffrey CR, Scory S, Steverding D. Cysteine proteinases of trypanosome parasites: novel targets for chemotherapy. Curr Drug Targets 2000; 1(2):155-162.
105. Troeberg L, Chen X, Flaherty TM et al. Chalcone, acyl hydrazide and related amides kill cultured Trypanosoma brucei brucei. Mol Med 2000; 6(8):660-669.
106. Vicik R, Hoerr V, Glaser M et al. Aziridine-2,3-dicarboxylate inhibitors targeting the major cysteine protease of Trypanosoma brucei as lead trypanocidal agents. Bioorg Med Chem Lett 2006; 16(10):2753-2757.
107. Mallari JP, Shelat A, Kosinski A et al. Discovery of trypanocidal thiosemicarbazone inhibitors of rhodesain and TbcatB. Bioorg Med Chem Lett 2008; 18(9):2883-2885.
108. Mallari JP, Shelat AA, Obrien T et al. Development of potent purine-derived nitrile inhibitors of the trypanosomal protease TbcatB. J Med Chem 2008; 51(3):545-552.
109. Watts KR, Ratnam J, Ang KH et al. Assessing the trypanocidal potential of natural and semi-synthetic diketopiperazines from two deep water marine-derived fungi. Bioorg Med Chem 2010; 18(7):2566-2574.
110. Agius B, Vogler B, Stokes SL et al. Inhibition of cruzain by triterpenoids isolated from a Salacia species from Monteverde, Costa Rica. Nat Prod Commun 2007; 2:1083-1084.
111. Setzer W, Stokes SL, Bansala A et al. Chemical composition and cruzain inhibitory activity of Croton draco bark essential oil from Monteverde, Costa Rica. Nat Prod Commun 2007; 2:685-689.
112. Mallari JP, Shelat AA, Kosinski A et al. Structure-guided development of selective TbcatB inhibitors. J Med Chem 2009; 52(20):6489-6493.
113. Nkemgu NJ, Grande R, Hansell E et al. Improved trypanocidal activities of cathepsin L inhibitors. Int J Antimicrob Agents 2003; 22(2):155-159.

114. Nikolskaia OV, de ALAP, Kim YV et al. Blood-brain barrier traversal by African trypanosomes requires calcium signaling induced by parasite cysteine protease. J Clin Invest 2006; 116(10):2739-2747.
115. Abdulla MH, O'Brien T, Mackey ZB et al. RNA interference of Trypanosoma brucei cathepsin B and L affects disease progression in a mouse model. PLoS Negl Trop Dis 2008; 2(9):e298.
116. Grab DJ, Garcia-Garcia JC, Nikolskaia OV et al. Protease activated receptor signaling is required for African trypanosome traversal of human brain microvascular endothelial cells. PLoS Negl Trop Dis 2009; 3(7):e479.
117. O'Brien TC, Mackey ZB, Fetter RD et al. A parasite cysteine protease is key to host protein degradation and iron acquisition. J Biol Chem 2008; 283(43):28934-28943.
118. Alexander J, Coombs GH, Mottram JC. Leishmania mexicana cysteine proteinase-deficient mutants have attenuated virulence for mice and potentiate a Th1 response. J Immunol 1998; 161(12):6794-6801.
119. Buxbaum LU, Denise H, Coombs GH et al. Cysteine protease B of Leishmania mexicana inhibits host Th1 responses and protective immunity. J Immunol 2003; 171(7):3711-3717.
120. Mottram JC, Coombs GH, Alexander J. Cysteine peptidases as virulence factors of Leishmania. Curr Opin Microbiol 2004; 7(4):375-381.
121. Pollock KG, McNeil KS, Mottram JC et al. The Leishmania mexicana cysteine protease, CPB2.8, induces potent Th2 responses. J Immunol 2003; 170(4):1746-1753.
122. Denise H, McNeil K, Brooks DR et al. Expression of multiple CPB genes encoding cysteine proteases is required for Leishmania mexicana virulence in vivo. Infect Immun 2003; 71(6):3190-3195.
123. Souza AE, Bates PA, Coombs GH et al. Null mutants for the lmcpa cysteine proteinase gene in Leishmania mexicana. Mol Biochem Parasitol 1994; 63(2):213-220.
124. Denise H, Poot J, Jimenez M et al. Studies on the CPA cysteine peptidase in the Leishmania infantum genome strain JPCM5. BMC Mol Biol 2006; 7:42.
125. Saravia NG, Escorcia B, Osorio Y et al. Pathogenicity and protective immunogenicity of cysteine proteinase-deficient mutants of Leishmania mexicana in nonmurine models. Vaccine 2006; 24(19):4247-4259.
126. Bart G, Frame MJ, Carter R et al. Cathepsin B-like cysteine proteinase-deficient mutants of Leishmania mexicana. Mol Biochem Parasitol 1997; 88(1-2):53-61.
127. Cameron P, McGachy A, Anderson M et al. Inhibition of lipopolysaccharide-induced macrophage IL-12 production by Leishmania mexicana amastigotes: the role of cysteine peptidases and the NF-kappaB signaling pathway. J Immunol 2004; 173(5):3297-3304.
128. Williams RA, Tetley L, Mottram JC et al. Cysteine peptidases CPA and CPB are vital for autophagy and differentiation in Leishmania mexicana. Mol Microbiol 2006; 61(3):655-674.

CHAPTER 7

CRUZAIN
The Path from Target Validation to the Clinic

Mohammed Sajid,[1] Stephanie A. Robertson,[2] Linda S. Brinen[2,3] and James H. McKerrow*,[2]

[1]Leiden University Medical Center, Afd. Parasitologie, Leiden, The Netherlands; [2]Sandler Center for Drug Discovery, [3]Department of Cellular and Molecular Pharmacology, University of California–San Francisco, San Francisco, California, USA

**Corresponding Author: J.H. McKerrow—Email: jmck@cgl.ucsf.edu*

Abstract: Cruzain is the major papain-like cysteine protease of *Trypanosoma cruzi*, the etiological agent causing Chagas' disease in humans in South America. Cruzain is indispensable for the survival and propagation of this protozoan parasite and therefore, it has attracted considerable interest as a potential drug target. This chapter charts the path from the initial identification of this proteases activity and its validation as a bone fide drug target to the arduous task of the discovery of an inhibitor targeting this protease and finally the path towards the clinic.

INTRODUCTION

In 1909, the Brazilian physician Carlos Chagas identified the euglenoid parasitic protozoan, *Trypanosoma cruzi* as the agent causing the potentially fatal infection, American trypanosomiasis; later to become known as Chagas' disease.[1] Chagas' disease is directly responsible for 50,000 deaths annually and is endemic in Mexico and many parts of South America. In 2010, the Center for Disease Control (CDC) estimated that there are between 8-11 million people infected with Chagas' disease in Latin America with over 100 million at risk, equating to about 25% of the population. The annual economic loss attributed to the disease is estimated to be in the order of 18 billion US dollars.[2]

The complex digenic lifecyle of *T. cruzi* involves shuttling between mammalian hosts, including humans, and an insect stage. Chagas' disease is transmitted by the hematophagous

Cysteine Proteases of Pathogenic Organisms, edited by Mark W. Robinson and John P. Dalton.

reduviid bug (also known as the kissing bug or assassin bug), a predatory blood-sucking insect. As it bites and feeds on mammalian blood, the kissing bug deposits infected feces into the skin and thus allows the transmission stage of the parasite, the trypomastigotes, to invade; invasion occurs through the bite wound and mucous membranes. The blood stream trypomastigote stage can invade a large number of cell types, a process that may or may not employ phagocytosis. Once in a cell the parasite transforms into the nonflagellar replicative amastigote. Infection of visceral organs causes a number of mega-syndromes, notably in the heart, colon and esophagus; heart failure is common in infected patients. Prior to host cell rupture, the amastigotes transform back into trypomastigotes that, once free, can either re-invade a new host cell or be taken up via a blood meal by the kissing bug. In the insect stage, the replicating form is the epimastigote; during movement into the insect rectum the parasite transforms into the transmissive metacyclic trypomastigote.

Transmission from infected blood and congenital transmission from mother to child have been well documented. The oral route is also a cause of transmission via ingestion of unwashed or undercooked food. In addition, due to increased international travel and immigration patterns, the last two decades have seen a significant spread of the disease spread to Southern US and Europe.

CURRENT TREATMENTS FOR CHAGAS' DISEASE

Although no prophylaxis currently exists, the nitroheterocyclics nifurtimox (Lampit, Bayer) and benznidazole (Radanil/Rochagan, Roche) are used in the clinic to cure acute phase Chagas' disease. However, these drugs are not available in the USA or Canada except through compassionate distribution by the CDC. These drugs have substantial side effects and are poorly effective (10-20%) against long-term (>10 years) chronic infections of Chagas' disease.[3] In addition, treatment with the aforementioned drugs have lengthy regimens of 2 to 3 pills per day for 60-120 days; the severe side-effects frequently result in patients abandoning treatment.

There is clearly an urgent and pressing need to develop new antiChagas' chemotherapies with improved patho-pharmacological properties compared to the currently available drugs. As the *T. cruzi* parasite undergoes cycles of intracellular replication, rupture and invasion, metabolic enzymes have become the focus of drug targeting. Key pathways of interest include polyamine metabolism, sterol and isoprenoid biosynthesis, redox sytems, transport pathways, trypanothione synthesis, trans-sialidase, pentose phosphate, arginase kinase and proteases.[4-6] Of the 200 or so proteases (MEROPS, see below) identified in *T. cruzi*, a papain family enzyme, termed cruzain, is critical in the pathobiology of Chagas' disease and has shown great promise as a novel drug target in the development of new antiChagasides.

CRUZAIN

In the mid to late 1970s, a number of protease activities, including cysteine protease (CP) activities, were described in crude *T. cruzi* extracts.[7,8] By the 1980s, CPs had been localized to the lysosome and purified to homogeneity from parasite extracts.[9-11] The last two decades have seen a progressive increase in the detailed molecular, biochemical, structural and drug-targeting of cruzain as evidenced by the publication output (Fig. 1). The enzyme was originally termed cruzipain as it was the major *T.* ***cruzi*** pa***pain*** family

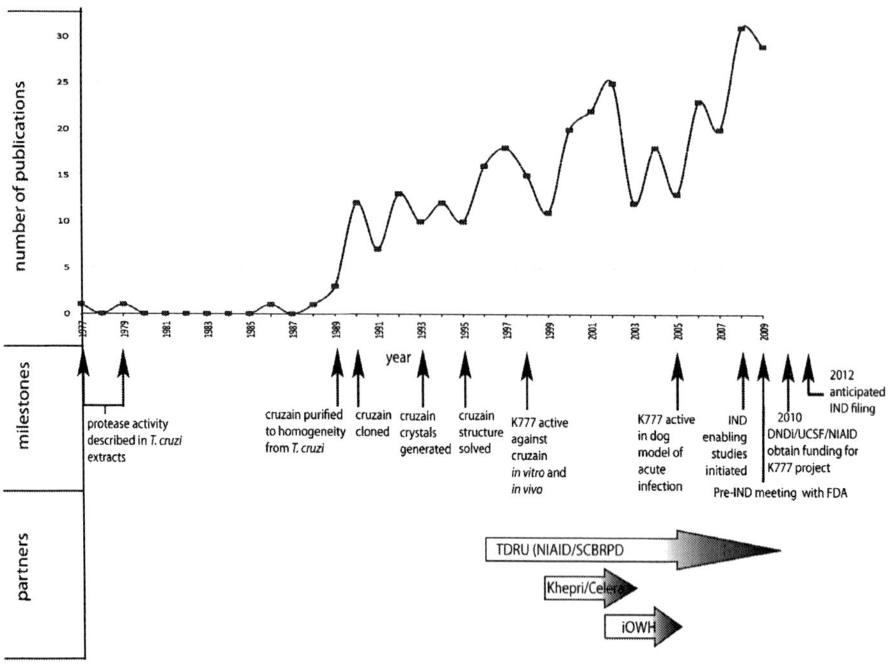

Figure 1. A temporal overview of publication output, key milestones and industry-academic partnerships involved with K777. The present day represents K777 progressing through IND-enabling studies in 2010. TDRU, Tropical Disease Research Unit; NIAID, National Institute for Allergies and Infections; SCBRPD, Sandler Center for Basic Research in Parasitic Disease (now Sandler Center for Drug Discovery).

CP. Cruzipain (a.k.a. Gp57/51; EC = 3.4.22.51; Swiss-Prot P25779 (CYSP_TRYCR)) referred to the native parasite-derived enzyme and 'cruzain' to the recombinantly-expressed protein (Fig. 2). However, in the literature, cruzain and cruzipain are used interchangeably and for purpose of this chapter, cruzain will be used to describe all forms.

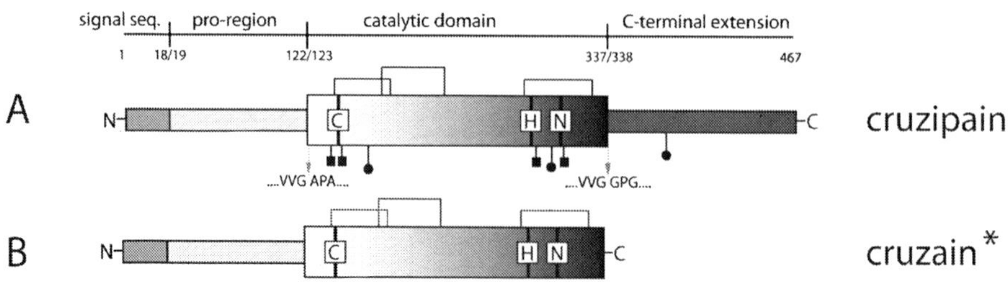

Figure 2. Domain structure of A) native cruzipain and B) *recombinant cruzain used in biochemical and structural studies. Key; the N- and C- are the amino and carboxyl ends of the protein; C, H and N are the catalytic Cys_{25}, His_{162} and Asn_{182} (cruzain numbering); squares above the catalytic domain represent internal disulfide bonds; N (filled squares) and O (filled circles) glycosylation sites are marked; the cleavage site between the pro-and catalytic domain and catalytic domain and the C-terminal extension is highlighted with a grey arrow and amino acids flanking the scissile bond are shown.

Cruzain has a typical domain architecture common to papain family enzymes. These comprise a signal peptide directing the protein to the ER, an indispensible multifunctional pro-region functioning first as an intramolecular chaperone responsible for the correct folding of the nascent enzyme, second as a potent inhibitor of cruzain enzymatic activity[12] and third as an essential element for intracellular trafficking.[13] In addition, cruzain contains a 205 amino acid core enzymatic domain that harbours all the catalytic residues, namely Cys_{25}, His_{162}, Asn_{182} and the oxyanion hole, Asn_{21} (cruzain numbering). Cruzain is heavily N- and O-glycosylated (up to 10% of the total mass) and the high mannose N-glycans have been shown to be sulfated.[14-16] Indeed, the glycosylation allowed affinity purification of native cruzain using ConA-sepharose.[17] Cruzain may also be sialycilated in the C-terminal domain.[18] In addition, *T. cruzi*, related trypanomastigotes and some plant CPs possess a unque C-terminal extension of unknown function (Fig. 2). The C-terminal domain contains considerable posttranslational modifications and has been shown to be an immunodominant protein, antibodies against which are found in patients with Chagas' disease.[19-21] A variant of the canonical C-terminal extension that is highly hydrophobic has also been described, however, there is no evidence that this variant is expressed.[22]

Cruzain is ubiquitously expressed in all life stages of the parasite (in all strains studied) with higher levels in the epimastigotes.[23] In *T. cruzi*, the cruzain activity represents a vastly-expanded number of parologous genes lacking introns arranged in tandem repeating arrays on a number of chromosomes. Depending on the strain and life-cycle stage, 37-130 copies have been described.[22,24] The number of genes and varied posttranslational modifications probably explains the heterogeneity observed in purified protein from parasites, with pI values between 6.4 to 8.1 and a molecular mass estimation between 40kDa to 60kDa.[25,26]

BIOCHEMISTRY OF CRUZAIN

Cruzain belongs to the Clan CA, family C1A cysteine peptidases as described by MEROPS, an online curated database of protease information, classification and nomenclature (http://merops.sanger.ac.uk/). As a member of the cathepsin L-superfamily, cruzain is described in the literature as cathepsin L-like. However, maybe a more accurate description would be to call cruzain cathepsin F-like, as it has greater sequence identity to (50.5%) and the least sequence divergence from, the human cathepsin F enzyme within the cathepsin L-superfamily (Fig. 3A). All papain family enzymes have a primary preference for amino acid residues that correspond to the P2 of the substrate and interact with the S2 subsite of the enzyme active site.[27] The amino acid at the base of the S2 substrate-binding pocket of papain family enzymes plays a significant role in the physicochemical preference for substrate residues at that position; typically amino acids with a noncharged aliphatic or aromatic side group. Sequence pileup analysis and modeling using the available papain atomic structure first revealed that the critical amino acid residue at the base of the S2 pocket of cruzain was Glu_{205} (Fig. 3B), which was unusual because at that time acidic S2 residues were only characterized in cathepsin B-like enzymes. All natural variations of cruzain that have been characterized have a Glu at this position.[24,28-30] The presence of Glu at position 205 in the base of the S2 pocket was confirmed when the molecular structure of cruzain was determined.[31] Because cruzain possesses Glu_{205}, it can accommodate Arg at the P2 position.

While hydrophobic side chains at P2 are preferred at the pH optimum of cruzain (pH 5.5), substrate specificity has been shown to be influenced by the pH of the milieu.[70] The carboxylic acid moiety of Glu_{205} swings into the S2 pocket at neutral pH but is

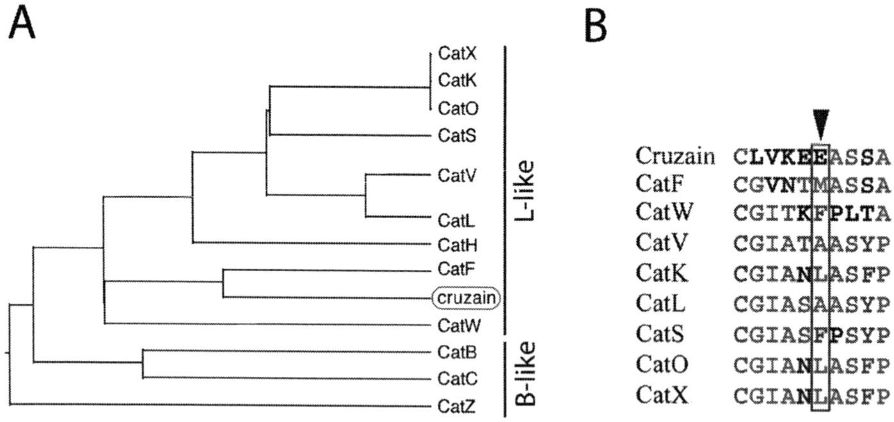

Figure 3. A) Phylogenetic tree using the core catalytic domain generated using ClustalW, reveals that cruzain is most similar to human cathepsin F. B) Comparison of the residue at the base of the S2 sub-site generated using ClustalW (at www.sacs.ucsf.edu) shows that, unlike all human cathepsin L-superfamily members, cruzain possesses an anionic (Glu_{205}) residue at that position.

directed away at acid pH.[70] This becomes biologically significant in the amastigote stage where cruzain is found on the parasite surface in contact with host cell cytoplasm. For recombinant cruzain, the Km values for the fluorescent peptide substrates, Z-Arg-Arg-AMC and Z-Phe-Arg-AMC, are typically around 120μM and 5-10μM, respectively, at the acid pH optimum.[32,33] The preference for hydrophobic residues is corroborated by studies using a positional scanning combinatorial peptide substrate library with a fixed P1 Arg,[34] where an S2 preference for Leu>Phe>Tyr>Val>Trp>Ile was obtained. Studies mapping preference for macromolecular substrates revealed that of the 49 characterized cleavage sites, 33 had a Leu at the P2 position (MEROPS). It is noteworthy, that the 2 autocatalytic *cis*-processing sites of the native form of cruzain, namely between the prodomain and catalytic domain and also between the catalytic domain and the C-terminal extension, both cleave after the tripeptide sequence -Val-Val-Gly (corresponding to P3-P2-P1; Fig. 2A). Further details of the contribution of Glu_{205} and chemical nature of the hydrophobic S2 pocket are discussed in the structure section below.

The gene encoding cruzain was originally cloned and sequenced in the early 1990s.[30] The primary amino acid sequence of cruzain is most closely related to the cysteine protease of *T. brucei* (59.3%) and murine cathepsin L (44.2%). The enzyme was first expressed in *E. coli* as a fusion polypeptide, with proteolytic processing events required to obtain the mature, active cruzain (this form was subsequently crystallized).[36] For the purpose of obtaining homogenous enzyme for crystallization and structural studies, a C-terminal truncated form (cruzain-Δc) was engineered by insertion of a stop codon in the gene at a position corresponding to the site of auto-proteolysis observed with the native enzyme purified from epimastigotes (Fig. 2B). The first X-ray crystallographic structure was solved from this construct of cruzain, bound to a small-molecule inhibitor, Z-Phe-Ala-fluoromethyl ketone.[31] This 2.35 Å atomic structure elucidated the key binding and specificity elements of cruzain's active site region and was the starting point for all future structure-guided drug design studies of this enzyme.

BIOLOGICAL ROLES OF CRUZAIN

Cruzain is expressed in all life-cycle stages and has been localized to the lysosome, the prelysosomal 'reservasome', the flagellar pocket and to the plasma membrane in both epimastigotes and amastogites.[22,30,37-40] In addition to contributing to general protein turnover as well as nutrient processing, cruzain may play a number of additional essential roles. Cruzain's role in invasion was proposed as Fab fragments of antibodies that targeted cruzain markedly reduced the invasion of trypomastigotes into macrophages.[37] In addition, strains with higher cruzain activities were shown to be more invasive, a process that was sensitive to CP inhibitors.[41] Plasma membrane resident cruzain has been shown to degrade the Fc region of antibodies, a mechanism that may thwart complement activation.[42,43]

Cruzain has also been shown to have a kininogenase activity and liberates the pro-inflammatory peptide Lys-bradykinin as well as trans-activating prekallikrein, which in turn may generate more Lys-bradykinin.[35,44,45] Lys-bradykinin (a.k.a. kallidin) is proinflammatory and also can activate host immune cells and therefore stimulate uptake of the parasite and contribute to propagating the disease.[46] Cruzain can hydrolyse human kininogen at Leu-Gly↑Met, Leu-Met↑Lys and Phe-Arg↑Ser (corresponding to the respective P3-P2↑P1').[35] Consistent with these extracellular roles, cruzain has also been shown to be released into the growth medium during in vitro culture of *T. cruzi* trypomastigotes.[47] Studies employing CP inhibitors have revealed a role for cruzain in the differentiation of trypomastigotes to amastigotes and amastigotes back to trypomastigotes[48,49] as well as in the proliferation of epimastigotes and amastigotes.[50-52] Consistent with the varied localization of cruzain, it has been shown to have a broad pH profile ranging from pH 4.5 to pH 9.[53,54]

In *T. cruzi* infection in humans there are a number of de novo naturally-occurring macromolecular inhibitors of cruzain. In addition to the cognate pro-domain (Ki at pH 7.0 is 0.018 nM)[55] parasites also express a cystatin-family inhibitor, chagasin, which significantly affects invasion and differentiation of *T. cruzi*.[56,57] Cruzain is also inhibited by plasma alpha-2-macroglobulin and kininogens as well as the intracellular stefin A and stefin B.[48,58-60] In a natural infection a strong humoral immune response against cruzain is elicited. Although the response is protective in that it limits the chronic infection, it cannot clear the parasite.[61,62] Vaccination with cruzain as an antigen has generated varying degrees of protection, nonetheless, parasite eradication in patients has not been achieved. Taken together, these results suggest that using cruzain as an antigen in a vaccine formulation is not a viable option at this point.[15,63-65] In light of this, chemotherapeutic targeting of cruzain is an appealing approach for the development of both a prophylaxis and cure for Chagas' disease.

CHEMICAL INHIBITION OF CRUZAIN—FIRST STUDIES AND BIOLOGICAL DATA

Early studies on chemical inhibition of cruzain activity in the parasite indicated parasite death was, at least in part, due to accumulation of the proteases precursor within the Golgi.[50,66] This mechanism of drug action reflects the autocatalytic activation of cruzain during, or shortly after, transit through the Golgi of the parasite. In the absence of protease activity, the prodomain remains attached to the catalytic domain of cruzain leading to unprocessed cruzain accumulation within the Golgi compartment. Eventually this protein accumulation leads to shock of the parasite Golgi and the entire endoplasmic reticulum and finally, parasite death.[50] Several classes of small-molecule inhibitors have been evaluated for their activity against cruzain. Many of these have been visualized *via* X-ray crystallographic structure determination in complex with cruzain. While most of these compounds have been peptidyl in their backbone[31,67-70] some recent potent inhibitors are nonpeptidyl.[71,72] The warheads that have been most examined structurally are based on vinyl sulfone, epoxysuccinate or ketone-based inhibitors.

THE STRUCTURAL BASIS OF CRUZAIN INHIBITION

As described above, cruzain is a papain-family cysteine protease that is folded into two domains. One is predominantly alpha-helical and the other consists of extensive antiparallel beta sheet interactions. The active site is found in the cleft at the interface of the two domains and within this cleft is, expectedly, the catalytic triad of residues Cys_{25}, His_{159} and Asn_{175} and extended substrate binding sites. [Note: cruzain numbering adheres to the numbering used in the original cruzain X-ray crystal structure, even though convention has changed and the most recently solved structures employ a slightly different numbering scheme.] The specificity of cruzain, as with other proteases of this family, is largely determined by the composition of the S2 pocket of the substrate binding cleft. Cruzain's S2 pocket is essentially hydrophobic in nature and is lined with residues Met_{68}, Ala_{133}, Leu_{157} and Gly_{160}. At the base of the pocket is Glu_{205}. There is ample room in the pocket for hydrophobic moieties at the P2 position of inhibitors and cruzain favors Phe and Leu at this position.[34,70,73] However, as mentioned previously, because Glu_{205} presents the opportunity for interactions with amines as well, Arg is also accepted at the P2 position. Cruzain prefers Phe over Arg, however, with a kcat/Km ratio of 90 at the acid pH optimum.[31] Comparison of the X-ray crystal structures of cruzain bound to three small molecule inhibitors, benzyol-Arg-Ala-fluoromethyl ketone (Cbz-RA-FMK), benzyol-Tyr-Ala-fluoromethyl ketone (Cbz-YA-FMK) and benzyol-Phe-Ala-fluoromethyl ketone (Cbz-FA-FMK) illustrates cruzain's ability to accommodate either hydrophobic or basic residues in the S2 pocket.[31,70] This flexibility is largely based on the ability of Glu_{205} at the base of the S2 pocket to swing into a position pointing into the pocket to interact with an incoming P2 moiety when that is favorable or to swing out of the pocket, orienting toward the outside of the protein when it is not. When Cbz-RA-FMK is bound to cruzain (PDB ID: 1AIM), Glu_{205} points into the S2 pocket forming a salt-bridge with the Arg side chain at P2. A similar orientation of Glu_{205} pointing inward to bind with an Arg moiety was observed in the structure of cruzain bound to NMePip-R-hPhe-vinylsulphonephenyl (PDB ID: 2LXS). In contrast, when Cbz-FA-FMK is bound, placing Phe into the S2 pocket, Glu_{205} is oriented out of the pocket, pointing towards the outside of the protein. When Cbz-YA-FMK is the inhibitor (PDB ID: 2AIM), the Tyr at P2 adopts the same rotomer conformation as that found in the Cbz-FA-FMK structure and once again, Glu_{205} orients itself to interact with the tyrosine hydroxyl. Peptide-based small molecule inhibitors in complex with cruzain have been extensively visualized *via* X-ray crystallography. Several of these complexes employ inhibitors containing a vinylsulfone warhead, which interacts with cruzain's active site Cys_{25} *via* an irreversible Michael addition.[72,74,75,87] A series of nonpolar interactions between a Phe moiety in the P2 position and the richly hydrophobic S2 subsite and a constellation of strong hydrogen bonds between the sulfone moiety and the S1′ residues of cruzain offer a consistent mode of binding for this class of inhibitors.

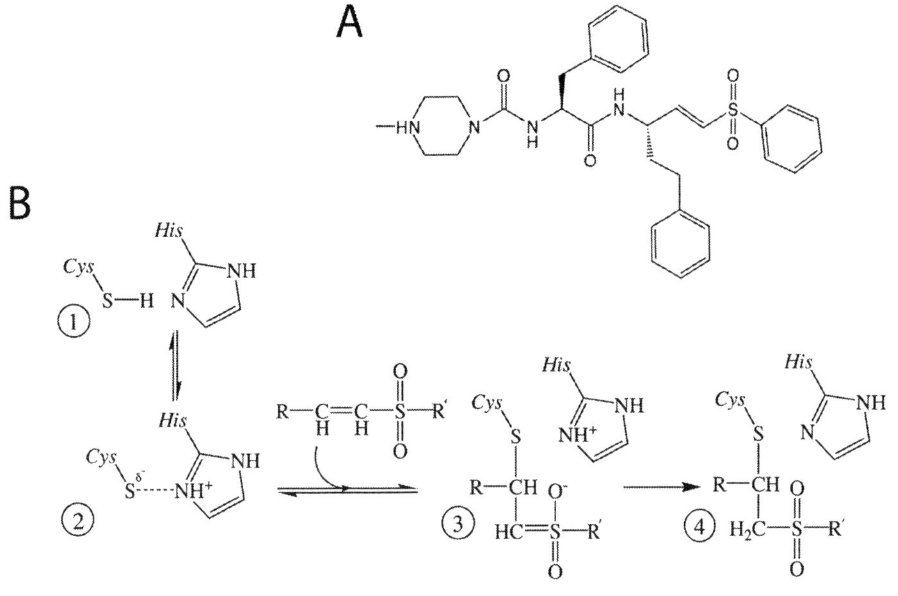

Figure 4. A) Structure of K777. B) Chemical mechanism of cruzain inhibition by K777. 1) The active site Cys and His are in equilibrium with (2) the thiolate-imidazolium charge relay dyad. In the activated form, the sulfur of the active site Cys initiates (3) a Michael addition by attacking the β-carbon of the γ-amino vinyl sulfone, K777. 4) Protonation of the α-carbon in the intermediate complex results in formation of a stable thiother. N.B. in solution the β-carbon of the vinyl sulfone is not polarized and so cannot undergo nucleophilic attack. However, once in correct register in the active site of a cysteine protease, the β-carbon is activated by the protonated imidazole, of the active site histidine. R, N-methyl piperazine-urea-phenyalanyl-homophenylalanyl of K777; R′, phenyl.

A vinylsulfone-cruzain complex of keen interest is that of K777 (Fig. 4A) bound to cruzain (Figs. 5A,5B), as this inhibitor is currently in preclinical development as a therapeutic against Chagas' disease (discussed in detail later; chemical mechanism Fig. 4B).[75,87] The crystallographic structure of this complex (PDB ID: 2OZ2) was solved and refined to 1.95 Å resolution. It shows that K777 spans subsites S3-S1′ of cruzain and forms an irreversible covalent adduct with the sulfur of Cys_{25}. The network of polar interactions between Gln_{19}, Gly_{66}, Asp_{158}, His_{159} and Trp_{177} observed is consistent with those found in all cruzain-vinylsulfone complexes that have been solved. These interactions serve to anchor the peptidyl backbone of the inhibitor into the protease active site. Since there is a hydrophobic Phe moiety in the P2 position of K777, Glu_{205} at the base of the S2 pocket is observed to point out towards the surface of the enzyme in order to avoid potentially unfavorable interactions (Figs. 5A,B).

The structures of two nonpeptidyl small molecule cruzain inhibitors have recently been solved in complex with the enzyme. First, a 1,2,3-triazole-based flurorophenoxymethyl ketone was visualized *via* X-ray crystallography to 1.2 Å resolution (PDB ID: 3IUT).[71] This compound (full chemical name: (3S)-3-(4-{(1S)-1,2-dimethyl-1-[(quinolin-6lmethyl)amino]propyl}-1H-1,2,3-triazol- 1-yl)heptan-2-one) completely eradicates the *T. cruzi* parasite in cell culture. It also represents a very promising mechanism-based pharmacophore due to its high selectivity for cysteine protease inhibition.[76-78] In addition, this class has shown to be well tolerated in animal studies and a tetrafluorophenoxy methyl

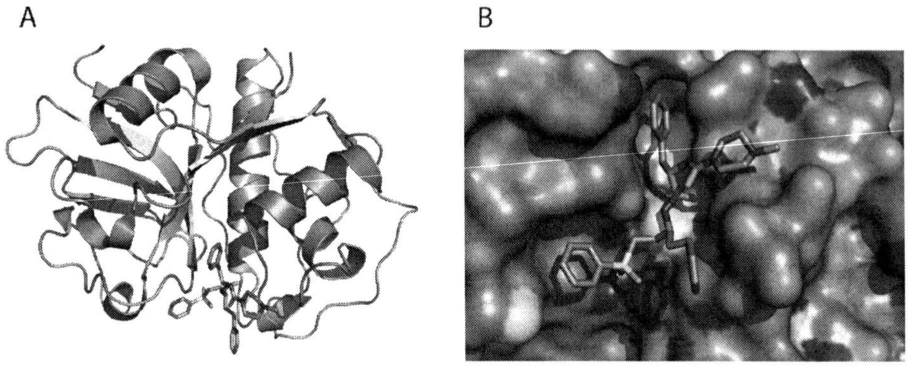

Figure 5. A) Cartoon representation of cruzain, with bound K777 inhibitor in the active site cleft, shown in stick representation. B) Cruzain's active site cleft with bound K777 inhibitor. Cruzain's surface is shown with K777 in stick representation. The Phe moiety of K777 points into the enzyme's S2 pocket. Figures prepared with PyMol [A: DeLano, W.L. The PyMOL Molecular Graphics System. (2008) DeLano Scientific LLC, Palo Alto, CA, USA. http://www.pymol.org].

ketone-based caspase inhibitor is currently in Phase II clinical trials.[79] Aryloxymethyl ketone inhibitors are well-studied as caspase inhibitors,[80,81] but this pharmacophore had not previously been structurally visualized with cruzain or any other member of the papain superfamily. The proposed mechanism of inhibition of cysteine proteases by activated ketones involves nucleophilic attack of the cysteine thiolate on the carbonyl carbon, leading to the formation of a thiohemiketal stabilized by the oxyanion hole.[78,81] Breakdown of the tetrahedral intermediate then results in a displacement of the leaving group and hence irreversible inhibition. The high-resolution crystal structure of the cruzain-inhibitor complex reveals that the thiol nucleophile of the active site Cys_{25} has effectively displaced the tetrafluorophenoxy moiety of the inhibitor and a 1.83 Å bond is formed between Cys_{25} and inhibitor. The pharmacophore is further stabilized in the S1′ subsite through the hydrogen bonds with the peptide amide of Cys_{25} and with Nε2 of Gln_{19}. Glu_{205} again demonstrates its flexibility and is found in dual conformations with one conformer pointing towards the inhibitor and interacting with the amine functionality *via* two water molecules. A second conformer adopts a solvent-exposed orientation and points out of the S2 pocket. The triazole moiety of this nonpeptidic inhibitor provides stabilizing interactions with the enzyme's active site similar to what has been seen with the amide bond in traditional peptidic inhibitors. This first structure of this pharmacophore complexed with cruzain enabled enhanced design of a new generation of inhibitors that showed a 4-fold increase in inhibitory activity and, for select inhibitors, improved physicochemical properties such as reduced molecular weight, lower hydrophobicity and a reduction in the number of rotatable bonds. In addition, several of these newer inhibitors exhibited comparable or modestly improved potency in cell culture evaluation.

The second nonpeptidic cruzain inhibitor visualized in complex *via* X-ray crystallography—full chemical name (1R,2R)-2-[(4-chlorophenyl)carbonyl]-N-{(1S)-1-[2-(phenylsulfonyl)ethyl]pentyl}cyclohexanecarboxamide—contains a vinylsulfone warhead.[72] The 1.75 Å structure (PDB ID: 3HD3) illustrates that the nonpeptidic P2/P3 moiety in such analogs bind the S2 and S3 subsites of cruzain, effectively recapitulating important binding interactions found in more traditional peptide-based protease inhibitors and natural substrates. Comparison of the nonpeptidic vinylsulfone inhibitor's interactions with cruzain to those found between inhibitor K777 and cruzain reveals several conserved interactions at the S1, S1′ and S2 subsites. These include the formation of two hydrogen bonds to the inhibitor backbone and another two with the sulfone moiety in the S1′ subsite. Conversely, the presence of a nonpeptidic group at P2 results in the loss of a hydrogen bonding interaction to the inhibitor backbone that is present in the cruzain-K777 complex. As is typically found in subsites S3-S1′ in the peptidyl inhibitors that have been visualized, nonpolar residues also contribute to the successful binding of enzyme and inhibitor in the nonpeptidic inhibitor complex. With a nonpolar P2 moiety placed in the S2 subsite, Glu_{205} is once again seen to point out from the S2 pocket and toward the surface of the protein.

The structures of cruzain bound to two potent reversible small molecule inhibitors have also been solved to 1.2 Å (PDB IDs: 1ME3 and 1ME4) and these compounds are based on an hydroxymethyl ketone scaffold.[69] N-Cbz-Phe-Phe hydroxymethyl ketone and 3-pyridinylmethoxycarbonyl-Phe-hPhe hydroxymethyl ketone were not covalently bound at Cys_{25} but were anchored by a series of hydrophobic interactions with the P2 Phe moiety in the S2 pocket as well as through stabilization with a series of hydrogen bonds with Gly_{66}. Additionally, both compounds exhibit a strong hydrogen bond between His_{159}, part of the canonical catalytic triad and the hydroxyl group of the inhibitor.

K777: THE PATH TO THE CLINIC

Khepri Pharmaceuticals showed that peptidomimetic scaffolds containing a vinyl sulfone moiety could selectively inhibit cysteine proteases such as cruzain over other protease classes (Fig. 4A). In addition, these vinyl sulfones also appear to be selective for cysteine proteases over other classes of proteins containing activated sulfers.[82] Treatment with this class of inhibitor not only cleared parasites from infected mammalian cells but also cured acutely infected mice.[83] Additionally, it appeared that development of resistance to this class of inhibitor would be slow to occur.[66] The results with this class of compound were compelling enough to convince the National Institute of Allergies and Infectious Diseases (NIAID) to fund preclinical safety studies to assess K777, one of the most active compounds of this class against *T. cruzi* in vitro and in vivo. SRI International performed in vitro mutagenicity tests as well as single-dose escalation exploratory toxicity studies in rats and dogs. These data indicated that K777 was not mutagenic and clinical observations indicated that K777 was well tolerated in both species. Based on these data, more detailed pharmacokinetics and a 7-day preliminary toxicity study (100 mg/kg, daily oral dosing) were performed using dogs. Severe emesis was noted in the dogs as well as significant elevations in liver enzymes, indicating that target organ toxicity for K777 would likely be the liver. Dogs are prone to emesis, so this was not necessarily an issue specific to the compound; however the emesis issue confounded clear interpretation of exposure data in the dog, so a second investigational study using monkeys was carried out to determine if this species would be more appropriate for use in more detailed toxicity and safety studies. A 7-day study in cynomolgus monkeys, evaluating oral daily dosing of K777 at 200 mg/kg showed again, significantly elevated liver enzymes, confirming that the liver was the likely target organ for toxicity. Pharmacokinetics in the monkey indicated that K777 had a reasonable exposure and half-life of 5 hours, suggesting K777 could be orally dosed in humans.

Historically the Institute of One World Health (iOWH) had an active Chagas program. There had been very little success in identifying compounds effectively targeting *T. cruzi* but the efficacy studies and early preclinical data for K777 were promising. In 2002, iOWH licensed K777 from Celera (formerly Khepri) with the hope of accelerating preclinical development efforts. iOWH performed several in vitro studies as well as in vivo central nervous system safety testing in rats, which showed K777 was well tolerated neurologically with only modest symptoms occurring at significantly high doses. While data generated by iOWH did not uncover any other concerns with K777, iOWH later determined that the risks associated with potential liver toxicities and the possible challenges in large-scale production of K777 warranted dropping this compound from their portfolio. In 2005, iOWH returned ownership of K777 and associated manufacturing processes to Celera and in 2007 dropped their Chagas program. The Sandler Center at UCSF along with the NIAID continued their interest in this compound and eventually UCSF was granted rights to K777 by Celera. During this time the Sandler Center performed numerous studies on K777 that have added to the evidence that the targeting of cruzain can be an effective means for eliminating *T. cruzi* in acute models of infection in mice. In addition, a study of acute *T. cruzi* infection in dogs showed that oral treatment with K777 at 50 mg/kg twice daily for 14 days was sufficient to protect against cardiac damage, as assessed by histopathology and troponin I levels.[84] The Sandler Center has also verified the efficacy of K777 against various *T. cruzi* strains (e.g., Y, Tulahuén, CL, CA-I/72, PSD-1 and PSD-2 isolates), that represent a spectrum of *T. cruzi* with various tissue tropisms, as well as against nifurtimox- and benznidazole-resistant *T. cruzi*.[85] Importantly, K777 also produced an additive effect on parasite killing when used in combination with benznidazole.[85]

To determine at what exposures elevated liver enzymes are detected and to evaluate other potential target organ effects that may need to be monitored in the clinic, a 14-day toxicity study in rats was performed. This study was performed under Good Laboratory Practices (GLP) and thus was of sufficient quality to submit as part of the K777 Investigational New Drug (IND) application to the FDA for approval of use of K777 in humans in clinical trials. A recovery arm was included in this study to evaluate whether any toxic effects, including those in the liver, were reversible upon cessation of drug treatment. This 14-day dosing study in rodents (SRI International) confirmed that modest (2-3x) ALT elevation occurs only at doses at or above 150 mg/kg and were reversible upon cessation of treatment. No treatment-related histopathological abnormalities were noted in any tissues at any doses, including the liver. Using allometric scaling to estimate the human equivalent dose from Chagas' infected mouse and dog model studies, it is predicted that a human oral dose of approximately 4 mg/kg for 14-30 days may be an effective therapeutic regimen for Chagas' disease intervention with this protease inhibitor. In comparison, the no observable adverse effect level (NOAEL) is anticipated to be approximately 24 mg/kg with a maximum tolerated dose of 160 mg/kg. Thus, there appears to be a 6 fold window between efficacious doses and drug-induced clinical morbidities.

On September 15th 2009, representatives from the Sandler Center and NIAID met with the FDA in a type B pre-IND meeting and agreed on what is needed to complete the IND package for K777. A significant portion of the IND package is already in place; outstanding elements as of this write-up include a 28-day nonrodent toxicity study with cardiopulmonary safety assessments and manufacture of the drug for use in the clinic. The Drugs for Neglected Disease Initiative (DNDi) joined forces with the Sandler Center and NIAID in the fall of 2009 to obtain funds to complete the IND package. Efforts are currently under way with a projected date for filing of the IND with the FDA in late 2012.

One of the first challenges in the clinic for K777 is going to be selecting the appropriate patient population for Phase II trials. The acute population has easily-detectable levels of parasite in blood; however, at the same time treatment is occurring the immune system is also at work clearing the parasite, likely making it difficult to distinguish drug effects from effects due to normal clearance by the body's natural defenses. In addition, this patient population responds well to current therapies. If a patient is fortunate enough to be diagnosed at this early stage, there is an ethical dilemma in treating patients with a drug of unproven human efficacy during the narrow window when they have a known effective drug (albeit with severe side-effects). Finally, aside from children, patients are rarely diagnosed in the acute stage, making recruitment for such a study difficult. Since the initial goals of a Phase II clinical trial will focus on changes in parasite levels, rather than progression to cardiac disease or mortality, the logical population will likely be patients in the indeterminate stage with no to few cardiac manifestations. This patient population is becoming diagnosed more frequently due, in part, to testing of blood donor samples for *T. cruzi* infection and represents a group where effectiveness of current therapies is questionable. In addition, as only 30% of these patients are projected to develop infection-related morbidities, it is questionable whether to subject patients to the severe side-effects of the current therapies available. One of the biggest challenges when using this patient population, where parasite levels are very low and sequestered in inaccessible tissues such as cardiac muscle, will be in defining cure. Currently patients are diagnosed serologically; however, this method only determines if patients have been exposed to infection but cannot confirm current infections. Patients which have gone from sera-positive to sera-negative are presumed cured but this transition can take years to occur. The current standard for determination of significant changes parasite burden is via assessment by polymerase chain reaction (PCR) -based techniques. Although this method is highly sensitive, a negative result does not necessarily indicate parasite is no longer present, as there are typically very small total numbers of parasites in chronically-infected patients and these parasites are sequestered in cells, limiting the amount of parasite detectable in blood sample. Thus it is necessary to examine blood samples several times, over the course of months, with numerous repeat negative results to suggest that the parasites may in fact be cleared. Currently there are clinical trials on-going, such as the BENEFIT trial (Benznidazole Evaluation for Interrupting Trypanosomiasis) which is incorporating long-term monitoring of progression and mortality coupled to detection of parasite using PCR.[86] This on-going international, multi-center, double-blind, placebo-controlled trial of trypanocidal treatment with benznidazole in patients with chronic Chagas' heart disease may well provide the needed evidence whether PCR-determined reduction of parasite load is predictive of improved outcome in chronically-infected patients.

CONCLUSION

The cysteine endoprotease cruzain from *T. cruzi* represents one the first examples of a protease from a parasitic organism that has had an inhibitor progress through the tropical diseases drug pipeline. Once validated as a bona fide drug target, the active enzyme was heterologously overexpressed, the structure derived and with aid of high throughput biochemical screening the peptide-based vinyl sulfone K777 was identified. This lead compound was shown to possess potent activity against the parasite in disease models. K777 is now in IND-enabling studies, poising this inhibitor for use in the clinic in humans. Thus, K777 may soon enter the clinic as one of the first protease targeted drugs to be used in treatment of a neglected tropical disease.

REFERENCES

1. Chagas C. Nova entidade morbida do homen. Resumo greal dos estudos estiologicos e clinicos. Mem Inst Oswaldo Cruz 1911; 3:219-275.
2. Wilson LS, Strosberg AM, Barrio K. Cost-effectiveness of Chagas disease interventions in latin america and the Caribbean: Markov models. Am J Trop Med Hyg 2005; 73:901-910.
3. Rodriques Coura J, de Castro SL. CoMFA and HQSAR of acylhydrazide cruzain inhibitors. Mem Inst Oswaldo Cruz 2002; 97:3-24.
4. Duschak VG, Couto AS. An insight on targets and patented drugs for chemotherapy of Chagas disease. Recent Pat Antiinfect Drug Discov 2007; 2:19-51.
5. Rivera G, Bocanegra-Garcia V, Ordaz-Pichardo C et al. New therapeutic targets for drug design against Trypanosoma cruzi, advances and perspectives. Curr Med Chem 2009; 16:3286-3293.
6. Moreira DR, Leite AC, dos Santos RR et al. Approaches for the development of new anti-Trypanosoma cruzi agents. Curr Drug Targets 2009; 10:212-231.
7. Itow S, Camargo EP. Proteolytic activites in cell extracts of Trypanosoma cruzi. J Protozool 1977; 24:591-595.
8. Avila JL, Casanova MA, Avila A et al. Acid and neutral hydrolases in Trypanosoma cruzi. Characterization and assay J Protozool 1979; 26:304-311.
9. Bontempi E, Martinez J, Cazzulo JJ. Subcellular localization of a cysteine proteinase from Trypanosoma cruzi. Mol Biochem Parasitol 1989; 33:43-47.
10. Cazzulo JJ, Couso R, Raimondi A et al. Further characterization and partial amino acid sequence of a cysteine proteinase from Trypanosoma cruzi. Mol Biochem Parasitol 1989; 33:33-41.
11. Rangel HA, Araujo PM, Repka D et al. Trypanosoma cruzi: isolation and characterization of a proteinase. Exp Parasitol 1981; 52:199-209.
12. Lalmanach G, Lecaille F, Chagas JR et al. Inhibition of trypanosomal cysteine proteinases by their propeptides. J Biol Chem 1998; 273:25112-25116.
13. Huete-Perez JA, Engel JC, Brinen LS et al. Protease trafficking in two primitive eukaryotes is mediated by a prodomain protein motif. J Biol Chem 1999; 274:16249-16256.
14. Cazzulo JJ, Hellman U, Couso R et al. Some kinetic properties of a cysteine proteinase (cruzipain) from Trypanosoma cruzi. Mol Biochem Parasitol 1990; 38:41-48.
15. Duschak VG, Couto AS. Cruzipain, the major cysteine protease of Trypanosoma cruzi: a sulfated glycoprotein antigen as relevant candidate for vaccine development and drug target. A review. Curr Med Chem 2009; 16:3174-3202.
16. Barboza M, Duschak VG, Fukuyama Y et al. Structural analysis of the N-glycans of the major cysteine proteinase of Trypanosoma cruzi. Identification of sulfated high-mannose type oligosaccharides. Febs J 2005; 272:3803-3815.
17. Labriola C, Sousa M, Cazzulo JJ. Purification of the major cysteine proteinase (cruzipain) from Trypanosoma cruzi by affinity chromatography. Biol Res 1993; 26:101-107.
18. Barboza M, Duschak VG, Cazzulo JJ et al. Presence of sialic acid in N-linked oligosaccharide chains and O-linked N-acetylglucosamine in cruzipain, the major cysteine proteinase of Trypanosoma cruzi. Mol Biochem Parasitol 2003; 127:69-72.
19. Cazzulo JJ, Martinez J, Parodi AJ et al. On the posttranslational modifications at the C-terminal domain of the major cysteine proteinase (cruzipain) from Trypanosoma cruzi. FEMS Microbiol Lett 1992; 79:411-416.
20. Parodi AJ, Labriola C, Cazzulo JJ. The presence of complex-type oligosaccharides at the C-terminal domain glycosylation site of some molecules of cruzipain. Mol Biochem Parasitol 1995; 69:247-255.
21. Martinez J, Campetella O, Frasch AC et al. The reactivity of sera from chagasic patients against different fragments of cruzipain, the major cysteine proteinase from Trypanosoma cruzi, suggests the presence of defined antigenic and catalytic domains. Immunol Lett 1993; 35:191-196.
22. Tomas AM, Kelly JM. Stage-regulated expression of cruzipain, the major cysteine protease of Trypanosoma cruzi is independent of the level of RNA1. Mol Biochem Parasitol 1996; 76:91-103.
23. Fampa P, Lisboa CV, Jansen AM et al. Protease expression analysis in recently field-isolated strains of Trypanosoma cruzi: a heterogeneous profile of cysteine protease activities between TC I and TC II major phylogenetic groups. Parasitology 2008; 135:1093-1100.
24. Campetella O, Henriksson J, Aslund L et al. The major cysteine proteinase (cruzipain) from Trypanosoma cruzi is encoded by multiple polymorphic tandemly organized genes located on different chromosomes. Mol Biochem Parasitol 1992; 50:225-234.

25. Martinez J, Cazzulo JJ. Anomalous electrophoretic behaviour of the major cysteine proteinase (cruzipain) from Trypanosoma cruzi in relation to its apparent molecular mass. FEMS Microbiol Lett 1992; 74:225-229.
26. Cazzulo JJ. Proteinases of Trypanosoma cruzi: patential targets for the chemotherapy of Changas desease. Curr Top Med Chem 2002; 2:1261-1271.
27. Sajid M, McKerrow JH. Cysteine proteases of parasitic organisms. Mol Biochem Parasitol 2002; 120:1-21.
28. Martinez J, Henriksson J, Rydaker M et al. Genes for cysteine proteinases from Trypanosoma rangeli. FEMS Microbiol Lett 1995; 129:135-141.
29. Lima AP, Tessier DC, Thomas DY et al. Identification of new cysteine protease gene isoforms in Trypanosoma cruzi. Mol Biochem Parasitol 1994; 67:333-338.
30. Eakin AE, Mills AA, Harth G et al. The sequence, organization and expression of the major cysteine protease (cruzain) from Trypanosoma cruzi. J Biol Chem 1992; 267:7411-7420.
31. McGrath ME, Eakin AE, Engel JC et al. The crystal structure of cruzain: a therapeutic target for Chagas' disease. J Mol Biol 1995; 247:251-259.
32. Judice WA, Cezari MH, Lima AP et al. Comparison of the specificity, stability and individual rate constants with respective activation parameters for the peptidase activity of cruzipain and its recombinant form, cruzain, from Trypanosoma cruzi. Eur J Biochem 2001; 268:6578-6586.
33. Alves LC, Melo RL, Cezari MH et al. Analysis of the S(2) subsite specificities of the recombinant cysteine proteinases CPB of Leishmania mexicana and cruzain of Trypanosoma cruzi, using fluorescent substrates containing nonnatural basic amino acids. Mol Biochem Parasitol 2001; 117:137-143.
34. Harris JL, Backes BJ, Leonetti F et al. Rapid and general profiling of protease specificity by using combinatorial fluorogenic substrate libraries. Proc Natl Acad Sci USA 2000; 97:7754-7759.
35. Del Nery E, Juliano MA, Lima AP et al. Kininogenase activity by the major cysteinyl proteinase (cruzipain) from Trypanosoma cruzi. J Biol Chem 1997; 272:25713-25718.
36. Eakin AE, McGrath ME, McKerrow JH et al. Production of crystallizable cruzain, the major cysteine protease from Trypanosoma cruzi. J Biol Chem 1993; 268:6115-6118.
37. Souto-Padron T, Campetella OE, Cazzulo JJ et al. Cysteine proteinase in Trypanosoma cruzi: immunocytochemical localization and involvement in parasite-host cell interaction. J Cell Sci 1990; 96:485-490.
38. Soares MJ, Souto-Padron T, De Souza W. Identification of a large prelysosomal compartment in the pathogenic protozoon Trypanosoma cruzi. J Cell Sci 1992; 102:157-167.
39. Parussini F, Duschak VG, Cazzulo JJ. Membrane-bound cysteine proteinase isoforms in different developmental stages of Trypanosoma cruzi. Cell Mol Biol 1998; 44:513-519.
40. Nascimento AE, de Souza W. High resolution localization of cruzipain and Ssp4 in Trypanosoma cruzi by replica staining label fracture. Biol Cell 1996; 86:53-58.
41. Aparicio IM, Scharfstein J, Lima AP. A new cruzipain-mediated pathway of human cell invasion by Trypanosoma cruzi requires trypomastigote membranes. Infect Immun 2004; 72:5892-5902.
42. Bontempi E, Cazzulo JJ. Digestion of human immunoglobulin G by the major cysteine proteinase (cruzipain) from Trypanosoma cruzi. FEMS Microbiol Lett 1990; 58:337-341.
43. Krettli AU, Weisz-Carrington P, Nussenzweig RS. Membrane-bound antibodies to bloodstream Trypanosoma cruzi in mice: strain differences in susceptibility to complement-mediated lysis. Clin Exp Immunol 1979; 37:416-423.
44. Scharfstein J, Monteiro AC, Schmitz V et al. Angiotensin-converting enzyme limits inflammation elicited by Trypanosoma cruzi cysteine proteases: a peripheral mechanism regulating adaptive immunity via the innate kinin pathway. Biol Chem 1998; 389:1015-1024.
45. Scharfstein J, Lima AP. Roles of naturally occurring protease inhibitors in the modulation of host cell signaling and cellular invasion by Trypanosoma cruzi. Subcell Biochem 2008; 47:140-154.
46. Scharfstein J. Parasite cysteine proteinase interactions with alpha 2-macroglobulin or kininogens: differential pathways modulating inflammation and innate immunity in infection by pathogenic trypanosomatids. Immunobiology 2006; 211:117-125.
47. Yokoyama-Yasunaka JK, Pral EM, Oliveira Junior OC et al. Trypanosoma cruzi: identification of proteinases in shed components of trypomastigote forms. Acta Trop 1994; 57:307-315.
48. Harth G, Andrews N, Mills AA et al. Peptide-fluoromethyl ketones arrest intracellular replication and intercellular transmission of Trypanosoma cruzi. Mol Biochem Parasitol 1993; 58:17-24.
49. Meirelles MN, Juliano L, Carmona E et al. Inhibitors of the major cysteinyl proteinase (GP57/51) impair host cell invasion and arrest the intracellular development of Trypanosoma cruzi in vitro. Mol Biochem Parasitol 1992; 52:175-184.
50. Engel JC, Doyle PS, Hsieh I et al. Cysteine protease inhibitors alter Golgi complex ultrastructure and function in Trypanosoma cruzi. J Exp Med 1998; 188:725-734.
51. Caffrey CR, Scory S, Steverding D. Cysteine proteinases of trypanosome parasites: novel targets for chemotherapy. Curr Drug Targets 2000; 1:155-162.

52. Campetella O, Martinez J, Cazzulo JJ. A major cysteine proteinase is developmentally regulated in Trypanosoma cruzi. FEMS Microbiol Lett 1990; 55:145-149.
53. Judice WA, Puzer L, Cotrin SS et al. Carboxydipeptidase activities of recombinant cysteine peptidases. Cruzain of Trypanosoma cruzi and CPB of Leishmania mexicana. Eur J Biochem 2004; 271:1046-1053.
54. Cazzulo JJ, Cazzulo Franke MC, Martinez J et al. Some kinetic properties of a cysteine proteinase (cruzipain) from Trypanosoma cruzi. Biochim Biophys Acta 1990; 1037:186-191.
55. Reis FC, Costa TF, Sulea T et al. The propeptide of cruzipain—a potent selective inhibitor of the trypanosomal enzymes cruzipain and brucipain and of the human enzyme cathepsin F. Febs J 2007; 274:1224-1234.
56. Santos CC, Sant'anna C, Terres A et al. Use of proteolytic enzymes as an additional tool for trypanosomatid identification. J Cell Sci 2005; 1 118:901-915.
57. Santos CC, Scharfstein J, Lima AP. Phytomonas serpens: cysteine peptidase inhibitors interfere with growth, ultrastructure and host adhesion. Parasitol Res 2006; 99:323-324.
58. Ramos AM, Duschak VG, Gerez de Burgos NM et al. Trypanosoma cruzi: cruzipain and membrane-bound cysteine proteinase isoform(s) interacts with human alpha(2)-macroglobulin and pregnancy zone protein. Exp Parasitol 2002; 100:121-130.
59. Cazzulo JJ, Stoka V, Turk V. Cruzipain, the major cysteine proteinase from the protozoan parasite Trypanosoma cruzi. Biol Chem 1997; 378:1-10.
60. Lima AP, Almeida PC, Tersariol IL et al. Heparan sulfate modulates kinin release by Trypanosoma cruzi through the activity of cruzipain. J Biol Chem 2002; 277:5875-5881.
61. Guinazu N, Pellegrini A, Carrera-Silva EA et al. Immunisation with a major Trypanosoma cruzi antigen promotes pro-inflammatory cytokines, nitric oxide production and increases TLR2 expression. Int J Parasitol 2007; 37:1243-1254.
62. Schnapp AR, Eickhoff CS, Scharfstein J et al. Cruzipain induces both mucosal and systemic protection against Trypanosoma cruzi in mice. Microbes Infect 2002; 4:805-813.
63. Cazorla SI, Becker PD, Frank FM et al. Oral vaccination with Salmonella enterica as a cruzipain-DNA delivery system confers protective immunity against Trypanosoma cruzi. Infect Immun 2008; 76:324-333.
64. Schnapp AR, Eickhoff CS, Sizemore D et al. Cruzipain induces both mucosal and systemic protection against Trypanosoma cruzi in mice. Infect Immun 2002; 70:5065-5074.
65. Cazorla SI, Frank FM, Malchiodi EL. Vaccination approaches against Trypanosoma cruzi infection. Expert Rev Vaccines 2009; 8:921-935.
66. Engel JC, Doyle PS, Palmer J et al. Cysteine protease inhibitors cure an experimental Trypanosoma cruzi infection. J Cell Sci 1998; 111:597-606.
67. Kerr ID, Lee JH, Farady CJ et al. Vinyl sulfones as antiparasitic agents and a structural basis for drug design. J Biol Chem 2009; 284:25697-25703.
68. Choe Y, Brinen LS, Price MS et al. Development of alpha-keto-based inhibitors of cruzain, a cysteine protease implicated in Chagas disease. Bioorg Med Chem 2005; 13:2141-2156.
69. Huang L, Brinen LS, Ellman JA. Crystal structures of reversible ketone-Based inhibitors of the cysteine protease cruzain. Bioorg Med Chem 2003; 11:21-29.
70. Gillmor SA, Craik CS, Fletterick RJ. Structural determinants of specificity in the cysteine protease cruzain. Protein Sci 1997; 6:1603-1611.
71. Brak K, Kerr ID, Barrett KT et al. Nonpeptidic tetrafluorophenoxymethyl ketone cruzain inhibitors as promising new leads for chagas disease chemotherapy. J Med Chem 2010; 53:1763-1773.
72. Bryant C, Kerr ID, Debnath M et al. Novel nonpeptidic vinylsulfones targeting the S2 and S3 subsites of parasite cysteine proteases. Bioorg Med Chem Lett 2009; 19:6218-6221.
73. Choe Y, Leonetti F, Greenbaum DC et al. Substrate profiling of cysteine proteases using a combinatorial peptide library identifies functionally unique specificities. J Biol Chem 2006; 281:12824-12832.
74. Brinen LS, Hansell E, Cheng J et al. A target within the target: probing cruzain's P1' site to define structural determinants for the Chagas' disease protease. Structure 2000; 8:831-840.
75. Kerr ID, Lee JH, Pandey KC et al. Structures of falcipain-2 and falcipain-3 bound to small molecule inhibitors: implications for substrate specificity. J Med Chem 2009; 52:852-857.
76. Krantz A, Copp LJ, Coles PJ et al. Peptidyl (acyloxy)methyl ketones and the quiescent affinity label concept: the departing group as a variable structural element in the design of inactivators of cysteine proteinases. Biochemistry 1991; 30:4678-4687.
77. Smith RA, Coles PJ, Spencer RW et al. Peptidyl O-acyl hydroxamates: potent new inactivators of cathepsin B. Biochem Biophys Res Commun 1998; 155:1201-1206.
78. Powers JC, Asgian JL, Ekici OD et al. Irreversible inhibitors of serine, cysteine and threonine proteases. Chem Rev 2002; 102:4639-4750.
79. Baskin-Bey ES, Washburn K, Feng S et al. Clinical Trial of the Pan-Caspase Inhibitor, IDN-6556, in Human Liver Preservation Injury. Am J Transplant 2007; 7:218-225.
80. Brady KD. Bimodal inhibition of caspase-1 by aryloxymethyl and acyloxymethyl ketones. Biochemistry 1998; 37:8508-8515.

81. Brady KD, Giegel DA, Grinnell C et al. A catalytic mechanism for caspase-1 and for bimodal inhibition of caspase-1 by activated aspartic ketones. Bioorg Med Chem 1999; 7:621-631.
82. Palmer JT, Rasnick D, Klaus JL et al. Vinyl sulfones as mechanism-based cysteine protease inhibitors. J Med Chem 1995; 38:3193-3196.
83. Engel JC, Doyle PS, McKerrow JH. Trypanocidal effect of cysteine protease inhibitors in vitro and in vivo in experimental Chagas disease. Medicina (B Aires) 1999; 59:171-175.
84. Barr SC, Warner KL, Kornreic BG et al. A cysteine protease inhibitor protects dogs from cardiac damage during infection by Trypanosoma cruzi. Antimicrob Agents Chemother 2005; 49:5160-5161.
85. McKerrow JH, Doyle PS, Engel JC et al. Two approaches to discovering and developing new drugs for Chagas disease. Mem Inst Oswaldo Cruz 2009; 104:263-269.
86. Marin-Neto JA, Rassi A, Morillo CA et al. BENEFIT Investigators Am Heart J 2009; 156:37-43.
87. Sterverding D, Caffrey C, Sajid M. Cysteine Proteinase Inhibitors as Therapy for Parasitic Diseases: Advances in Inhibitor Design. Mini Rev Med Chem 2006; 6:1025-1032.

CHAPTER 8

THE PHYLOGENY, STRUCTURE AND FUNCTION OF TREMATODE CYSTEINE PROTEASES, WITH PARTICULAR EMPHASIS ON THE *FASCIOLA HEPATICA* CATHEPSIN L FAMILY

Colin Stack,[1] John P. Dalton[2] and Mark W. Robinson*,[3]

[1]School of Biomedical and Health Sciences, University of Western Sydney (UWS), Narellan Road, Campbelltown, NSW, Australia; [2]Institute of Parasitology, McDonald Campus, McGill University, St. Anne de Bellevue, Quebec, Canada; [3]i3 Institute (IBID), University of Technology Sydney (UTS), Ultimo, Sydney, Australia

**Corresponding Author: Mark W. Robinson—Email: mark.robinson@uts.edu.au*

Abstract: Helminth parasites (nematodes, flatworms and cestodes) infect over 1 billion of the world's population causing high morbidity and mortality. The large tissue-dwelling worms express papain-like cysteine peptidases, termed cathepsins that play important roles in virulence including host entry, tissue migration and the suppression of host immune responses. Much of our knowledge of helminth cathepsins comes from studies using flatworms or trematode (fluke) parasites. The developmentally-regulated expression of these proteases correlates with the passage of parasites through host tissues and their encounters with different host macromolecules. Recent phylogenetic, biochemical and structural studies indicate that trematode cathepsins exhibit overlapping but distinct substrate specificities due to divergence within the protease active site. Here we provide an overview of the evolution, biochemistry and structure of these important enzymes and highlight how recent advances in proteomics and gene silencing techniques are allowing researchers to probe their biological functions. We focus mainly on members of the cathepsin L gene family of the animal and human pathogen, *Fasciola hepatica*, because of our deep understanding of their function, biochemistry and structure.

Cysteine Proteases of Pathogenic Organisms, edited by Mark W. Robinson and John P. Dalton.

INTRODUCTION

Helminth parasites include the nematodes, flatworms and cestodes. Human diseases caused by these large tissue-dwelling parasites are amongst the most prevalent on earth with over one billion people infected worldwide, predominantly in poverty-stricken regions in developing countries.[1]

The flatworms (trematodes or flukes), are of major importance since many species are important human pathogens.[2] Medically important trematode genera include *Schistosoma, Paragonimus, Opisthorchis, Clonorchis* and *Fasciola*. Blood flukes of the genus *Schistosoma* reside in the vascular system of humans and cause a disease known as Bilharzia or schistosomiasis. This disease afflicts >250 million people in over 70 tropical countries and results in >280,000 deaths annually in sub-Saharan Africa alone and is by far the most important trematode infection.[3,4] Tissue flukes reside in the lungs (e.g., *P. westermani*) or liver (*C. sinensis, Opisthorchis spp.*) placing >600 millon people at risk of infection across Asia.[5,6] Liver flukes of the genus *Fasciola* infect domestic animals (sheep, cattle and water buffalo) causing a disease known as fasciolosis. Although traditionally regarded as a disease of livestock, fasciolosis has recently emerged as an important human zoonosis in South America, Egypt, Iran and Vietnam. Estimates suggest that 2.4 to 17 million people are infected worldwide, with a further 91.1 million people currently living at risk of infection.[2,5,7,8]

TREMATODE CYSTEINE PROTEASES

Trematodes express several types of peptidases that are involved in many aspects of the host-parasite relationship.[9] Of particular significance are those belonging to the papain superfamily (clan CA, family C1: CA1 peptidases) referred to as cathepsin L, B, F and C proteases.[10-12] Cysteine proteases make up a large proportion of the total transcripts of each of trematode parasites studied to date. For example, nearly 15% of the transcripts derived from adult *F. hepatica* (http://www.sanger.ac.uk/Projects/Helminths/),[13] 10% from adult *C. sinensis*,[14,15] as well as 18% and 27% from adult diploid and triploid *P. westermani*, respectively,[16] encode cysteine proteases. Cathepsin L, F and B proteases are often found among the profile of molecules secreted by trematodes, termed their secretome (also known as excretory-secretory proteins; ES), which allows them to perform a number of critical extracellular roles in parasite-host interactions.[17,18] The ability of secreted trematode cathepsin proteases to modulate the host immune response is well documented and will be covered in the chapter by Donnelly et al.[19] By contrast, cathepsin C, functions alongside aminopeptidases in the hydrolysis of ingested host macromolecules that have been absorbed into the gastrodermis from the gut lumen and is therefore not found in fluke secretions.[10,13] While the types of cathepsin proteases expressed by each trematode species is similar, their developmental expression within each parasite may vary considerably.[10,13] For example, *Fasciola* express both cathepsin L and cathepsin B-like proteases but, as discussed later, the expression of the latter is restricted to early invasive-stage parasites. On the other hand, *Schistosoma* express cathepsin L-like, cathepsin F-like and cathepsin B-like proteases in the early and late stages of infection. Interestingly, cathepsin F-like proteases are the predominant cysteine proteases expressed by the Asian flukes including *Clonorchis spp., Paragonimus spp.* and *Opisthorchis spp.*[20-22] and may be related to their use of fish as intermediate hosts. A detailed classification and characterisation of the trematode cysteine proteases can be found in Kasny et al.[11]

EVOLUTIONARY RELATIONSHIPS OF TREMATODE CYSTEINE PROTEASES

Phylogenetic studies indicate that at the time of helminths emergence 480 to 540 million years ago, several cathepsin classes had evolved:[9] two with endopeptidase activity (cathepsin L, F), one with both exo- and endopeptidase activity (cathepsin B) and one with dipeptidylpeptidase activity (cathepsin C).[10,12,13] Sequence homologies of trematode cysteine protease mature domains reveal clustering into to three subfamilies: cathepsin L-like, cathepsin B-like and cathepsin F-like indicating that these enzymes diverged early during eukaryotic evolution (Fig. 1). In contrast to the other cysteine proteases which are endopeptidases, cathepsin Cs are exopeptidases and are involved in removing dipeptides from the N-terminus of oligopeptides and macromolecules. To date, trematode cathepsin C proteases have only been identified in *S. mansoni* and *S. japonicum*, termed SmCC and SjCC and share more than 60% sequence similarity. In terms of phylogeny, cathepsin C genes do not cluster with cathepsin Ls instead they are more related to those of cathepsin B.[23]

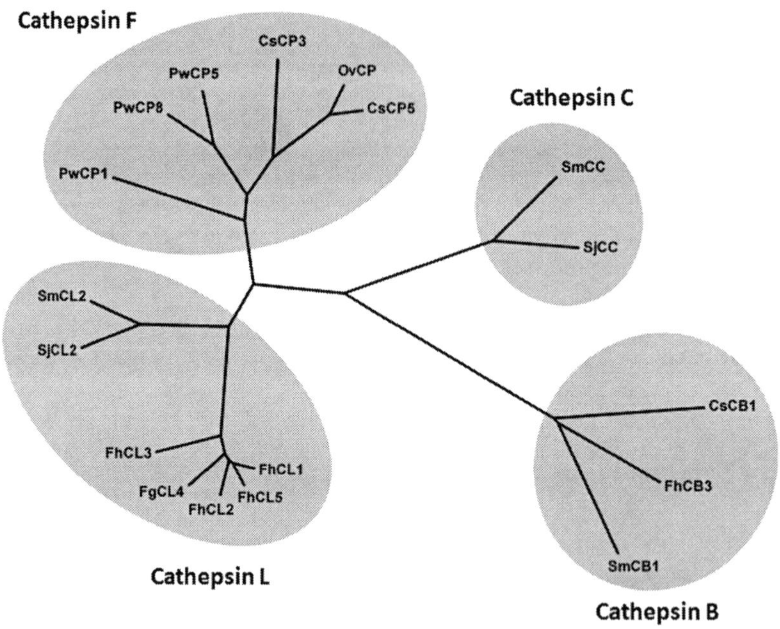

Figure 1. Papain-like cysteine proteases of trematode pathogens. Trematode pathogens express 4 major classes of cathepsin proteases (L, B, F and C) that have functions in parasite virulence including tissue invasion and feeding. Bootstrapped (1000 trials) neighbour-joining phylogenetic tree showing the evolutionary relationship of the trematode cathepsin superfamily. The tree was constructed using trematode cDNA sequences (including the prosegment region); adapted from references 12 and 13. The branches are coloured according to the 4 major enzyme classes. Fh, *Fasciola hepatica*; Fg, *Fasciola gigantica*; Pw, *Paragonimus westermani*; Sj, *Schistosoma japonicum*; Sm, *Schistosoma mansoni*; Ov, *Opisthorchis viverrini*, Cs, *Clonorchis sinensis*; CP, cysteine protease; CB, cathepsin B; CC, cathepsin C; CL, cathepsin L. Reprinted from Robinson MW, Dalton JP, Donnelly S. Pathogen cathepsin proteases: it's a family affair.Trends Biochem Sci 2008; 33:601-608, with permission from Elsevier.

Gene duplication followed by divergence is one of the primary means by which organisms generate proteins with new functions[24] and there is evidence that this process has created the diversity of cathepsin proteases expressed by *F. hepatica*.[17,25] Phylogenetic analysis of trematode CA1 peptidases reveals that the cathepsin Ls of *Fasciola* and *Schistosoma* species separate into two lineages. The *Fasciola* cathepsins L1, L2 and L3 and cathepsins L2 and L3 of schistosomes (SmCL2, SmCL3, SjCL2 [*S. japonicum* CL2], SjCL3) belong to a lineage that includes human cathepsin L (cathepsin L-like group). Sequence analysis by Caffrey et al of a protease previously known as SmCL1, revealed that this protease is actually more homologous to cathepsin F of *P. westermani* cathepsin F (61%) than to SmCL1 (44%).[10] Accordingly, both SmCL1 and SjCL1 were recently renamed cathepsin F. Phylogenetic analysis of *C. sinensis* cathepsin Fs (CsCF) demonstrates the existence of a large gene family, comprising of at least 12 CsCF genes grouping into three different subfamilies.[26]

FASCIOLA AS AN EXAMPLE OF A LARGE CYSTEINE PROTEASE GENE FAMILY

The various developmental stages of *F. hepatica* express and secrete cathepsin L and B proteases, but not cathepsin F. In fact >80% of proteins secreted by *F. hepatica* adults are cathepsin L cysteine proteases.[17,27] No other class of endoprotease of exoprotease have been identified in fluke secretions, demonstrating an exclusive reliance by adult parasites on cathepsin Ls. The cathepsin L-like protease genes of *F. hepatica* constitute a large and well-characterised multi-gene family that belong to a clade that includes mammalian cathepsin L, S and K proteases (parasite enzymes show 40-55% similarity with their mammalian homologues).[12]

The functional diversity of the various members of the *Fasciola* cathepsin L gene family, their relationship to pathogen virulence and host adaptation are of particular interest from an evolutionary view point as they reflect both adaptation and speciation of this parasite.[9,12,17,25] Irving et al[23] showed that the duplication and subsequent divergence of this family was estimated to have occurred over the last 135 million years. Interestingly the timing of these duplications correlates with the evolution of rodents, ruminants and higher mammals. However, most of these duplications took place relatively recently, approximately 25 million years ago, at about the same time climatic conditions favoured the development of grasslands and the expansion of common hosts of *F. hepatica*. This suggests that the divergence of the cathepsin L protease family was important in the evolution and adaptation of this parasite to a wider host range.[25]

Phylogenetic analysis revealed that *F. hepatica* expresses the largest monophyletic cysteine protease family described to date.[17,25] The family can be grouped into five distinct phylogenetic clades; cDNAs for two of these are found only in the early infective larvae (clades FhCL3 and FhCL4) while the remaining three clades (clades FhCL1, FhCL2 and FhCL5) were derived from adult parasites taken from bile ducts (Fig. 2).[17,25] FhCL3 is believed to be the most ancient member of this gene family. Adult FhCL2 arose and subsequently evolved following duplication of the ancestral FhCL3 gene and likewise, a subsequent duplication event in FhCL2 gave rise to the FhCL1 gene. Thus, a family of proteases with subtle, but important, amino acid changes within the enzyme`s active site cleft evolved (see below).[28-30] It has been suggested that these residue changes conferred

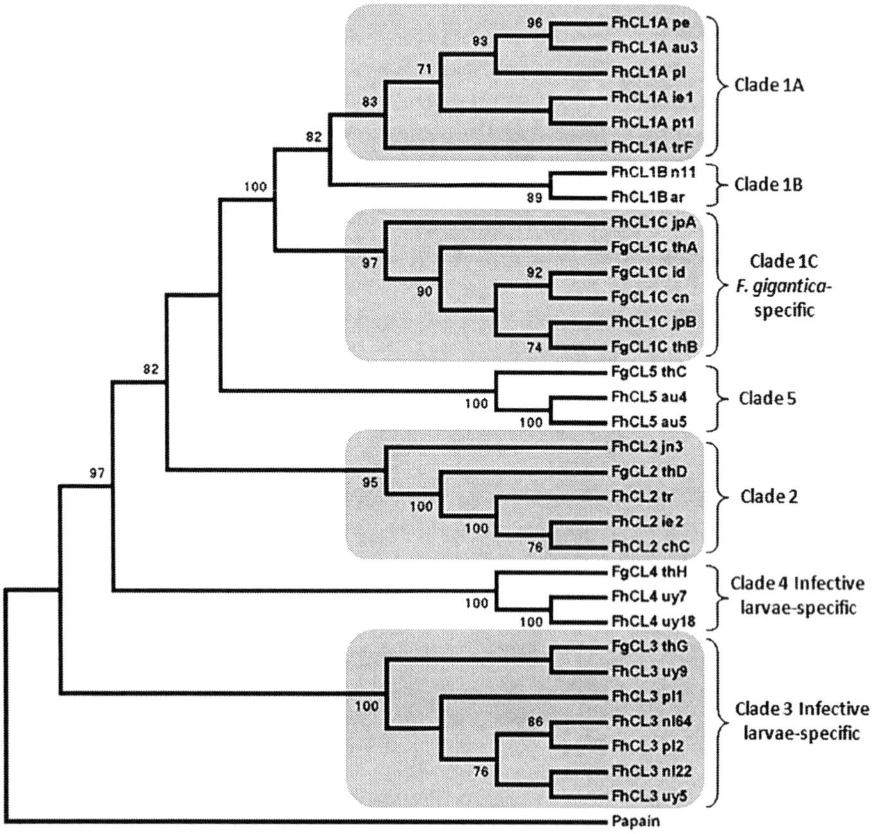

Figure 2. Phylogenetic analysis of the *Fasciola* cathepsin L gene family (adapted from Robinson et al[17]). The *Fasciola* cathepsin L gene family expanded by a series of gene duplications, followed by divergence of residues in the active site, which gave rise to three clades associated with mature adult worms and two clades specific to infective juvenile stages. These changes resulted in repertoire of cathepsin L proteases with overlapping and complementary substrate specificities. Bootstrapped (1000 trials) neighbour-joining tree constructed using *Fasciola hepatica* and *Fasciola gigantica* cathepsin L cDNA sequences. The tree is rooted to *Carica papaya* papain. This research was originally published in Molecular and Cellular Proteomics. Robinson MW, Tort JF, Wong E et al. Proteomics and phylogenetic analysis of the cathepsin L protease family of the helminth pathogen, Fasciola hepatica: expansion of a repertoire of virulence-associated factors. Mol Cell Proteomics 2008; 7:1111-1123. © 2008 the American Society for Biochemistry and Molecular Biology.

overlapping and yet complementary specificities enabling the parasite to degrade a wider variety of host macromolecules.[25,29]

Our recent proteomics analysis of the secretome of mature flukes was in agreement with the phylogenetic studies and confirmed that adult flukes residing within the bile ducts rely solely on cathepsin Ls for feeding on host blood.[12,17,25,27] Clade FhCL1 and clade FhCL2 proteases were by far the most predominantly expressed proteins, accounting for 67.39% and 27.63% of total secreted cathepsin Ls respectively.[17] Furthermore, neither FhCL3 or cathepsin B (FhCB) proteases were secreted by adult flukes, supporting a specific role for these enzymes in the infective larval stages that initiate and establish infection through the host intestinal wall.[31]

VARIATION IN THE S2 SUBSITE OF THE ACTIVE SITE AND ITS INFLUENCE ON ENZYME SPECIFICITY

It is well established that substrate specificity of papain-like cysteine proteases is primarily determined by the composition and arrangement of amino acids that create the S2 subsite within the active site.[29-34] The deep S2 subsite is composed of residues occupying positions 67, 68, 133, 157, 160 and 205 (papain numbering) within the active site which interact with the P2 amino acid of the substrate. It is these positions which exhibit most variation among members of the papain superfamily. An analysis of these residues in the various *Fasciola* cathepsin L clade members reveals divergence within the S2 subsite, in particular at the three positions that have the greatest influence on P2 binding i.e., residues 67, 157 and 205.[29]

F. hepatica cathepsins L1 and L2 have partly overlapping specificities, often making it difficult to discriminate between them biochemically. However, our biochemical data using both native and recombinant forms of FhCL1 and FhCL2 show just how significant changes within the S2 subsite can be. Overall, FhCL1 (S2 subsite: Leu_{67}, Val_{157} and Leu_{205}) showed an affinity (k_{cat}/K_m) for small, fluorogenic substrates in the following order: Z-Leu-Arg-NHMec > Z-Phe-Arg-NHMec >> Z-Pro-Arg-NHMec whereas for FhCL2 (S2 subsite: Tyr_{67}, Leu_{157} and Leu_{205}) the order of affinity was Z-Leu-Arg-NHMec >> Z-Phe-Arg-NHMec ≈ Z-Pro-Arg-NHMec. While both enzymes efficiently cleaved substrates with hydrophobic residues (Phe and Leu) in the P2 position, the catalytic rates (k_{cat}/K_m) are 25- and 8-fold greater, respectively, for FhCL1 than FhCL2. Another key difference was that while FhCL1 showed a lower affinity for substrates containing proline at the P2 position, FhCL2 displayed a greater preference for Z-Pro-Arg-NHMec with an approximately 6-fold greater affinity for this substrate at pH 5.5 and 3-fold greater affinity at pH 7.3 than FhCL1.[29]

Recently, we correlated the ability of FhCL2 to accommodate Pro in the S2 subsite with the capacity to cleave native collagen, which contains the repeat motif Gly-Pro-Xaa (where Xaa is any amino acid).[29] Since collagen is a major component of interstitial matrices, this property is likely to be important for disrupting cellular integrity making it easier for the parasite to migrate through host tissues. We believe the evolution of this biochemical property by trematode proteases has been critical to the expansion of parasites into various hosts. The ability of papain-like cysteine proteases to degrade collagen is quite rare, having been previously restricted to a small number of bacterial collagenases, matrix metalloproteases and, more relevantly, human cathepsin K but not human cathepsin L.[35] Recently, Corvo et al demonstrated that recombinant FhCL3 displayed a restricted substrate specificity, with a 70-fold preference for Tos-Gly-Pro-Arg-AMC as well as an ability to cleave native collagen.[30] FhCL3, secreted by the newly excysted juvenile stages (NEJs), is known to play a pivotal in the initial steps of host penetration.[31]

The overall similarity between trematode cathepsin L-like peptidases and their human homologs ranges between 35-47%. A comparison of residues within the S2 pocket of the various FhCLs family members, human cathepsin L and cathepsin K is shown in

Table 1. Residues forming the S2 active site of human and *F. hepatica* cathepsin L proteases. Comparison of the residues from the S2 active site that contribute to differential substrate-binding in *Fasciola hepatica* cathepsin Ls (clades 1-5) and human cathepsin L. (Residues were identified using primary sequence alignments and analysis of the atomic structure of *F. hepatica* cathepsin L1;[29] PDB ID: 2O6X)

	Residues						
	67	68	133	157	158	160	205
Human cathepsin L	Leu	Met	Ala	Met	Asp	Gly	Ala
Human cathepsin K	Tyr	Met	Ala	Leu	Asn	Ala	Leu
Adult: FhCL 1A	Leu	Met	Ala	Val	Asn	Ala	Leu
FhCL 1B	Leu	Met	Ala	Leu	Asn	Ala	Leu
FhCL 2	Tyr	Met	Ala	Leu	Thr	Ala	Leu
FhCL 5	Leu	Met	Ala	Leu	Asn	Gly	Leu
Juvenile: FhCL 3	Trp	Met	Ala	Val	Thr	Ala	Val
FhCL 4	Phe	Met	Ala	Leu	Asn	Ala	Phe

Table 1. Most variation occurs in residues at positions 67 (Leu, Tyr, Trp or Phe) and 205 (Leu, Val or Phe), however, variation can also observed at position 157 which lies at the opening of the S2 pocket.[29] Interestingly, all cathepsins with the ability to degrade collagen posses a Tyr residue at position 67. Using site-directed mutagenesis, Lecaille et al substituted Tyr_{67} in human cathepsin K to Leu_{67} (creating a similar S2 subsite to that of human cathepsin L) and showed that the resulting enzyme lacked the ability to degrade collagen.[35] However, we found that replacing Leu_{67} of FhCL1 with Tyr, thus mimicking the S2 subsite of FhCL2 and human cathepsin K, neither enhanced the ability of variant FhCL1Tyr_{67} to accept Pro at the P2 position nor did it endow the engineered enzyme with an ability to degrade collagen.[29] These results clearly suggest the contribution of other key residues within the S2 pocket regarding collagenase activity.

FhCL5 which diverged from FhCL2 prior to its divergence from FhCL1 exhibits an intermediate S2 subsite (Leu_{67}, Leu_{157}, Leu_{205}). Smooker et al demonstrated that despite sharing 80% sequence identity to FhCL2, FhCL5 did not exhibit substantial activity against substrates containing Pro in the P2 position.[28] However, by mutating Tyr_{67} for Leu, the ability of FhCL5Tyr_{67} to cleave substrates containing P2 Pro was greatly enhanced. Taken together these results indicate that separation of FhCL1 and FhCL5 from FhCL2 was accompanied by the loss of the ability to cleave substrates with a P2 proline, like collagen. Recently Lowther et al demonstrated that this alteration enabled the S2 subsite of FhCL1 to "open up" so that it could accommodate hydrophobic amino acids, such as Leu, Ala, Phe and Val.[36] Interestingly, these residues are the most common residues found in host haemoglobin (accounting for 42% of the protein), the major source of nutrients for adult flukes.

ATOMIC STRUCTURE AND SUBSTRATE SPECIFICITY

Although eukaryotic cathepsins are diverse at the primary sequence level, their atomic structures are remarkably conserved. We have recently solved the 3-D atomic structure of the *F. hepatica* FhproCL1 zymogen at 1.4 Å, using an active site mutant FhproCL1Gly$_{25}$ (PDB ID 206X).[29] This structure is almost identical to that of human cathepsin L (PDB ID 1CJL), despite their primary structures exhibiting only 35% identity (and 71% similarity). The mature domain of FhproCL1Gly$_{25}$ is bi-lobed: the left-hand lobe (L) is predominantly composed of α–helices, whereas the right-hand lobe (R) contains several elements of β–sheet. The L and R lobes fold together to give the typical "V"-shaped active site cleft configuration at the interface between the two domains.[34] The N-terminal extension, or prosegment, runs through the active site cleft, in a reverse direction to a normal protein substrate and prevents uncontrolled proteolysis (preventing premature activation of the catalytic domain).[29,37]

As mentioned previously, the ability to accept or exclude particular substrate moieties is highly dependent on the size, shape and volume of the S2 pocket, as well as an ability to form stabilizing interactions. In the atomic structure of FhproCL1Gly$_{25}$, Leu$_{67}$ and Val$_{157}$ are situated at the entrance to the pocket and act as "gatekeepers", while Met$_{68}$, Ala$_{133}$ and Ala$_{160}$ sit below them deeper within the pocket and Leu$_{205}$ lies at the floor of the pocket (Fig. 3). The crystal structure of the FhproCL1Gly$_{25}$ was used to create a robust homology model of FhCL2 that was compared to the atomic structure of human cathepsin K (PDB ID 1ATK) to investigate which residue(s) are important in determining the ability of both cathepsin K and FhCL2 to degrade collagen. Although the S2 pockets

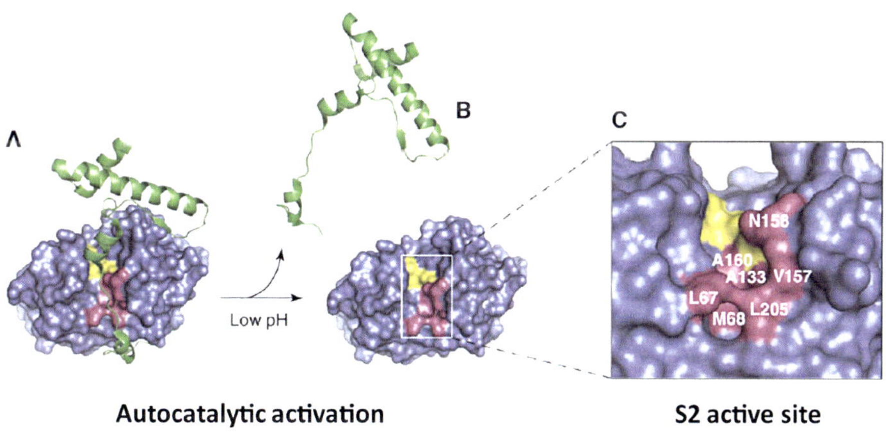

Figure 3. Structural analysis of *F. hepatica* cathepsin L proteases. *Fasciola* cathepsin Ls are stored in specialised secretory vesicles within the parasite's gut epithelial cells as inactive zymogens consisting of a prosegment and mature enzyme domain. The prosegment is removed by catalytic cleavage following secretion into the parasite intestine to reveal an active mature protease. A) Surface representation of the major *F. hepatica* secreted cathepsin L protease (FhCL1) (PDB ID: 2O6X). The prosegment is shown as a green ribbon and the catalytic machinery is shown in yellow (P1 residues) and magenta (P2 residues). B) At low pH (4.0-4.5) the prosegment is auto-catalytically removed (B) to produce the mature active enzyme. C) Detail from the active site of the enzyme. S2 residues that determine substrate-binding specificity are labelled. Reprinted from Robinson MW et al. Trends Biochem Sci 2008; 33:601-608;[23] ©2008, with permission from Elsevier.

of human cathepsin K and FhCL2 are very similar, there are some noteworthy differences within a 5 Å radius of the active site. Most significant is the 'gatekeeper' residue 157, which sits near the opening of the S2 pocket. Residue 157 is a Val in FhCL1 and is one carbon shorter than the Leu found in FhCL2 and human cathepsin K and accordingly its extension into the pocket is approximately 1.5 Å less and thus does not allow it to extend far enough into the available space to participate in aliphatic interactions. Hence the absence of both stabilizing Tyr and Leu residues would account for the lack of favour for P2 proline in the S2 pocket of FhCL1 (and human cathepsin L).[29]

Based simply on spatial constraints, a Pro residue would be accommodated in the S2 area of FhCL2. Our analyses suggest that acceptance of P2 Pro is achieved by providing opportunities for stabilizing interactions with the 5-membered proline ring of the substrate at the entrance to the pocket and that such stabilization involves ring-ring interactions between Tyr_{67} and the P2 Pro. The location and positioning of Leu_{157} in FhCL2 and human cathepsin K suggests its availability to further stabilize the presence of a P2 Pro, perhaps with constructive aliphatic interactions.[29]

Our biochemical and structural data on FhCL2 and that of Corvo et al on FhCL3 clearly demonstrate that both enzymes, like cathepsin K, cleave collagen within the triple helical regions.[30] However, both FhCL2 and FhCL3 can digest native collagen at neutral pH while degradation of collagen by cathepsin K requires acidic conditions. In humans, cleavage of collagen by cathepsin K is essential for bone remodelling. In contrast, FhCL3 activity is vital for penetrating the host gut wall, while adult parasites which must penetrate and migrate through large host organs, including the liver, require FhCL2 activity for degrading collagen-rich interstitial matrices, both likely occurring under physiological pH.[29-31,35]

THE ROLE OF THE PROSEGMENT IN REGULATING THE FUNCTION OF TREMATODE CATHEPSINS

Cathepsin proteases are produced as inactive zymogens requiring the cleavage of prosegments to become functionally active.[33,35] Prosegments function as regulators of enzymatic activity by binding to the substrate cleft,[29,31,37,38] as intermolecular chaperones that are important for correct protein folding,[39,40] and are involved in intracellular trafficking of human cathepsins to the lysozome.[41] Activation of cathepsin proteases, to generate active mature enzyme, occurs via a specific cleavage event between the prosegment region and the catalytic domain. Processing is believed to be facilitated by a low pH environment which disrupts the salt bridge involving an Asp contained within a conserved G*X*N*X*F*X*D motif located between residues –42 and –36 (papain numbering). In mammalian cells, this environment is provided by the acidic lysosome. Subsequent removal of the prosegment involves possible intramolecular and/or intermolecular cleavages that take place within the central portion of the prosegment and then at the C-terminal region close to the N-terminus of the mature enzyme.[37,41,42]

We and others have proposed that auto-activation of trematode zymogens to the mature active form occurs in the slightly acidic milieu of the parasite gut. Experimental evidence for auto-activation comes from in vitro studies using recombinant FhproCL1 and have demonstrated that autocatalytic activation can occur within the pH range 4.5 to 7.3.[29,43] Recently Lowther et al have shown that activation occurs far more rapidly

A

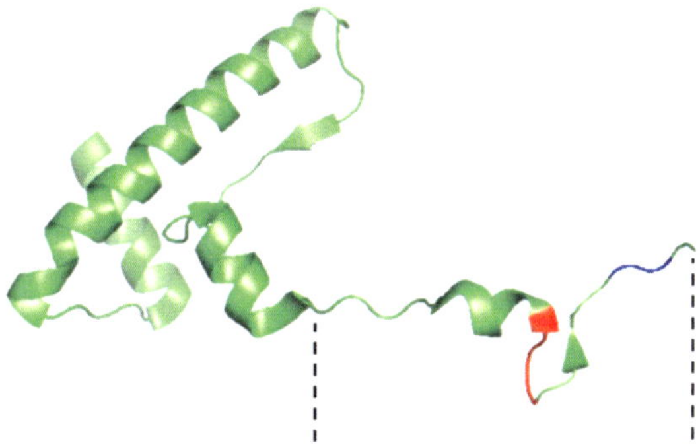

B

```
FhCL consensus    -E----S--LSHG--Y------
FhCL 1A_ie1       T.MSRA.DI..H.VP.EANNRa
FhCL 1A_au3       T.MSRA.DI..H.VP.EANNRa

FhCL 1B_nl1       T.MPRA.DI..H.IP.EANNRa
FhCL 1B_ar        T.MPRA.DI..H.IP.EANNRa

FhCL 2_ie2        I.IPRS.EL..R.IP.KANKLa
FgCL 2_thD        I.IPRS.EL..R.IP.KANKPa

FhCL 5_au4        T.MPRA.EL..H.IP.KANKRa
FhCL 5_au5        T.MPRA.EL..H.IP.KANKRa

FgCL 3_thG        I.MSPE.ES..D.IS.EAEGNd
FhCL 3_uy3        M.MSPE.ES..D.IS.EAEGNd

FhCL 4_uy4        R.IPRA.DM..H.IP.EANDRa
FgCL 4_thH        R.IPRA.DIH.H.IP.EANDRa
```

Figure 4. Structural and sequence analysis of *Fasciola* cathepsin L prosegments. Cathepsin L prosegments regulate enzyme activity by binding to the substrate cleft and act as a molecular chaperones to ensure correct folding of the enzyme. Prosegment removal is mediated by cleavage at two conserved motifs towards the C-terminal end of the molecule. A) The atomic structure of the *F. hepatica* major secreted cathepsin L protease prosegment region (PDB ID: 2O6X). The conserved Leu-Ser↓His motif required for zymogen auto-catalytic processing is highlighted in red and the asparagine residues required for *trans*-activation by apsaraginyl endopeptidases are shown in blue. B) Primary sequence alignment of the nonconserved prosegment C-terminal regions from selected *Fasciola* cathepsin L proteases. A consensus sequence for all *Fasciola* cathepsin Ls (FhCL consensus) is shown at the top of the alignment. Gaps in the alignment are represented by a dash (-) and amino acids that are conserved in all sequences are indicated by a dot (.). Prosegment residues are in capital letters whereas the first amino acids of the mature enzymes are in lowercase letters. The colouring of residues follows that of (A). Reprinted from Robinson MW et al. Trends Biochem Sci 2008; 33:601-608.[23] ©2008, with permission from Elsevier.

at lower pH values (activation at pH 4.0 was 40-times faster than at pH 7.0).[36] The precise nature of the initial events that transform the inactive zymogen to active enzyme still remains a mystery. The presence of a functional active site cysteine (Cys_{25}) is essential for activation of both human and *F. hepatica* cathepsin Ls.[37,41,43] Using the same active site variant as used in our crystallographic studies ($FhproCL1Gly_{25}$), we have shown that this variant cannot auto-catalytically process. Processing of this variant could however be achieved via *trans*-processing using pre-activated wildtype FhCL1. This intermolecular processing occurred as a result of direct cleavage within the Leu_{12}-Ser_{11}↓His_{10} motif located at the nonconserved C-terminal region of the propeptide. Mutations that alter this motif to a Pro_{12}-$Ser_{11}\downarrow His_{10}$ prevent or slow down FhproCL1 activation.[44]

Robinson et al recently demonstrated that phylogenetic analyses using only the prosegment domains of the *Fasciola* cathepsin Ls produced a tree similar to that produced by the mature domains indicating parallel adaptation of both regions.[17] An alignment of the prosegments (residues P1 to P91) shows that the N-terminal and intermediate regions (residues P1 to P70) of these enzymes are remarkably conserved across all cathepsin L clades. It is noteworthy that although the C-terminal portion of the *F. hepatica* cathepsin L prosegments (21 residues, P70 to P91) show striking variability between the phylogenetic clades it is conserved within each clade. This is particularly evident in the final five residues which form the boundary between the prosegment and mature enzyme and gives each clade its signature sequence as shown in Figure 4. In addition to the conserved Leu_{12}-$Ser_{11}\downarrow His_{10}$ motif (described above) that occurs within this region, cleavage of the prosegment after a conserved asparagine residue is believed to be required for *trans*-activation by asparaginyl endopeptidases (see below).[17] The FhCL1 atomic structure suggests that this region does not make contacts with the main body of the enzyme, but rather moves freely in space creating a 'protease-accessible' region that facilitates prosegment removal.[29]

Since the stability of the propeptide-protease complex is dependent on electrostatic interactions, reduction of the environmental pH likely weakens the bonds between the propeptide and the catalytic site. As a consequence, the proenzyme possibly adopts a looser conformation, in which the propeptide is bound less tightly into the active site making it more susceptible to proteolysis.[45,46] However, circular dichroism studies reveal that activation does not involve significant conformational changes in the structure of FhproCL1, procathepsin L or procathepsin B.[36,41,45] These data suggest two distinct possibilities: (1) the initial event may involve an active proenzyme, possibly created by the reduced pH, *trans*-processing another proenzyme setting off a chain reaction or (2) *trans*-processing by an enzyme of a different class.

TRANS-ACTIVATION OF TREMATODE CATHEPSINS BY ASPARAGINYL ENDOPEPTIDASE?

Schistosome asparaginyl endopeptidase (SmAE; also known as Sm32 or schistosome legumain) is a clan CD cysteine endoprotease that cleaves C-terminal to asparginyl (Asn) residues and was originally thought to have a direct role in the hydrolytic degradation of host haemoglobin. In support of exogenous cleavage of trematode cathepsins by a different type of enzyme, Dalton and Brindley, proposed that SmAE was responsible *trans*-processing and activating the cysteine proteases involved in the haemoglobin degradation cascade.[47] They also proposed that the enzyme's primary role was the *trans*-processing and activation of other schistosome proteases after noting that zymogens of Clan AA aspartic proteases (cathepsin D) and Clan CA cysteine proteases cathepsins L, F, B1 and C each possessed an asparaginyl endopeptidase cleavage site at the juncture between the prosegment and mature enzyme domain. Immunolocalisation studies have shown that asparaginyl endopeptidases are co-expressed with the haemoglobin-degrading cathepsin proteases in the gastrodermis surrounding the gut lumen of several trematode species including *F. gigantica* and *F. hepatica*,[48-50] *O. viverrini*,[51] and *C. sinensis*.[52] Sajid et al provided experimental support for this hypothesis by showing that recombinant

SmAE could convert the schistosome cathepsin B1 zymogen to a mature enzyme in vitro.[47,53] Furthermore, *Fasciola* cathepsin L and B cysteine proteases also have preserved the asparaginyl endopeptidase-processing site at the prosegment-mature domain junction and SmAE was also shown to *trans*-process *F. hepatica* cathepsin B.[17,54] If exogenous cleavage is the primary mechanism for cathepsin activation, then inhibiting asparaginyl endopeptidase activity could be an attractive target for pharmacological intervention. Gene silencing studies by Delcroix et al demonstrated that 20% of cathepsin B activity was lost when SmAE expression in 3-week old male and female schistosomes was knocked down using RNAi.[55] However, cathepsins are capable of autoactivation which begs the question, why do trematodes require two different mechanisms for the activation of their cysteine proteases?

To answer this question, we have proposed that the cleavage events leading to cathepsin prosegment removal take place in two steps.[17] Initially in a bimolecular process, a small number of cathepsin L zymogens either autoactivate at low pH or are *trans*-activated by asparaginyl endopeptidase. These enzymes are then able to *trans*-process another cathepsin L zymogens through specific cleavage at the Leu-Ser↓His motif and thus the rate increases exponentially.[29] Interestingly, the prosegment of *O. viverrini* cathepsin F (Ov-CF-1) lacks the conserved asparagine residues found in other trematodes and could not be *trans*-processed by *O. viverrini* asparaginyl endopeptidase that is also localised in the gut cells of adult worms.[22] However, we have recently shown that *O. viverrini* cathepsin B (Ov-CB-1) is capable of *trans*-processing Ov-CF-1 via a specific cleavage between the prosegment and mature enzyme domain.[56]

BIOLOGICAL ROLES OF TREMATODE CATHEPSIN PROTEASES

The phylogenetic and biochemical studies described above collectively show how divergence within the *F. hepatica* cathepsin L family produced a repertoire of enzymes with overlapping and complementary substrate specificities. This parasite also expresses cathepsin B proteases that exhibit independent regulation to the cathepsin L proteases but function in concert with these to facilitate parasite infection and migration.

Following ingestion by a mammalian host, *Fasciola* NEJs must penetrate and traverse the wall of the duodenum, move through the peritoneum and penetrate the liver. After migrating through and feeding on the liver, the parasites move to their final destination within the bile ducts where they mature and produce vast quantities of eggs (30-50,000 eggs/day/worm).[13] Using an integrated transcriptomics and proteomics analysis of the molecules secreted by various developmental stages of *F. hepatica*, Robinson et al demonstrated that cysteine protease expression/secretion is highly regulated during the development of the parasite and correlates with its penetration and migration through the tissues of the host.[48] By combining the temporal expression of individual cysteine proteases with our understanding of fluke biochemistry and recent insights from RNAi studies, we now have a complete picture of how the developing *F. hepatica* migrates through host tissues and sustains itself during its maturation. (Fig. 5). Dormant larvae emerge from cysts in the duodenum of their mammalian hosts as NEJs and directly penetrate the gut wall and NEJ-specific proteases such as FhCL3 and cathepsin B have specialised functions in relation to these tasks.[31] Using inhibitors of cathepsin-like proteases we have previously shown that excystment of the infective larvae is dependent on both FhCL3 and FhCB. Together these enzymes account for over 80% of the total protease activity

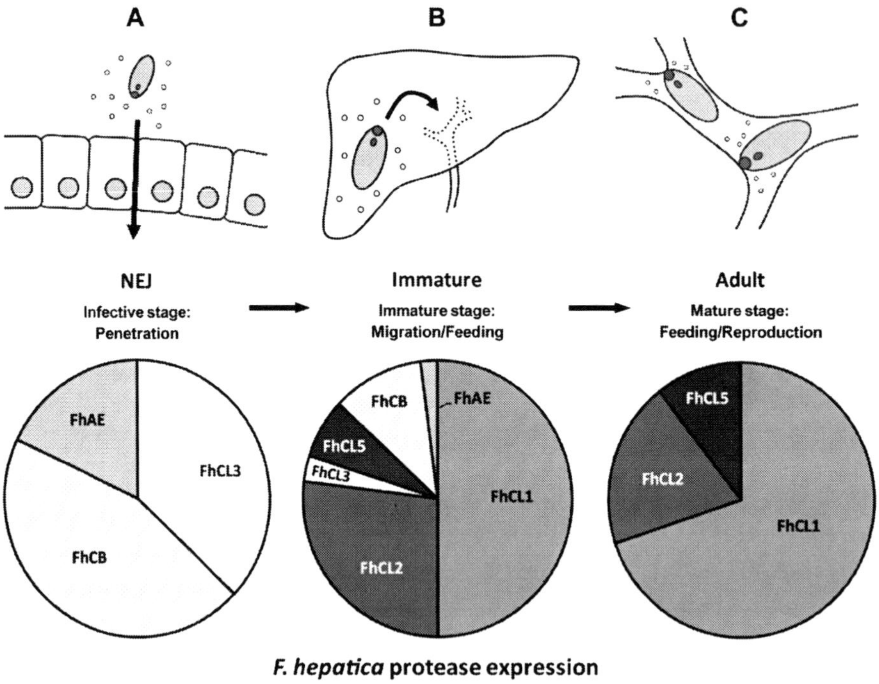

Figure 5. Developmental regulation of *Fasciola hepatica* cathepsin proteases. Trematode pathogens such as *F. hepatica* undergo complex life-cycles involving transitions between host organ systems and tissues. Movement from one site to another, as the life-cycle progresses, is associated with the differential expression of cathepsin enzymes. A) Penetration of the host intestinal wall by newly excysted juvenile (NEJ) *F. hepatica* is facilitated by the secretion of proteolytic enzymes. This process involves cathepsin B and clade 3 cathepsin Ls. B) The migration of immature flukes through host liver tissue corresponds to a period of rapid growth and development of the parasite. At this stage, host protein degradation is achieved primarily by *Fasciola* clade 1 and 2 cathepsin Ls although some cathepsin B and clade 3 and 4 cathepsin L activity may remain. C) Adult *F. hepatica* reside within the bile ducts where they feed on host red blood cells. Proteomics analyses have shown that clade 1, 2 and 5 cathepsin Ls are the only proteases secreted by adult flukes, thus suggesting crucial roles for these enzymes in penetrating the bile duct wall and digesting host haemoglobin. Pie charts represent protease expression levels determined by quantitative mass spectrometry analysis of *F. hepatica* secretory proteins.[48] Reprinted from Robinson MW et al. Trends Biochem Sci 2008; 33:601-608.[23] Copyright 2008, with permission from Elsevier. This research was originally published in Robinson MW et al. Mol Cell Proteomics 2009; 8:1891-1907;[48] Copyright 2009, with permission of the American Society for Biochemistry and Molecular Biology.

detectible in NEJ stage.[48] Further experimental evidence has been provided by McGonigle et al who have shown that RNAi-mediated silencing of either cathepsin B or cathepsin L-like transcripts in infective juveniles blocks their ability to penetrate the intestinal wall of experimentally-infected rats.[31] Together this evidence clearly demonstrates that both enzyme types are required for successful host penetration. Similar studies on the infective larvae of other medically important trematodes also demonstrate a reliance of cysteine proteases for successful infection.[11,12,57]

As the immature flukes enter the liver parenchyma and begin feeding on the tissue, the secretion of FhCL3 and FhCB is much reduced and account for only 3% and 2%

of total secreted protease activity respectively. Concurrently, the expression of other cathepsin L proteases is up-regulated: FhCL1 (50%), FhCL2 (27%) and FhCL5 (7%) become more highly represented in the secretome. This striking modification of protease activities represents a clear shift in the requirements of the parasite for a different subset of proteases during this developmental phase.[48]

Adult flukes residing in the bile ducts become obligate blood-feeders and draw blood through punctures they make in the duct wall. The nutrients obtained from this process are used to support the production of an enormous number of progeny.[13] Biochemical and immunological studies indicate that by the time the parasites have moved into the immunologically safe environment of the bile ducts, they become reliant solely on cathepsin L proteases to degrade host haemoglobin and serum proteins; expression profiles: FhCL1 (69%), FhCL2 (22%) and FhCL5 (9%).[17] Accordingly, FhCL1, FhCL2 and FhCL5 proteases must collectively possess the hydrolytic machinery capable of digesting the macromolecules in the bloodmeal to peptides that are sufficiently small to be absorbed into the parasite gastrodermis. Recently, we showed that one of these enzymes, FhCL1, can efficiently degrade haemoglobin to peptides of between 3-14 amino acids that are presumably absorbed into the gastrodermis for further catabolism to amino acids by intracellular cathepsin C and aminopeptidases.[36] Thus, the developmentally-regulated expression of trematode cathepsins correlates directly with the passage of the parasite through the various host tissues and macromolecules its encounters, implying that individual enzymes have evolved very specific functions.

Studies on the infective larvae of other trematodes, including those of *Schistosoma spp., Trichobilharzia spp., Paragonimus spp.* and *Clonorchis spp.* point to the universal use of cysteine proteases in host infection as well as haemoglobin degradation.[12,57,58] There is strong experimental evidence supporting the critical roles played by these enzymes in the establishment and maintenance of infection; a) blocking proteolysis of host haemoglobin with cysteine protease inhibitors results in significant antiparasitic and antipathology effects in schistosomasis;[59,60] b) administration of cysteine protease inhibitors such as K11777 to *S. mansoni*-infected mice reduces worm burden and egg production;[59] and c) vaccination of sheep and cattle with purified *F. hepatica* cathepsin L elicits protection (50-73%) and antifecundity effects against challenge infection. It has been proposed that the effects of inhibitors and vaccines on parasite development may be mediated through disruption of parasite feeding and migratory behaviour, while the antifecundity effects may be related to the inability of starved parasites to synthesise viable eggs.[61,62] New insights into the fundamental roles of other trematode cysteine proteases have recently been provided by a number of RNAi and transgenesis studies.[63,64] Skelly et al were the first to successfully perform knockdown of protease expression in schistosomes.[65] Specific silencing of SmCB1 activity was demonstrated by immunofluorescence, RT-PCR and more importantly proteolytic activity. SmCB1 had been proposed to play a central role in haemoglobin digestion in the schistosome gut. While long-term suppression of SmCB1 did retard schistosome growth, parasites were still viable and capable of degrading host haemoglobin. In contrast to *Fasciola* parasites, the picture of haemoglobin degradation is somewhat more complicated in schistosomes as they also possess an aspartic protease, SmCD, which acts as part of a multi-enzyme cascade to perform this task. One possible reason for the lack of any notable deleterious phenotype may have been due to proteolytic compensation by SmCD in this pathway. These results clearly indicate that although SmCB1 is necessary for normal parasite development, its function is not essential in the process of haemoglobin catabolism.

In a recent transcriptional study analysing temporal transcriptional changes in *S. mansoni* during early schistosomula development, Gobert et al found that genes encoding cathepsin proteases were highly up-regulated by day five and included SmCB (65-fold), cathepsin L (37-fold), cathepsin D (13-fold) and cathepsin C (11-fold).[66] This strong temporal control of protease expression, similar to that seen in *Fasciola*, is again indicative of a complex array of functionality. Delcroix et al used RNAi to dissect the roles of a number of proteases believed to be involved in haemoglobin degradation and found the primary role of SmCD to be in haemoglobin proteolysis, as was previously hypothesised by Brindley et al.[55,67] The same study demonstrated the primary role of SmCB1 to be albumin degradation. Recently, Morales et al also using RNAi demonstrated that SmCD is essential for schistosome survival due to its critical role in haemoglobin catabolism.[68]

LOCALISATION PATTERNS AND SECRETION OF TREMATODE CATHEPSINS

Immunolocalisation and in situ hybridisation studies on *F. hepatica* have shown that cathepsin L proteases are synthesised within the gastrodermis that lines the parasite gut.[61] These cells have both a secretory and absorptive function and spread extended lamellae into the gut lumen.[43] Liver flukes possess a blind-ended gut, requiring the contents of the gut to be emptied by frequent regurgitation (approximately every 3 h) and refilled with fresh host blood.[69] Using antibodies directed against the propeptide of FhCL1 we have demonstrated that it is the inactive procathepsin L that is packaged into secretory vesicles within the gut epithelial cells.[43] Electron microscopy revealed the specific location of these cathepsin L-containing secretory bodies to be at the apex of the gastrodermal cells ready for secretion into the gut.[43,61] Collectively these observations provide important insights into the regulation of *F. hepatica* protease activity. First, the accumulation of inactive procathepsin L molecules, rather than active mature proteases reduces the risk of potential internal damage due to uncontrolled proteolysis. Second, protease activation takes place following secretion of proteases into the low pH environment of the parasite gut lumen. Once activated, proteases are in direct contact with the bloodmeal and thus are ready to digest host macromolecules for protein catabolism. In addition, active mature cysteine proteases are released regularly into the surrounding host tissues where they participate in functions outside the parasite gut.

Immunohistochemical studies have also localised a number of schistosome cysteine proteases to the parasite gut.[70-72] One of the best characterised schistosome proteases is SmCB. SmCB has two distinct forms, SmCB1 and SmCB2, with similar biochemical properties. SmCB1 is expressed in the gut where it was originally thought to participate in haemoglobin degradation but it has since been shown to be involved in the digestion of albumin.[55] In contrast, SmCB2 has been localised to the tegument and parenchyma but its function is still unknown.[10] SmCF (previously known as SmCL1) is also expressed in the schistosome gut but there is some conjecture as to the exact localisation of SmCL2. SmCF and SmCL2 have been localised within the reproductive system of female worms and the gynecophoral canal of males,[73] and in the gastrodermal surface of both males and females.[60] In contrast to the findings of Michel et al[73] Bogitsh et al did not find SmCL2 in the reproductive organs.[60] However, the recently-discovered SmCL3 has been detected in the gastrodermis of both males and females but also within the female vitellaria.[74]

Immunolocalisation studies have shown that the major cathepsin F proteases of adult *C. sinsensis* (Cs-CF-4 and Cs-CF-9) and *P. westermani* are expressed exclusively in the cells lining the gut.[21,26,75] In contrast, the cathepsin F expressed by *O. viverrini* (Ov-CF-1) has been localised within the gut, vitellaria, testes and eggs of adult worms[22] suggesting a wider role for this protease. Interestingly, the cathepsin Fs expressed and secreted from *P. westermani* newly excysted metacercariae (PwMc28a and PwMc28b) have been localised to the tegument and anterior border of the oral sucker. This suggests that proteases are secreted trans-tegumentally or exit via the excretory canal network in the juvenile fluke that have not yet developed a functional gut.[76]

THE ROLE OF PH IN REGULATING CATHEPSIN FUNCTION IN TREMATODES

Once released from the gut, the extracorporeal roles of trematode proteases are performed at physiological pH, i.e., between two and three pH units higher than the microenvironment in which the proteases function in the parasite gut. The gut lumen of *F. hepatica*, like that of other trematodes, is believed to be slightly acidic (around pH 5.5) however, the precise pH is unknown.[77] Estimates suggest the gut lumen of *S. mansoni* is in the range pH 5.0-6.0[78] to 6.84,[53] although studies by Delcroix et al suggest the existence of micro-compartments between the lamellae of the gut with pH as low as 3.9.[55] Our studies show that FhCLs are active over a broad pH range (3.0-9.0) and are highly stable at neutral pH. This points to a specific adaptation of parasite cysteine proteases to carry out functions over a wide pH range (recombinant FhCL1 retained 100% of its activity when incubated at 37 °C for 24 h and ~45% activity when incubated at 37 °C for 10 days at pH 4.5). It is interesting to note that the pH optimum of FhCL1, pH 6.2, is approximately mid-point between the pH values at which it works inside and outside the parasite.[36]

Mammalian lysosomal cathepsin Ls are active only at approximately pH 4.5, in keeping with the environment in which they function and are inherently unstable at neutral pH so that cellular damage due to leakage from the lysosome is avoided.[37] Interestingly, Corvo et al found the pH profile of FhCL3 to be in the neutral range with an optimum pH of 7.0.[30] In contrast to FhCL1 that displays impressive stability at low pH, FhCL3 demonstrated only 35% of maximum activity at pH 5.5.[36] In terms of stability, FhCL3 retained 50% activity after 8 h at neutral pH and 37 °C. FhCL3 is not secreted by adult parasites and therefore does not play any role in the acquisition of nutrients from haemoglobin or serum proteins. These findings are consistent with the secretion and function of this enzyme in the environment encountered by the NEJ stages and hence it's proposed role in penetration of the host duodenum.[30,31]

CONCLUSION

While there have been numerous comprehensive studies characterising the biochemical properties of trematode cathepsins, their exact biological roles are often merely inferred and at best poorly understood. For instance, although many cysteine proteases are expressed in the parasite gut, the transcription levels of some of these enzymes are 50- to 1000-fold less than those of other gut-associated proteases. Prime examples are FhCL5 and SmCL3 the expression levels of which are relatively minor when compared to FhCL1 or SmCB1.[48,74]

Proteases by their nature are promiscuous, functioning both within the parasite and externally at the host-parasite interface. The exact proteolytic contribution to parasite biology of such proteases remains ambiguous. The ability to isolate and purify native enzymes and/or produce recombinant forms of parasite enzymes has been fundamental in terms of both our understanding cysteine protease biochemistry and in inferring their biological roles.[12,53,43] Using both native and recombinant forms of *F. hepatica* cysteine proteases we have shown that these enzymes efficiently degrade a variety of host macromolecules. We have proposed that by producing proteases with overlapping specificity, *F. hepatica* parasites could more efficiently infect and penetrate host tissue and organs. Our integrated transcriptomics and proteomics analysis has provided an in-depth overview of protein secretion by *F. hepatica* parasites.[17,48] However, it is only when these data are combined with the recent functional RNAi studies that we now are able to definitively state that these enzymes are crucial to successful establishment and maintenance of *F. hepatica* infection.[31]

Trematode cysteine proteases are promising targets against which novel intervention strategies can be targeted. New insights into fundamental trematode biology are continually emerging through various genomics, transcriptomics and proteomics projects and due to the application of genetic manipulation technologies such as RNAi and transgenesis.[64,79,80] Integration of these functional analyses will be invaluable in furthering our understanding of parasite pathogenesis and represents a significant step toward a comprehensive understanding of host-parasite interplay. Although RNAi and transfection are powerful tools in validating the function of target molecules these technologies are still in their infancy when it comes to their use in trematode biology. However, their continued development and integration with other postgenomic technologies will allow us to answer many biologically-relevant questions and aid both antiparasite vaccine and drug discovery programs.

ACKNOWLEDGEMENTS

JPD holds a Canada Research Chair (Tier 1) funded by the Canadian Institutes of Health Research (CIHR). MWR holds a University of Technology, Sydney (UTS) Chancellor's Postdoctoral Research Fellowship.

REFERENCES

1. Hotez PJ, Molyneux DH, Stillwaggon E et al. Neglected tropical diseases and HIV/AIDS. Lancet 2006; 368:1865-1866.
2. Robinson M, Dalton JP. Zoonotic helminth infections with particular emphasis on fasciolosis and other trematodes. Philos Trans R Soc B Biol Sci 2009; 364:2763-2776.
3. Engels D, Chitsulo L, Montresor A et al. The global epidemiological situation of schistosomiasis and new approaches to control and research. Acta Trop 2002; 82:139-146.
4. Gryseels B, Polman K, Clerinx J et al. Human schistosomiasis. Lancet 2006; 368:1106-1118.
5. Keiser J, Utzinger J. Emerging foodborne trematodiasis. Emerg Infect Dis 2005; 11:1507-1514.
6. Lun ZR, Gasser RB, Lai DH et al. Clonorchiasis: a key foodborne zoonosis in China. Lancet. Infect Dis 2005; 5:31-41.
7. Mas-Coma S, Bargues MD, Valero MA. Fascioliasis and other plant-borne trematode zoonoses. Int J Parasitol 2005; 35:1255-1278.
8. Parkinson M, O'Neill SM, Dalton JP. Endemic human fasciolosis in the Bolivian Altiplano. Epidemiol Infect 2007; 135:669-674.
9. Littlewood DTJ. The evolution of parasitism in flatworms. In: Maule A.G. and Marks N.J., eds. Parasitic Flatworms: Molecular Biology, Biochemistry, Immunology and Physiology, Oxford:CAB International, 2006:1-36.

10. Caffrey CR, McKerrow JH, Salter JP et al. Blood 'n' guts: an update on schistosome digestive peptidases. Trends Parasitol 2004; 20:241-248.
11. Kašný M, Mikes L, Hampl V et al. Chapter 4. Peptidases of trematodes. Adv Parasitol 2009; 69:205-297.
12. Tort J, Brindley PJ, Knox D et al (1999). Proteinases and associated genes of parasitic helminths. Adv Parasitol 1999; 43:161-266.
13. Dalton JP, Caffrey CR, Sajid M et al. Proteases in trematode biology. In: Maule A.G. and Marks N.J., eds. Parasitic Flatworms: Molecular Biology, Biochemistry, Immunology and Physiology, Oxford: CAB International, 2006.
14. Cho PY, Lee MJ, Kim TI et al (2006). Expressed sequence tag analysis of adult Clonorchis sinensis, the Chinese liver fluke. Parasitol Res 2006; 99:602-608.
15. Cho PY, Kim TI, Whang SM et al. Gene expression profile of Clonorchis sinensis metacercariae. Parasitol Res 2008; 102:277-282.
16. Kim TS, de Guzman JV, Kong HH et al. Comparison of gene representation between diploid and triploid Paragonimus westermani by expressed sequence tag analyses. J Parasitol 2006; 92:803-816.
17. Robinson MW, Tort JF, Lowther J et al. Proteomic and phylogenetic analysis of the cathepsin L protease family of the helminth pathogen, Fasciola hepatica: expansion of a repertoire of virulence-associated factors. Mol Cell Proteomics 2008; 7:1111-1123.
18. Curwen RS, Ashton PD, Sundaralingham S et al. Identification of novel proteases and immunomodulators in the secretions of schistosome cercariae that facilitate host entry. Mol Cell Proteomics 2006; 5:835-844.
19. Donnelly SM, Dalton JP, Robinson MW. How pathogen-derived cysteine proteases modulate host immune responses. In Robinson MW, Dalton JP, eds. Cysteine Proteases of Pathogenic Organisms. Austin: Landes Bioscience, 2011.
20. Park H, Hong, KM, Sakanari JA et al. Paragonimus westermani: cloning of a cathepsin F-like cysteine proteinase from the adult worm. Exp Parasitol 2001; 98:223-227.
21. Na BK, Kang JM, Sohn WM. CsCF-6, a novel cathepsin F-like cysteine protease for nutrient uptake of Clonorchis sinensis. Int J Parasitol 2008; 38:493-502.
22. Pinlaor P, Kaewpitoon N, Laha T et al. Cathepsin F cysteine protease of the human liver fluke, Opisthorchis viverrini. PLoS Negl Trop Dis 2009; 3:e398.
23. Robinson MW, Dalton JP, Donnelly S. Helminth pathogen cathepsin proteases: it's a family affair. Trends Biochem Sci 2008; 33:601-608.
24. Lynch M. (2002) Gene duplication and evolution. Science 2009; 297:945-947.
25. Irving JA, Spithill TW, Pike RN et al. The evolution of enzyme specificity in Fasciola spp. J Mol Evol 2003; 57:1-15.
26. Kang JM, Bahk YY, Cho PY et al. A family of cathepsin F cysteine proteases of Clonorchis sinensis is the major secreted proteins that are expressed in the intestine of the parasite. Mol Biochem Parasitol 2010; 170:7-16.
27. Morphew RM, Wright HA, LaCourse EJ et al. Comparative proteomics of excretory-secretory proteins released by the liver fluke Fasciola hepatica in sheep host bile and during in vitro culture ex host. Mol Cell Proteomics 2007; 6:963-972.
28. Smooker PM, Whisstock JC, Irving JA et al. A single amino acid substitution affects substrate specificity in cysteine proteinases from Fasciola hepatica. Protein Sci 2000; 9:2567-2572.
29. Stack CM, Caffrey CR, Donnelly SM et al. Structural and functional relationships in the virulence-associated cathepsin L proteases of the parasitic liver fluke, Fasciola hepatica. J Biol Chem 2008; 283:9896-9908.
30. Corvo I, Cancela M, Cappetta M et al. The major cathepsin L secreted by the invasive juvenile Fasciola hepatica prefers proline in the S2 subsite and can cleave collagen. Mol Biochem Parasitol 2009; 167:41-47.
31. McGonigle L, Mousley A, Marks NJ et al. The silencing of cysteine proteases in Fasciola hepatica newly excysted juveniles using RNA interference reduces gut penetration. Int J Parasitol 2008; 38:149-155.
32. Turk D, Guncar G. Lysosomal cysteine proteases (cathepsins): promising drug targets. Acta Crystallogr D Biol Crystallogr 2003; 59:203-213.
33. Turk D, Guncar G, Podobnik M et al. Revised definition of substrate binding sites of papain-like cysteine proteases. Biol Chem 1998; 379:137-147.
34. Lecaille F, Broome D, Lalmanach G. Biochemical properties and regulation of cathepsin K activity. Biochimie 2008; 90:208-226.
35. Lecaille F, Choe Y, Brandt W et al. Selective inhibition of the collagenolytic activity of human cathepsin K by altering its subsite specificity. Biochemistry 2002; 41:8447-8454.
36. Lowther J, Robinson MW, Donnelly SM et al. The importance of pH in regulating the function of Fasciola hepatica cathepsin L1 cysteine protease. PLoS Negl Trop Dis 2009; 3:e369.
37. Coulombe R, Grochulski P, Sivaraman J et al. Structure of human procathepsin L reveals the molecular basis of inhibition by the prosegment. EMBO J 1996; 15:5492-5503.
38. Wiederanders B, Kaulmann G, Shilling K. Functions of propeptide parts in cysteine proteases. Curr Protein Pept Sci 2003; 4:309-326.

39. Baker D, Silen JL, Agard DA. Protease pro region required for folding is a potent inhibitor of the mature enzyme. Proteins 1992; 12:339-344.
40. Shinde UP, Liu JJ, Inouye M. Protein memory through altered folding mediated by intramolecular chaperones. Nature 1997; 389:520-522.
41. Menard R, Carmona E, Takebe S et al. Autocatalytic processing of recombinant human procathepsin L. Contribution of both intermolecular and unimolecular events in the processing of procathepsin L in vitro. J Bio Chem 1998; 273:4478-4484.
42. Vernet T, Berti, PJ, de Montigny C et al. Processing of the papain precursor. The ionization state of a conserved amino acid motif within the Pro region participates in the regulation of intramolecular processing. J Biol Chem 1995; 270:10838-10846.
43. Collins PR, Stack CM, O'Neill SM et al. Cathepsin L1, the major protease involved in liver fluke (Fasciola hepatica) virulence: propeptide cleavage sites and autoactivation of the zymogen secreted from gastrodrmal cells. J Biol Chem 2004; 279:17038-17046.
44. Stack CM, Donnelly SM, Lowther J et al. The major secreted cathepsin L1 protease of the liver fluke, Fasciola hepatica: a Leu-12 to Pro-12 replacement in the nonconserved C-terminal region of the prosegment prevents complete enzyme autoactivation and allows definition of the molecular events in prosegment removal. J Biol Chem 2007; 282:16532-16543.
45. Rozman J, Stojan J, Kuhelj R et al. Autocatalytic processing of recombinant human procathepsin B is a bimolecular process. FEBS Lett 1999; 459:358-362.
46. Jerala R, Zerovnik E, Kidric J et al. pH-induced conformational transitions of the propeptide of human cathepsin L. A role for a molten globule state in zymogen activation. J Bio Chem 1998; 273:11498-11504.
47. Dalton JP, Brindley PJ. Schistosome asparginyl endopeptidase SM32 in haemoglobin digestion. Parsitol Today 1996; 12:125.
48. Robinson MW, Menon R, Donnelly SM et al. An integrated transcriptomics and proteomics analysis of the secretome of the helminth pathogen Fasciola hepatica: proteins associated with invasion and infection of the mammalian host. Mol Cell Proteomics 2009; 8:1891-1907.
49. Tkalcevic J, Ashman K, Meeusen E. Fasciola hepatica: rapid identification of newly excysted juvenile proteins. Biochem Biophys Res Commun 1995; 213:169-174.
50. Adiskwattana P, Viyanant V, Chaicumpa W et al. Comparative molecular analysis of two asparaginyl endopeptidases and encoding genes from Fasciola gigantica. Mol Biochem Parasitol 2007; 156:102-116.
51. Laha T, Sripa J, Sipra B et al. Asparaginyl endopeptidase from the carcinogenic liver fluke, Opisthorchis viverrini and its potential for serodiagnosis. Int J Infect Dis 2008; 12:e49-59.
52. Ju JW, Joo HN, Lee MR et al. Identification of a serodiagnostic antigen, legumain, by immunoproteomic analysis of excretory-secretory products by Clonorchis sinensis adult worms. Proteomics 2009; 9:3066-3078.
53. Sajid M, McKerrow JH, Hansell E et al. Functional expression and characterisation of Schistosoma mansoni cathepsin B and its trans-activation by an endogenous asparginyl endopeptidase. Mol Biochem Parasitol 2003; 131:65-75.
54. Krautz-Peterson G, Skelly PJ. Schistosome asparaginyl endopeptidase (legumain) is not essential for cathepsin B1 autoactivation in vivo. Mol Biochem Parasitol 2008; 159:54-58.
55. Delcroix M, Sajid M, Caffrey CR et al. A multienzyme network functions in intestinal protein digestion by a platyhelminth parasite. J Biol Chem 2006; 281:39316-39329.
56. Sripa J, Laha T, To J et al. Secreted cysteine proteases of the carcinogenic liver fluke, Opisthorchis viverrini: regulation of cathepsin F activation by autocatalysis and trans-processing by cathepsin B. Cell Microbiol 2010; 12:781-795.
57. Dalton JP, Clough KA, Jones MK et al. The cysteine proteases of Schistosoma mansoni cercariae. Parasitology 1997; 114:105-112.
58. Kašný, M, Mikes L, Dalton JP et al. Comparison of cysteine protidase activities in Tricjobilharzia regenti and Schistosoma mansoni cercariae. Parasitology 2007; 134:1599-1609.
59. Wasilewski MM, Lim KC, Philips J et al. Cysteine protease inhibitors block schistosome hemoglobin degradation in vitro and decrease worm burden and egg production in vivo. Mol Biochem Parasitol 1996; 81:179-189.
60. Bogitsh BJ, Kirschner KF, Rotmans JP. Schistosoma japonicum: immunoinhibitory studies on hemoglobin digestion using heterologous antiserum to bovine cathepsin D. J Parasitol 1992; 78:454-459.
61. Dalton JP, Neill SO, Stack C et al. Fasciola hepatica cathepsin L-like proteases: biology, function and potential in the development of first generation liver fluke vaccines. Int J Parasitol 2003; 33:1173-1181.
62. McManus DP, Dalton JP. Vaccines against the zoonotic trematodes Schistosoma japonicum, Fasciola hepatica and Fasciola gigantica. Parasitology 2006; 133:S43-S61.
63. Brindley PJ, Pearce EJ. Genetic manipulation of schistosomes. Int J Parasitol 2007; 37:465-473.
64. Kalinna BH, Brindley PJ. Manipulating the manipulators: advances in parasitic helminth transgenesis and RNAi. Trends Parasitol 2007; 23:197-204.
65. Skelly PJ, Da'dara A, Harn DA. Suppression of cathepsin B expression in Schistosoma mansoni by RNA interference. Int J Parasitol 2003; 33:363-369.

66. Gobert GN, Tran MH, Moertel L et al. Transcriptional changes in Schistosoma mansoni during early schistosomula development and in the presence of erythrocytes. PLoS Negl Trop Dis 2010; 4:e600.
67. Brindley PJ, Kalinna BH, Wong JY et al. Proteolysis of human haemoglobin by schistosome cathepsin D. Mol Biochem Parasitol 2001; 112:103-112.
68. Morales ME, Rinaldi G, Gobert GN et al. RNA interference of Schistosoma mansoni cathepsin D, the apical enzyme of the hemoglobin proteolysis cascade. Mol Biochem Parasitol 2008; 157:160-168.
69. Hanna RE, Ballway SS, Jura W. Methods for in vitro study of the invasive processes of Fasciola gigantica. Res Vet Sci 1975; 19:96-97.
70. Klinkert MQ, Felleisen R, Link G et al. Primary structures of Sm31/Sm32 diagnostic proteins of Schistosoma mansoni and their identification as proteases. Mol Biochem Parasitol 1989; 33:113-122.
71. Ruppel A, Shi YE, Wei DX et al. Sera of Schistosoma japonicum-infected patients cross-react with diagnostic 31/32 kD proteins of S. mansoni. Clin Exp Immunol 1987; 69:291-298.
72. Hola-Jamriska L, Tort J, Dalton JP et al. Cathepsin C from Schistosoma japonicum-cDNA encoding the preproenzyme and its phylogenetic relationships. Eur J Biochem 1998; 255:527-534.
73. Michel A, Ghoneim H, Resto M et al. Sequence, characterisation and localisation of a cysteine proteine protease cathepsin L in Schistosoma mansoni. Mol Biochem Parasitol 1995; 73:7-18.
74. Dovorak J, Mashiyama ST, Sajid M et al. SmCL3, a gastrodermal cysteine protease of the human blood fluke Schistosoma mansoni. PLoS Negl Trop Dis 2009; 3:e449.
75. Choi JH, Lee JH, Yu HS et al. Molecular and biochemical characterization of hemoglobinase, a cysteine proteinase, in Paragonimus westermani. Korean J Parasitol 2006; 44:187-196.
76. Na BK, Kim SH, Lee EG et al. Critical roles for excretory-secretory cysteine proteases during tissue invasion of Paragonimus westermani newly excysted metacercariae. Cell Microbiol 2006; 8:1034-1046.
77. Halton DW. Nutritional adaptations to parasitism within the platyhelminthes. Int J Parasitol 1997; 27:693-704.
78. Senft AW, Senft DG, Himlman GR et al. Influence of hycanthone on morphology and serotonin uptake of Schistosoma mansoni. Am J Trop Med Hyg 1976; 25:832-840.
79. Hoffmann KF, Fitzpatrick JM. Gene expression studies using self-fabricated parasite cDNA microarrays. Methods Mol Biol 2004; 270:219-236.
80. Brindley PJ, Mitreva M, Ghedin E et al. Helminth genomics: The implications for human health. PLos Negl Trop Dis 2009; 3:e538.

CHAPTER 9

CATHEPSINS B1 AND B2 OF *TRICHOBILHARZIA* SPP., BIRD SCHISTOSOMES CAUSING CERCARIAL DERMATITIS

Martin Kašný,*[,1] Libor Mikeš,[1] Kateřina Dolečková,[1] Vladimír Hampl,[1] Jan Dvořák,[2] Marian Novotný[3] and Petr Horák[1]

[1]Department of Parasitology, Faculty of Science, Charles University in Prague, Czech Republic; [2]Institute of Parasitology, Biology Centre of the Academy of Sciences of the Czech Republic, Budějovice, Czech Republic; [3]Department of Cell Biology, Faculty of Science, Charles University in Prague, Prague, Czech Republic
**Corresponding Author: Martin Kašný—Email: kasa@post.cz*

Abstract: *Trichobilharzia regenti* and *T. szidati* are schistosomes that infect birds. Although *T. regenti*/*T. szidati* can only complete their life cycle in specific bird hosts (waterfowl), their larvae—cercariae are able to penetrate, transform and then migrate as schistosomula in nonspecific hosts (e.g., mouse, man). Peptidases are among the key molecules produced by these schistosomes that enable parasite invasion and survival within the host and include cysteine peptidases such as cathepsins B1 and B2. These enzymes are indispensable bio-catalysts in a number of basal biological processes and host-parasite interactions, e.g., tissue invasion/migration, nutrition and immune evasion. Similar biochemical and functional characteristics were observed for cathepsins B1 and B2 in bird schistosomes (*T. regenti*, *T. szidati*) and also for their homologs in human schistosomes (*Schistosoma mansoni, S. japonicum*). Therefore, data obtained in the research of bird schistosomes can also be exploited for the control of human schistosomes such as the search for targets of novel chemotherapeutic drugs and vaccines.

INTRODUCTION

Members of the genus *Schistosoma* are amongst the most important human parasites. They affect more than 200 million people in tropical and subtropical countries; research on these pathogens is supported by many institutions including WHO TDR (http://apps.who.int/tdr/). However, the importance of the remaining genera within the family

Cysteine Proteases of Pathogenic Organisms, edited by Mark W. Robinson and John P. Dalton.

Schistosomatidae for human/animal health seems to be neglected and these parasites are studied only by a few research teams. This applies to bird schistosomes in general and in particular the largest genus *Trichobilharzia* (consisting of more than 40 species; see web pages of the Schistosome Group Prague, www.schistosomes.cz).

As far as *Trichobilharzia* spp. is concerned, freshwater larvae (cercariae) of these trematodes are notorious due to their ability to penetrate human skin and cause cercarial dermatitis (swimmer's itch) (Fig. 1). Reports on the attacks of humans by cercariae of *Trichobilharzia* spp. (and other bird schistosomes) are increasing, e.g., the disease affects 15% of swimmers in the Savoy district, France,[1] and approximately one-half of Swedish municipalities (125 out of 248) with lakes that have reported problems with dermatitis.[2] As a probable consequence of recent global climate changes[3] or special conditions in water reservoirs (e.g., geothermally warmed lakes), cercarial dermatitis has also been reported from cold lake areas in higher latitudes.[4,5] Based on such data, cercarial dermatitis is regarded as an emerging disease.

The occurrence of cercarial dermatitis is always linked with transmission of schistosomes between water snails and birds. Studies have shown that members of Schistosomatidae belong to the most common parasites of birds; although under-reported in the past, they are now frequently detected in bird tissues. For example, visceral *Trichobilharzia* spp. was found in 78-100% of New Zealand scaup (*Aythya novaeseelandiae*) in New Zealand[6] and 94% of three anatid species in the United States,[7] and nasal *Trichobilharzia* spp. parasitized 75% of mallards in Iceland.[5] If heavily infected, birds may suffer from the infection; their internal organs and tissues (e.g., the lungs, liver, intestine and central nervous system) may be impaired and serious clinical symptoms such as enteritis, hepatitis, endophlebitis, leg paralysis and balance/orientation disorders may appear. Fatal infections of birds have also been reported (for a review see ref. 8).

Experimental infections proved that *Trichobilharzia* larvae not only penetrate the skin of mammals, but may escape and migrate via the circulatory system through the mammalian body. Therefore, juvenile parasites of visceral *Trichobilharzia* can regularly be found in the lungs of first-time infected mammals. In addition, our view on schistosome migration and pathogenicity was substantially changed in 1999 when neurotropic migration of *T. regenti* was discovered (see http://www.schistosomes.cz/regenti.htm).[9] To our knowledge, this is the only schistosome for which migration of schistosomula through the nerves and attack of the spinal cord and brain of birds (natural host) and experimental mammals (accidental host) has been described. The importance of these findings for assessment of human infections by bird schistosomes remains to be clarified.[10]

Thanks to their two-host life cycle and availability of experimental intermediate and definitive hosts, bird schistosomes represent convenient model organisms for studies of host-trematode (host-schistosome) interactions at the molecular level. Three research streams are noteworthy in this regard: (a) monitoring of immune and hormonal (neuroendocrine) status of *Lymnaea stagnalis* snails infected by *T. szidati*,[11,(a)] (b) identification of chemical stimuli produced by snail/vertebrate hosts that trigger specific behavioral responses of *T. szidati* miracidia and cercariae,[12,13] and (c) characterization of essential peptidases used by *T. regenti* and *T. szidati* for tissue lysis and digestion.[14-18] Data and hypotheses arising from the latter subject will be discussed in this chapter.

[(a)] Synonymy of German and Dutch isolates of *Trichobilharzia ocellata* and *T. szidati* was explained by Rudolfová et al.[66]

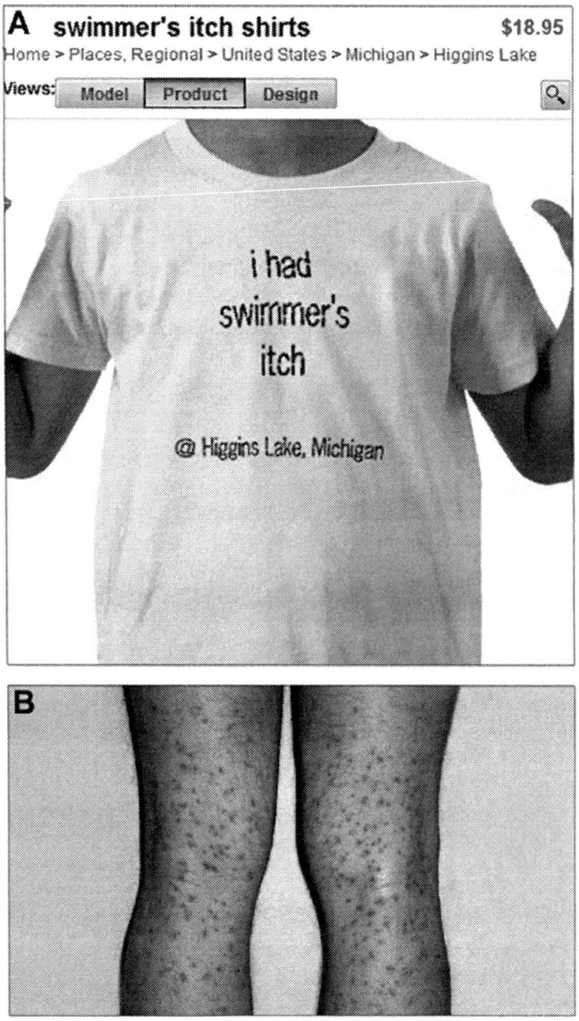

Figure 1. Cercarial dermatitis (swimmer's itch; B). It is a severe inflammatory reaction caused by penetration of the skin by schistosome larvae (cercariae) which frequently belong to the genus *Trichobilharzia*. Nowadays, cercarial dermatitis is classified as an emerging disease reported from a number of lakes in all over the world, with evident economic impact in recreational areas. Numerous cases of human cercarial dermatitis imply that precautions should be adopted in risk areas, e.g., application of protective cream against cercariae developed by Wulff et al (2007).[75] On the other hand the emergence of swimmer's itch can bring an economic profit to some entrepreneurs (picture A). Picture (A) was kindly provided by Zazzle.com (web sites http://www.zazzle.com/swimmers_itch_tshirt-235405014772762831), amendments by Kašný.

PEPTIDASES OF *TRICHOBILHARZIA*

Although members of the genus *Trichobilharzia* parasitize bird definitive hosts exclusively, their strategies of exploiting enzymatic equipment are, in general, similar to the well described human schistosomes, such as *Schistosoma mansoni*, *S. japonicum*

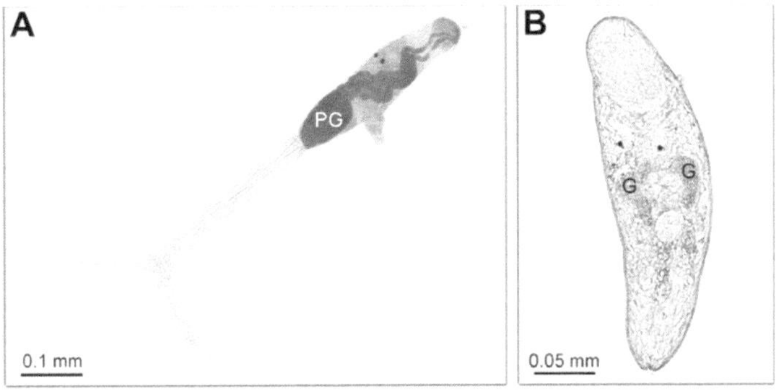

Figure 2. Cercaria (A) and schistosomulum (B) of *Trichobilharzia regenti*. PG—postacetabular glands (stained by lithium carmine). G—gut. Photo (B) taken by Kateřina Blažová, MSc.

and *S. haematobium*. From the biochemical/molecular point of view, *T. regenti* and *T. szidati* were frequently investigated to elucidate the proteolytic machinery linked with skin penetration (cercariae) and tissue migration (schistosomula).[14-17,19] It was confirmed that, for this purpose, these two schistosomes adopted typologically similar enzymes (peptidases).[15,16] The most active peptidases present in *T. regenti/T. szidati* penetration glands of cercariae and gut of schistosomula belong to the group of cysteine peptidases (CP) including also cathepsin B1 and B2 (Fig. 2).[14-17] These are now intensively studied, because cathepsins B1 or B2 might operate as potent penetration enzymes similar to those expressed by *S. mansoni and S. japonicum*.[16-18,20-22,(b)] The immunogenicity of *Trichobilharzia* cercarial whole extract and excretory-secretory products (ESP) has been recently demonstrated and secreted cathepsin Bs (CBs) have been considered as allergic factors too.[23,24] Their contribution to the development of cercarial dermatitis (swimmer´s itch) is currently under investigation.

T. regenti/T. szidati cathepsin B1/2 (TrCB1/2, TsCB1/2) genes were identified and cloned recently by Dvořák et al (TrCB1),[14] Dolečková et al (TrCB2)[25] and Kašný et al (unpublished) (TsCB1/2) (Table 2). TrCB1/2 and TsCB1/2 figure amongst other cca 25 orthologous trematode peptidases of CA peptidase clan, CA1 family, cloned from nine trematode species (MEROPS database 9.1, see Fig. 3).[18,26,(c)]

[(b)] In *S. mansoni* and some other schistosomes, serine peptidases (elastases) are believed to be responsible for degradation of skin proteins,[67,68] but these were not found in *T. regenti* and *T. szidati*.[15,16,22,25] Activity of the presumed *T. ocellata* CE (= *T. szidati*)[66] originating from excretory/secretory products was monitored by Bahgat and Ruppel.[19] It has been discussed by several authors that the activity reported by Bahgat and Ruppel[19] was probably of trypsin-like type, whereas the activity of *S. mansoni* CE is chymotrypsin-like.[18,67,68] Up-to-now no cercarial elastase gene has been identified in *T.regenti/T.szidati*.[18,25]

[(c)] Without broader contextual comments, phylogenetic relationships of 4 *Trichobilharzia* spp. cathepsins B and the other 30 trematode representatives analyzed by methods of molecular phylogenetics are shown in Figure 3. CB1/2 peptidases expectably split into two main phylogenetic clades of cathepsin B1 and B2 (blue and green branch, Fig. 3), although the clades are related not very distantly (Fig.3). According to this analysis, it is supposed that former submission of certain cathepsin genes into CB1 and CB2 group could be incorrect (e.g. *Fasciola hepatica* CB2, GB AN: AJ488928). Trematode cathepsin B sequences show demonstrable sequence similarity to enzymes of other metazoans. The presented phylogenetic tree reflects also the results of alignment analysis, showing that there is only 48-51% sequence identity between schistosome cathepsins B1 and B2 (Table 1, Fig. 4A,E).

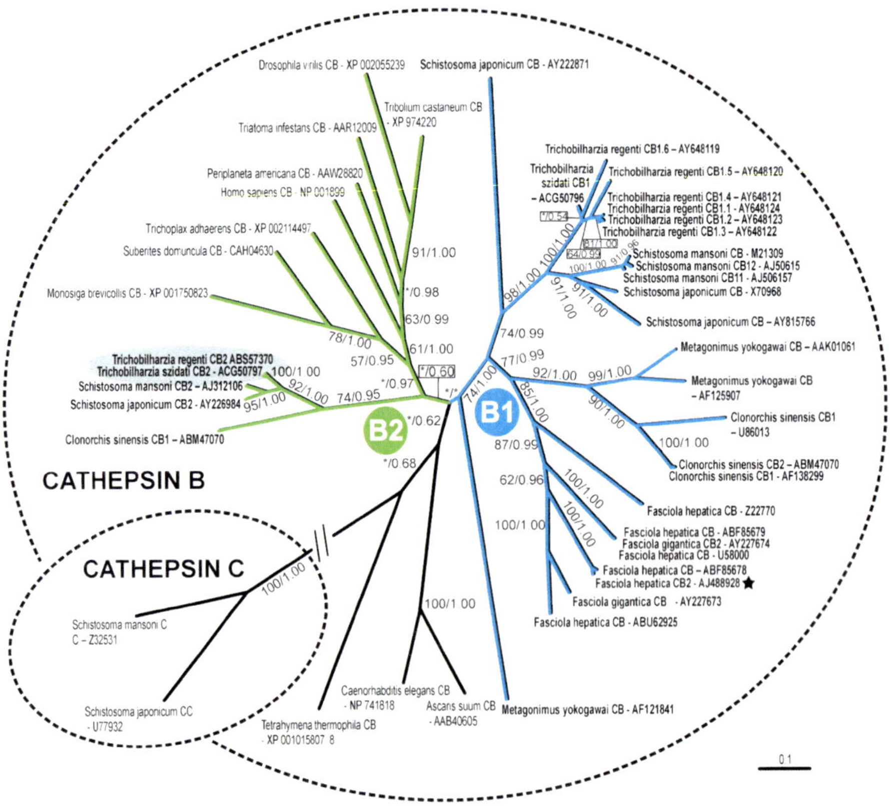

Figure 3. Phylogenetic tree of cathepsin B protein sequences. The cathepsin B peptidases split into two groups, cathepsins B1 and B2 (blue and green branch). Cathepsins B1/2 of trematodes are highlighted by bold labels and cathepsins of *Trichobilharzia* by grey ovals (*T. regenti*/*T. szidati* CB1—light gray oval and CB2—dark gray oval). All cathepsin B sequences are supplemented by GenBank accession number. The black star indicates cathepsin B1 of *Fasciola hepatica* which was previously incorrectly named as cathepsin B2. The tree was constructed using maximum likelihood method in RAxML (PROTGAMMAWAG model of amino acid change).[76] Values at internal branches represent maximum likelihood bootstraps/Bayesian posterior probabilities (asterisk designates values lower than 50% or 0.50). Cathepsin C sequences were chosen as outgroups to root the tree. To illustrate relationships to nontrematode peptidases the representative cathepsins B from other large groups of organisms were included in the tree (http://www.ncbi.nlm.nih.gov/Genbank/index.html).[77]

T. regenti/*T. szidati* Cathepsin B1

Cathepsins B1 are probably the best characterized peptidases among trematode enzymes (for review see refs. 18, 27 and 28). It was shown that CB1 is one of the pivotal peptidases of the worm gut, involved in blood protein digestion (hemoglobin etc.) by *S. mansoni*, *S. japonicum* and most likely also *T. regenti* schistosomula/adults.[14,29-36,(d)] These properties have made CB1 a potential target of novel antischistosomiasis drugs.[37-40]

CB1 is present in the gastro-intestinal content of *T. regenti* schistosomula (and probably also in *T. szidati*) and released outside the worm body, implying its interaction

Table 1. Percentage of cathepsin B1 and B2 AA sequence identities. Sequences of CB1 and CB2 included in the table: **TrCB1.1-6**; *Trichobilharzia regenti* CB1-1.6 (GB AN: AY648119-24), **TsCB1**; *T. szidati* CB1 (GB AN: ACG50796), **TrCB2**; *T. regenti* CB2 (GB AN: ABS57370), **TsCB2**; *T. szidati* CB2 (GB AN: ACG5077), **SmCB1**; *Schistosoma mansoni* CB1 (GB AN: AJ506157), **SmCB2**; *S. mansoni* CB2 (GB AN: AJ312106), **SjB1**; *S. japonicum* CB1 (GB AN: P43157), **SjCB2**; *S. japonicum* CB2 (GB AN: AY226984). Percentage of identity was recorded via ExPASy Proteomics Server using CLUSTALW alignment tool

%	TrCB1.1	TrCB1.2	TrCB1.3	TrCB1.4	TrCB1.5	TrCB1.6	TsCB1	SmCB1	SjCB1	TrCB2	TsCB2	SmCB2	SjCB2
TrCB1.1		99.12	99.12	97.95	88.30	84.50	91.81	69.88	70.47	51.16	51.45	50.28	50.86
TrCB1.2	99.12		98.25	97.37	88.30	84.21	91.81	69.88	70.47	51.16	51.45	50.28	50.86
TrCB1.3	99.12	98.25		97.08	88.01	84.80	91.81	70.18	70.76	51.45	51.73	50.00	50.86
TrCB1.4	97.95	97.37	97.08		89.47	84.21	90.94	68.71	70.18	50.87	51.16	50.00	50.57
TrCB1.5	88.30	88.30	88.01	89.47		81.29	86.26	66.67	68.13	49.13	49.42	48.58	49.14
TrCB1.6	84.50	84.21	84.80	84.21	81.29		86.26	66.37	66.08	49.42	50.00	47.16	48.00
TsCB1	91.81	91.81	91.81	90.94	86.26	86.26		70.47	71.35	51.73	52.02	50.57	50.86
SmCB1	69.88	69.88	70.18	68.71	66.67	66.37	70.47		73.98	49.13	49.42	47.01	47.71
SjCB1	70.47	70.47	70.76	70.18	68.13	66.08	71.35	73.98		50.58	51.16	49.29	50.00
TrCB2	51.16	51.16	51.45	50.87	49.13	49.42	51.73	49.13	50.58		97.67	80.52	81.03
TsCB2	51.45	51.45	51.73	51.16	49.42	50.00	52.02	49.42	51.16	97.67		76.50	81.61
SmCB2	50.28	50.28	50.00	50.00	48.58	47.16	50.57	47.01	49.29	80.52	76.50		86.53
SjCB2	50.86	50.86	50.86	50.57	49.14	48.00	50.86	47.71	50.00	81.03	81.61	86.53	

with the host immune system.[14] Therefore, it is hypothesized that cathepsin B1 could be potentially exploited for serodiagnosis of *Trichobilharzia* infection in humans with symptoms of cercarial dermatitis.[18,24]

Cysteine peptidase activity (possibly of cathepsin B1) in cercariae and schistosomula of *T. regenti*/*T. szidati* was identified, but at present, only *T. regenti* CB1 (isoforms TrCB1.1/TrCB1.4 see below) has been produced in a recombinant form (differently glycosylated recombinant pro-enzymes of 38-45 kDa).[14-16]

TrCB1 amino acid (AA) sequence shows 92% similarity to TsCB1 (see Table 1). TrCB1/TsCB1 possess equal number (342/342) of AAs, including a 21 AA signal sequence and a 68 residue pro-peptide region (Kašný et al unpublished, SignalP 3.0 software).[14] Dvořák et al[14] described six TrCB1 isoforms from schistosomula (TrCB1.1–6.) (Table 2).[(e)] TrCB1.1 was determined as the most abundant (68%) among all sequenced clones (the other isoforms were represented as follows: 10%—1.2, 5%—1.3, 13%—1.4, 3%—1.5/1.6; TrCB1.1-TrCB1.6 GenBank accession number; GB AN: AY648119-24).[14] Cathepsin B1 from *T. szidati* sporocysts with cercarial germ balls was amplified by Kašný et al (unpublished) employing the primers previously designed for TrCB1.1 full sequence (see Tab. 2, TsCB1: GB AN: EU877763, Kašný et al unpublished). No isoforms of TsCB1 were recognized (Kašný et al unpublished).

Both *T. regenti*/*T. szidati* cathepsins B1 possess three highly conserved catalytic residues Cys/His/Asn equally numbered for TrCB1.1 (GB AN: AY648119) and TsCB1 (GB AN: EU877763) as Cys-29/His-199/Asn-219 (Fig. 4E, Kašný et al unpublished, MEROPS 9.1).[14,26] TrCB1/TsCB1 sequences possess the other AAs such as Gln-23, Gly-72, Gly-73, Trp-221, which were previously defined as highly conserved with side chains which possibly interact with the residues of bound substrate.[14,41,42] Twelve Cys residues are present in the CB1 molecule and form intra-molecular disulfide bridges which are conserved for both members of the genus *Trichobilharzia*, as well as other schistosomes.[14,41,43]

Multiple alignment analysis of all TrCB1 isoforms (TrCB1.1-TrCB1.6) showed that in the AA sequences of TrCB1.5 and TrCB1.6, residue Cys-29 that occurs within the conserved sequence motif CGSCWAF is substituted by Gly-29 (Fig. 4D).[14,44,45] Expression of recombinant TrCB1.6 is currently in progress, but we hypothesize that substitution within the catalytic triad could lead to crucial biochemical changes such as the loss of enzymatic activity due to the lack of nucleophile.[14] Moreover, the synthetic dipeptidyl-7-amido-4-methylcoumarin (AMC) fluorescent substrate Z-Phe-Arg-AMC is not incorporated so "deep" into the active site cleft of TrCB1.5. This is supported also by a 3D model showing the binding of Z-Phe-Arg-AMC (see Fig. 4C,D). Substitution of AAs in the cathepsin B1 catalytic domain has previously been reported in *S. japonicum*

[(d)]Sequence of TrCB1/TsCB1 contains hemoglobinase motif (YWLIANSWxxDWGE, Fig. 4E) typical for blood-feeding helminths,[69] Asn219 of catalytic triad (Cys-29/His-199/Asn-219) is underlined (see Fig. 4E). It was recorded by Dvořák et al[14] that recombinant TrCB1.1 and TrCB1.4 isoforms can degrade duck, turkey and bovine hemoglobin during overnight incubation at pH 4.0-7.0, although the myelin basic protein was more effectively degraded substrate under these conditions.[5]TrCB1/TsCB1 of bird schistosomes occurred as highly active peptidases with low abundance.[14-16] Up to now TrCB1/TsCB1 have not been reliably identified in worm extracts even by sensitive mass-spectrometry methods (Kašný et al unpublished).

[(e)]Sequence analysis revealed 6 genes coding cathepsin B-like peptidase isoforms differing in amino acid code (85%—99%) and confirmed their orthology to the schistosomal gut-associated cathepsin B enzymes, SmCB1 and SjCB1, with 70% sequence identity (Table 1).[14] Therefore, the isoforms were named TrCB1.1—TrCB1.6 in accordance with the previously used terminology.[27,32]

Table 2. Cathepsins B1/B2 of *T. regenti*/*T. szidati*

	Species	Stage	Accession Number (GenBank, UniProtKB/ TrEMavBL, MEROPS)	Nucleotide/ Amino Acid Sequence Length (Bases/ Amino Acids)	pI (Theoretical)	MW (kDa) (Practical/ Theoretical)	Clan, Family	Properties
CB1	*T. regenti* (Dvořák et al 2005)[14] (Dolečková et al 2007)[25]	(C,Sp, Sc)	TrCB1.1: AY648119, Q4VRW9, MER049448	1148/342	6.91	33-35/39	CA,C1A	**Type of activity:** exopeptidase/endopeptidase **pH optimum:** between 4.5-6 **Specific substrate (50 μM):** Z-Arg-Arg-AMC **Common cysteine peptidase substrates (50 μM):** Z-Phe-Arg-AMC, Boc-Val-Leu-Lys-AMC **Specific inhibitor (10 μM):** CA-074 **Common effective cysteine peptidase inhibitors (10 μM):** E-64, Z-Phe-Ala-CHN_2, Z-Phe-Phe-CHN_2, Z-Phe-Tyr-CHN_2, elastatinal **Function and localization:** proteolysis, cathepsins B1 are associated with the gut and facilitate digestion of host proteins to absorbable nutrients, they may also facilitate penetration of cercariae through the host skin and migration of schistosomula through the host tissues **Primers:** **TrCB1/TsCB1fwd:** 5′-ATTCGTGACCAATCCCGATGTGGTTCAT-GTTGG-3′ **TrCB1/TsCB1rev:** 5′-CCAATCTTCATTCCACGAATTAGCAATC-3′
			TrCB1.2: AY648120, Q4VRW8, -	1144/342	7.14			
			TrCB1.3: AY648121, Q4VRW7, -	1139/342	7.16			
			TrCB1.4: AY648122, Q4VRW6, -	1139/342	6.85			
			TrCB1.5: AY648123, Q4VRW5, MER049449	1139/342	6.83			
			TrCB1.6: AY648124, Q4VRW4, MER049450	1139/342	8.29			

continued on next page

Table 2. Continued

	Species	Stage	Accession Number (GenBank, UniProtKB/ TrEMavBL, MEROPS)	Nucleotide/ Amino Acid Sequence Length (Bases/ Amino Acids)	pI (Theoretical)	MW (kDa) (Practical/ Theoretical)	Clan, Family	Properties
	T. szidati (Kašný et al 2008, unpublished)	(C,Sp)	TsCB1: EU877762, B5AXI3, MER158388	1026/342	8.59	–/39	CA,C1A	
CB2	*T. regenti* (Dolečková et al 2009)[17]	(A,E,MC, Sp,Sc)	TrCB2: EF682129, A7L844, MER127123	1035/344	6.16	30/39	CA,C1A	**Type of activity:** exopeptidase/endopeptidase **pH optimum:** between 4.5-6 **Specific substrate (50 μM):** Z-Arg-Arg-AMC **Common cysteine peptidase substrates (50 μM):** Z-Phe-Arg-AMC, Boc-Val-Leu-Lys-AMC **Specific inhibitor (10 μM):** CA-074 **Common effective cysteine peptidase inhibitors (10 μM):** E-64, Z-Phe-Ala-CHN_2, Z-Phe-Phe-CHN_2, Z-Phe-Tyr-CHN_2, elastatinal **Function and localization:** proteolysis, localization of TrCB2 in cercarial postacetabular penetration glands suggest its function as penetration facilitating peptidase **Primers:** **TrCB2/TsCB2fwd:** 5′-GAGGCTAATCGACA-CAAGTTTATG-3′ **TrCB2/TsCB2rev:** 5′-TTTAAGTTTCGGAATTC-CAGC-3′
	T. szidati (Kašný et al 2008, unpublished)	(C,Sp)	TsCB2: EU877763, B5AXI4, MER158389	1032/344	6.15	–/39	CA,C1A	

Legend: A—adults, E—eggs, M—miracidia, Sp—sporocysts, C—cercariae, Sc—schistosomula

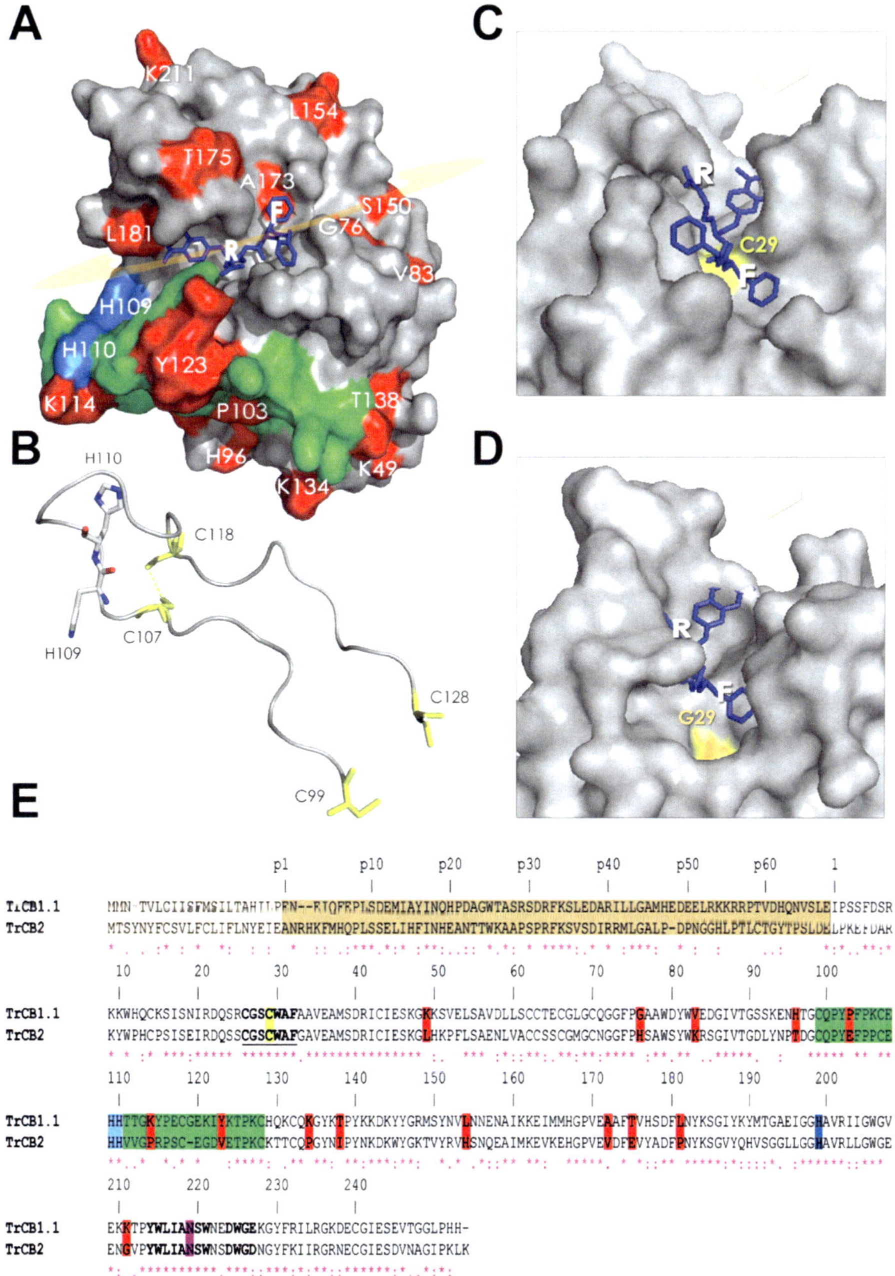

Figure 4. 3D modelling and sequence analysis of *Trichobilharzia* cathepsins B1/B2. 3D *Trichobilharzia regenti* CB1.1 protein structure was built using Modeller 9.7 software.[78] The model is based on two different template structures: human pro-cathepsin B (PDB ID: 3pbh)[79] and bovine cathepsin B complexed with the inhibitor NS-134 (PDB ID: 1sp4).[80] The fluorogenic peptide substrate Z-Phe-Arg-AMC was docked at the active site clefts of TrCB1.1 via PatchDock and SymmDock server.[81] Legend is continued on next page.

Figure 4, continued from previous page. Panel A, E; the red areas with white numbers of amino acids (e.g., Y-123) shown on the surface of 3D cathepsin B molecule (panel A) represent 3D positions of major visible AA sequence differences between *T. regenti* cathepsin B1.1. and B2. This also corresponds to differences of red labeled AA representatives shown in the "text alignment" of TrCB1.1/TrCB2 on panel E. Occluding loop is in green on panel (A) and two histidines responsible for CB1/2 exopeptidase activity are in light blue (panel A,E). Panel (E) shows sequence alignment of TrCB1.1 and TrCB2 cathepsins. Above the sequences, the letter "p" indicates numbering of the pro-peptide region (labeled by light brown together with dark brown), light brown color only shows the signal sequence, the number "1" indicates the start of mature enzyme AA sequence. Catalytic Cys-29 is marked in yellow, His-199 in dark blue and Asp-219 in purple. The conserved sequence motif CGSCWAF is underlined and the hemoglobinase motif YWLIANSWxxDWGE is dashed underlined. The zoomed occluding loop structure is shown on panel (B), cysteines which form the occluding loop (Cys-99—C-128, Cys-107—Cys-118) are in yellow. Position and orientation of exo-peptidolytically active His109—His-110 are shown on panel (B), too. The fluorogenic peptide substrate Z-Phe-Arg-AMC shown in blue (orientation labeled by white bold F and R) (panel A,C,D) is docked into at the active site cleft between two CB1 domains; it is labeled by orange oval (panel A) or orange arrow (panel C,D). The active site residue of TrCB1.1 isoform Cys-29 is in yellow (panel C,E) and its substitution in TrCB1.5 isoform by Gly-29 in orange (panel D). The alignment was performed via ExPASy Proteomics Server using CLUSTALW alignment tool.

and other examples can be found in schistosome EST databases.[46-48] Various substitutions in the catalytic Cys-His dyad are common and present in 8% of genes encoding helminth Clan CA family C1 peptidases (papain-like cysteine peptidases to which cathepsin B is assigned). Generally, enzymes with active site mutations are not rare in metazoan organisms,[49] including parasites.[50] Biological function of catalytically inactive isoforms could hypothetically lie in regulation of peptidase activity. An increased expression of inactive isoenzymes could induce a decrease in expression of active enzyme forms and in this way suppress the level of proteolytic activity (and vice versa).[14,50]

Cathepsin B-like activity was detected against the fluorescent AMC substrates Z–Phe-Arg–AMC and Z-Arg-Arg–AMC in samples (whole worm extracts and ESP) of all *Trichobilharzia* larval stages (*T. regenti* and *T. szidati* miracidia, cercariae and schistosomula) at pH optima between 4.5-6.0 (Kašný unpublished, see Table 2).[14-16] Similar enzymatic activity was observed with recombinant schistosomula TrCB1 enzymes (TrCB1.1/TrCB1.4).[14] The recombinants and cercarial whole worm extracts/ESP were also capable of effectively degrading native macromolecular proteins such as hemoglobin, myelin basic protein, collagen, keratin, fibronectin and immunoglobulin G (Kašný et al unpublished).[14,16,24,(f)] The ability of *Trichobilharzia* cysteine peptidases (CB1/2) to degrade skin macromolecular components, together with the localization of these enzymes in the gut of schistosomula and probably also in cercarial penetration glands, supports the theory that cysteine peptidases participate in cercarial penetration through the host skin and migration of schistosomula within the host tissues (Fig. 2A).[16-18] TrCB1.1/TrCB1.4 recombinants and cercarial whole worm extract/ESP activities are inhibited by a panel of general cysteine peptidase (CB1) inhibitors including E-64, CA-074, Z-Phe-Ala-CHN_2, Z-Phe-Phe-CHN_2, Z-Phe-Tyr-CHN_2 and also by elastatinal—an irreversible inhibitor of neutrophil/pancreatic elastases (see Table 2).[14-16] The data above (except for elastatinal) correspond with results observed in studies of human schistosomes.[16,18]

[(f)]Myelin basic protein is the major component of the nervous tissue and its degradation by both isoforms of *T. regenti* cathepsins TrCB1.1/TrCB1.4 probably corresponds with a unique migration route of schistosomula through the peripheral nerves and spinal cord to the brain and nasal cavity (see also note d). *T. regenti* schistosomula might use myelin as the main nutrient. In addition, less than 5% of schistosomula isolated from CNS contained hematin in the gut.[14,70] and hemoglobin probably becomes a major nutritive source first after schistosomula migration to the nasal cavity.[9]

S2 subsite of papain-like peptidases is a major determinant of their substrate specificity.[27] Substrate specificities of *Trichobilharzia* cathepsin B peptidases were revealed using a limited set of dipeptidyl fluorogenic substrates with varying P2 residues (Phe, Leu, Val and Arg). TrCB1.1 displayed a P2 preference in the order Val>Phe>Leu>Arg at acidic pH whilst the activity against substrates containing Arg at P2 was optimal at pH 7. The TrCB1.4 isoform has P2 preference for Val>>Leu>Phe with essentially no activity against substrates with Arg at P2 at any pH. This "cathepsin L-like preference" expressed by TrCB1.4 can be explained by replacement of negatively charged Glu-245 by a smaller neutral Ala-245 which causes a change in both the charge and the volume of S2 subsite binding pocket. This was verified using comparative 3D models of TrCB1.1 and TrCB1.4 atomic structures.[14]

T. regenti CB1.1 and CB1.4 isoforms differ in their in vitro processing from exo-peptidolytically active pro-enzyme (see below) to endo-peptidolytically active mature enzyme. Recombinant TrCB1.1 is incapable of auto-processing, but similar to *S. mansoni* CB1, it could be fully-processed by two other cysteine peptidases from the clans CD and CA, namely asparaginyl endopeptidase and cathepsin C, which cleave off the N-terminal pro-peptide.[14,27,32] On the other hand, the TrCB1.4 zymogen was shown to be resistant to the action of *S. mansoni* asparaginyl endopeptidase but it could be trans-processed to the mature enzyme following overnight incubation at pH 4.0.[14,42,(g)]

In addition to the endopeptidase activity of TrCB1 (provided by Cys-29/His-199/Asn-219), this enzyme would also be expected to display exo-peptidase activity due to a unique structural component—the occluding loop (see Fig. 4B). The occluding loop is present in all trematode cathepsin B1 genes characterized to date.[14,17,32,43,51-53] This 20-30 AA long sequence motif comprises a catalytic dyad (His-110—His-111) that is responsible for peptidyl C-terminus (exo) dipeptidase activity.[14,32,54,55,(h)] Besides exopeptidase activity, the occluding loop stabilizes the conformation of the pro-enzyme active site cleft at low pH (e.g., in the lysosome) via a His-110—Asp-22 salt bridge and consequently affects the processing of the endopeptidase and its activity (see Fig. 4).[41,54,56,57] Due to the deprotonation of His-110 (followed by breaking of the H-110—Asp-22 salt bridge) at higher pH (e.g., when the peptidase is released from the lysosome), the bond between the occluding loop and the enzyme is destabilized which theoretically leads to an increase of endopeptidase activity.[55,57] The primary action of CB1 as an exo-peptidase in the acidic environment of the lysosome could imply that endopeptidase activity may have arisen later in eukaryotic evolution.[54,57,58]

TrCB1.1 gene expression is probably developmentally regulated. Cathepsin B1 enzymes are highly expressed as pro-enzymes, predominantly by the adult worms and intra-vertebrate larval stages of *S. mansoni*, *S. japonicum* and *T. regenti*.[33,36,59,60] Low

[(g)]TrCB1.1 is in vivo probably fully-processed by *T. regenti* asparaginyl endopeptidase and cathepsin C. In vitro TrCB1.1 could be activated by orthologous enzymes of nonrelative species (e.g., *Ixodes ricinus* asparaginyl endopeptidase and rat cathepsin C). This mode of activation is called "trans-activation" or "trans-processing" and was previously used in context of pro-enzyme conversion experiments with *S. mansoni* CB1.[14,32,71] Whether SmAE action during the peptidase processing in schistosomal gut in vivo reflects the situation in vitro is subject of ongoing experiments and discussions.[32,72,73] By using peptidyl substrate libraries it was proved that *S. mansoni* asparaginyl peptidase (the same CA Clan of peptidases to which cathepsin B belongs) is strictly selective for asparagine at P1 subsite position. [32,73,74]

[(h)]It has been proposed that prolins and cysteins situated in both end parts of CB1 occluding loop may assist to keep the structure stability via interhydrogen bonds (Fig. 4B).[41]

level expression of the TrCB1.1 gene was detected in eggs, miracidia, sporocysts with cercarial germ balls and cercariae.[36] Its activity is significantly increased in schistosomula and adults, that are highly metabolically active, within their definitive bird hosts.[36] This developmentally-dependent switch likely correlates with the development of the parasite's intestine and initiation of digestion.[14,36] Levels of TrCB1.1 transcripts in schistosomula and adults support a central role for this peptidase in schistosome digestion.[14,33,34,36]

T. regenti/*T. szidati* Cathepsin B2

As mentioned above, cysteine peptidase activities were described in homogenates of cercariae, in the contents of their penetration glands and in schistosomular extracts of *T. regenti* and *T. szidati*. Fluorometric analyses using synthetic fluorogenic peptide substrates (e.g., Z-Phe-Arg-AMC, Z-Arg-Arg-AMC at pH optimum between 4.5-6) and a set of inhibitors (e.g., E-64 and CA-074) showed that cathepsin B-like activity predominated in these *T. regenti/T. szidati* samples (see Table 2).[14-17]

The TrCB2 gene was cloned from intra-molluscan stages of *T. regenti* (sporocysts containing developing cercariae—cercarial germ balls) and an active recombinant form of the enzyme was expressed in the yeast, *Pichia pastoris* (TrCB2: GB AN; EF682129).[17] For TsCB2, only sequence data is available (TsCB2: GB AN: EU877763, Kašný et al unpublished). Both TrCB2 and TsCB2 are 344 AAs in length and include a 21 AA signal sequence and a 68 AA long pro-peptide region (SignalP 3.0 software).[17] TrCB2 showed a high AA sequence similarity of 98% with TsCB2 and 81% with *S. japonicum* and *S. mansoni* cathepsin B2 (Table 1).[17] Therefore, the molecular characteristics of TrCB2 are very similar to those of SmCB2, SjCB2 and partially to TrCB1 as discussed above (Table 1, Fig. 4A,E).[(i)]

Recombinant TrCB2 migrated in SDS-PAGE gels as a 32 kDa band. After deglycosylation its size was reduced to 30 kDa which was slightly higher than the calculated theoretical molecular mass of 28.5 kDa. It is expressed as a pro-enzyme which auto-activates under the conditions of yeast expression medium. N-terminal sequencing revealed three residual AAs (Leu-68/Asp-69/Glu-70) of the pro-domain remaining after auto-activation (Fig. 4E) when compared to predicted mature domain processing site.[17] A similar situation was observed in the case of *S. mansoni* CB2 where six pro-domain residues (Gly-68/Ty-69/Ile-70/Ser-71/Asp-72/Glu-73) remained after auto-activation.[43,(j)] Analysis of subsite specificities of recombinant TrCB2 by means of positional scanning a synthetic combinatorial peptide substrate library[61] showed preferences for Lys or Arg at the P1 position of the substrate and for hydrophobic residues at P2 and P3 positions in accordance with TrCB1 and SmCB2.[14,17,43]

Using DCG-04, a biotinylated derivative of the inhibitor E-64,[62] cysteine peptidases (possibly CBs) migrated at 31-33 kDa on ligand blots of *T. szidati*/*T. regenti* cercarial extracts and 30-32 kDa on blots with recombinant TrCB2.[15,17] Immunoblots employing

[(i)]This includes catalytic AA triad Cys29/His199/Asn219, a highly conserved sequence motif CGSCWAF, residues of the papain superfamily endopeptidases interacting with the chain of the substrate—Gln-23, Gly-72, Gly-73, Trp-221, 12 cysteine residues conserved between schistosome and human enzymes which form 6 intramolecular disulfide bridges, an occluding loop (Cys-99—Cys-128) responsible for dipeptidyl peptidase activity (Fig. 4B)[41] and a slightly modified unique motif YWLIANSXxxDWGE ascribed to those cathepsins which are assumed to cleave hemoglobin in blood-feeding helminths (see Fig. 4E).[69]

[(j)]Presence of six extra AAs probably does not affect biochemical properties of recombinant enzymes.[43]

anti-TrCB2 and anti-SmCB2 polyclonal antibodies showed reaction in a corresponding area using protein extracts of *T. regenti* cercariae. The same antibodies revealed two bands in homogenates of schistosomula at 31 kDa and 28 kDa which may reflect either different posttranslational modifications or different phases of pro-enzyme activation in vivo.[17] Moreover, both immune sera reacted also with protein extracts of *S. mansoni* cercariae and adults, thus indicating immuno-cross-reactivity between cathepsins B2 of both species.

Besides confirming the presence of CB2 in cercariae of *T. regenti*, the immunoblots showed that CB2 is also present in cercariae of *S. mansoni*, a species for which this peptidase was known only from adults.[17,43] Histological sections of *T. regenti* cercariae showed that TrCB2 is immunolocalized in postacetabular penetration glands which is in agreement with the results of Dvořák et al[63] who demonstrated CB2 in the secretome of penetration glands of *S. japonicum* cercariae. In *S. mansoni* however, CB2 has so far been immunolocalized to the dorsal and lateral tubercles of the tegument of male worms and in the parenchymal cells of both sexes.[43] These observations suggest that cathepsin B2 peptidases may play different roles in the various developmental stages throughout schistosome life cycle.

Occurrence of CB2 near the body surface of blood-dwelling adult schistosomes suggests its participation in the turnover of tegumental proteins, degradation of endocytosed host proteins or protection against the host immune response.[43] However, the fact that *T. regenti* and *S. mansoni* cercariae contain CB2 in penetration glands implies its involvement in degradation of tissues during invasion of the host via the skin. This hypothesis was supported by Dolečková et al[17] who demonstrated the capability of recombinant TrCB2 to cleave skin proteins including keratin, collagen and elastin.[(k)] Thus, it seems that cercariae from different schistosome species employ distinct tools for histolysis and in this respect, bird schistosomes of the genus *Trichobilharzia* seem to be more similar to *S. japonicum* than to *S. mansoni*, regardless of phylogenetic relationships. Hydrolytic activity of *T. regenti* CB2 against other protein substrates (e.g., fibrinogen and myelin basic protein) could also be employed by schistosomula and adults to facilitate migration, feeding and immune evasion within the host. On the other hand, it may simply reflect an overall broad substrate specificity similar to other cysteine peptidases, including cathepsins B.[64,65]

The low activity of TrCB2 towards hemoglobin is noteworthy, illustrating the fact that CB2 probably is not an intestinal digestive peptidase involved in blood digestion by schistosomes (see note d).

The temporal expression of the CB2 gene has recently been evaluated by means of real-time PCR by Dolečková et al.[36] The highest level of TrCB2 expression was observed in schistosomula and adults (similar to TrCB1 transcript levels). On the other hand, the results show a TrCB2 expression pattern distinct from that of TrCB1.1, most significantly in the eggs, miracidia and sporocysts with cercarial germ balls, where a relatively high level of expression of TrCB2 was recorded, compared to cercariae. It has been speculated that TrCB2 could be involved in the invasion of cercariae into the skin.[17] Steady and relatively high expression of TrCB2 in intra-molluscan stages is, therefore, of particular interest. These stages are believed to be transcriptionally highly active because they reproduce asexually (mother and daughter sporocysts) giving rise to thousands of larvae

---

[(k)]Elastin is a structural protein of connective tissue. It is resistant to hydrolysis by most peptidases.

that must synthesize all the enzymatic equipment (including cathepsin B2) necessary to infect the next host.[59]

CONCLUSION

Investigations of schistosomes of the genus *Trichobilharzia* is becoming increasingly important. Its members deserve our interest because they are the causative agents of cercarial dermatitis of humans and particular attention should be paid to the neuropathogenic species *T. regenti*. Nowadays, an increasing risk of infections by bird schistosomes in some areas is attracting newly established research teams. Their ongoing studies may elucidate the unknown aspects of *Trichobilharzia* (neuro-) pathogenicity which should also provide relevant comparative data for human schistosome research.

Undoubtedly, parasite-derived peptidases including cathepsins B are the key factors in trematode survival, because they are involved in many physiological processes like tissue invasion/migration, nutrition, immune evasion and other host-parasite interactions. Although cathepsin B-like activities have been detected in several developmental stages of *T. regenti* and *T. szidati*, their origin is still not fully understood. Based on gene expression analyses, these activities could be attributed to both CB1 and CB2, at least in schistosomula and adult worms. Our knowledge about localization of both enzymes in particular life stages is still not complete, but the data presented above suggest multiple roles for cathepsin B during the development of bird schistosomes. From this point of view, only sporadic evidence of their involvement in the biology of intra-molluscan stages has been published compared to those developmental stages invading vertebrate hosts. It is generally accepted that cathepsin B1 plays a crucial role in protein digestion within the schistosome gut. The existence of multiple isoforms in case of *T. regenti* cathepsin B1 with significant differences in their biochemical properties, despite high sequence similarities, are worthy of further investigation. The intra-vertebrate stages of this species encounter various protein substrates during their migration and therefore a wide range of proteolytic activities may be advantageous in terms of parasite adaptability. Whether the low activity of *T. regenti* cathepsins B towards hemoglobin and high activity on myelin basic protein reflects the unique neurotropism of this species among other schistosomes still remains a matter of ongoing research. We can only speculate on the role of cathepsin B2 in schistosomes. Its presence in different compartments of particular life stages (the tegument of adult *S. mansoni* and the penetration glands of *T. regenti* cercariae) implies a role in household biological processes. Besides host invasion, it might play a role in nutrition or parasite-host interactions such as immune evasion.

The eventual involvement of schistosome cathepsin B peptidases in pathology of disease caused by human schistosomes is only partially understood. In *Trichobilharzia*, preliminary data suggest that cathepsins B may be involved in the development of cercarial dermatitis. In *T. regenti*, particularly, their histolytic effect on the nervous tissue may contribute to host neuromotor disorders and paralyses. The ability of *T. regenti* to invade the CNS of primary-infected nonspecific mammalian hosts again raises a question about the plasticity of the parasite which may be based also on the variability of proteolytic equipment. Further characterization of peptidases of bird schistosomes will contribute to a better understanding of multiple peptidase functions in trematode biology and parasite-host interactions based on proteolysis.

ACKNOWLEDGEMENTS

The research of Martin Kašný, Libor Mikeš, Vladimír Hampl and Petr Horák and their students has recently been supported by the Czech Science Foundation (Grants Nos. 206/07/0233 and 206/09/H026), the Czech Ministry of Education (Grants Nos. MSM 0021620828 and MSM LC06009) and Charles University in Prague (Grant No. SVV-261212/2010). Marian Novotný was supported by grant No. LC07032 from the Czech Ministry of Education, Youth and Sports. Jan Dvořák is a holder of FP7-People-Marie Curie Reintegration Grant (No. 248642). We thank to Dr. Andrea Rossi (University of California, San Francisco, CA, USA) who kindly provided TrCB1.1 3D template model.

REFERENCES

1. Dubois JP, Buet A, Cusin I et al. Epidemiological studies related to cercarial dermatitis in lakes of the Savoy District (France). Helminthologia 2001; 38:244.
2. Thors C, Linder E. Swimmer's itch in Sweden. Helminthologia 2001; 38:244.
3. Mas-Coma S, Valero MA, Bargues MD. Climate change effects on trematodiases, with emphasis on zoonotic fascioliasis and schistosomiasis. Vet Parasitol 2009; 163:264-280.
4. Larsen AH, Bresciani J, Buchmann K. Increasing frequency of cercarial dermatitis in higher latitudes. Acta Parasitol 2004; 49:217-221.
5. Skírnisson K, Aldhoun JA, Kolářová L. A review on swimmers itch and the occurrence of bird schistosomes in Iceland. J Helminthol 2009; 83:165-171.
6. Davis NE. A survey of waterfowl for echinostomes and schistosomes from Lake Wanaka and the Waitaki River watershed. N Z J Helminthol 2006; 80:33-40.
7. Brant SV, Loker ES. Molecular systematics of the avian schistosome genus Trichobilharzia (Trematoda: Schistosomatidae) in North America. J Parasitol 2009; 95:941-963.
8. Horák P, Kolářová L, Adema C. Biology of the schistosome genus Trichobilharzia. Adv Parasitol 2002; 52:155-233.
9. Horák P, Dvořák J, Kolářová L et al. Trichobilharzia regenti, a pathogen of the avian and mammalian central nervous system. Parasitology 1999; 119:577-581.
10. Horák P, Mikeš L, Rudolfová J et al. Penetration of Trichobilharzia cercariae into mammals: dangerous or negligible event? Parasite 2008; 15:299-303.
11. De Jong-Brink M, Bergamin-Sassen M, Solis SM. Multiple strategies of schistosomes to meet their requirements in the intermediate host. Parasitology 2002; 123:S129-S141.
12. Haas W. Parasitic worms: strategies of host finding, recognition and invasion. Zoology 2003; 34:927-934.
13. Hertel J, Holweg A, Haberl B et al. Snail odour-clouds: spreading and contribution to the transmission success of Trichobilharzia ocellata (Trematoda, Digenea) miracidia. Oecologia 2006; 147:173-180.
14. Dvořák J, Delcroix M, Rossi A et al. Multiple cathepsin B isoforms in schistosomula of Trichobilharzia regenti: identification, characterization and putative role in migration and nutrition. Int J Parasitol 2005; 35:895-910.
15. Mikeš L, Zídková L, Kašný M et al. In vitro stimulation of penetration gland emptying by Trichobilharzia szidati and T. regenti (Schistosomatidae) cercariae. Quantitative collection and partial characterization of the products. Parasitol Res 2005; 96:230-241.
16. Kašný M, Mikeš L, Dalton JP et al. Comparison of cysteine peptidase activities in Trichobilharzia regenti and Schistosoma mansoni cercariae. Parasitology 2007; 134:1599-1609.
17. Dolečková K, Kašný M, Mikeš L et al. The functional expression and characterization of a cysteine peptidase from the invasive stage of the neuropathogenic schistosome Trichobilharzia regenti. Int J Parasitol 2009; 39:201-211.
18. Kašný M, Mikeš L, Hampl V et al. Peptidases of trematodes. Adv Parasitol 2009; 69:205-297.
19. Bahgat M and Ruppel A. Biochemical comparison of the serine protease (elastase) activities in cercarial secretions from Trichobilharzia ocellata and Schistosoma mansoni. Parasitol Res 2002; 88:495-500.
20. Dalton JP, Clough KA, Jones MK et al. Characterization of the cathepsin-like cysteine proteinases of Schistosoma mansoni. Infect Immun 1996; 64:1328–1334.
21. Dalton JP, Clough KA, Jones MK et al. The cysteine proteinases of Schistosoma mansoni cercariae. Parasitology 1997; 114;105-112.

22. Dvořák J, Mashiyama ST, Braschi S et al. Differential use of protease families for invasion by schistosome cercariae. Biochimie 2008; 90:345-348.
23. Kouřilová P, Hogg KG, Kolářová L et al. Cercarial dermatitis caused by bird schistosomes comprises both immediate and late phase cutaneous hypersensitivity reactions. J Immunol 2004; 172:3766-3774.
24. Lichtenbergová L, Kolbeková P, Kouřilová P et al. Antibody responses induced by Trichobilharzia regenti antigens in murine and human hosts exhibiting cercarial dermatitis. Parasite Immunol 2008; 30:585-595.
25. Dolečková K, Kašný M, Mikeš L et al. Peptidases of Trichobilharzia regenti (Schistosomatidae) and its molluscan host Radix peregra s. lat. (Lymnaeidae): construction and screening of cDNA library from intramolluscan stages of the parasite. Folia Parasitol 2007; 54:94-98.
26. Rawlings ND, Morton FR, Kok CY et al. MEROPS: the peptidase database. Nucleic Acids Res 2008; 36:320-325.
27. Sajid M., McKerrow JH. Cysteine proteases of parasitic organisms. Mol Biochem Parasitol 2002; 120:1-21.
28. McKerrow JH, Caffrey C, Kelly B et al. Proteases in Parasitic Diseases Annu Rev Pathog Mech Dis 2006; 1:497–536.
29. Caffrey CR, Rheinberg CE, Moné H et al. Schistosoma japonicum, S mansoni, S haematobium, S intercalatum and S rodhaini: cysteine-class cathepsin activities in the vomitus of adult worms. Parasitol Res 1997; 83:37-41.
30. Caffrey CR, Ruppel A. Affinity isolation and characterization of the cathepsin B-like proteinase Sj31 from Schistosoma japonicum. The J Parasitol 1997a; 82:1112-1118.
31. Caffrey CR, Ruppel A. Cathepsin B-like activity predominates over cathepsin L-like activity in adult Schistosoma mansoni and S. japonicum. Parasitol Res 1997b; 83:632-635.
32. Sajid M, McKerrow JH, Hansell E et al. Functional expression and characterization of Schistosoma mansoni cathepsin B and its trans-activation by an endogenous asparaginyl endopeptidase. Mol Biochem Parasitol 2003; 131:65-75.
33. Caffrey CR, McKerrow JH, Salter JP et al. Blood 'n' guts: an update on schistosome digestive peptidases. Trends Parasitol 2004; 20:241-248.
34. Delcroix M, Sajid M, Caffrey CR et al. A multienzyme network functions in intestinal protein digestion by a platyhelminth parasite. J Biol Chem 2006; 282:39316–39329.
35. Delcroix M, Medzihradsky K, Caffrey CR et al. Proteomic analysis of adult S. mansoni gut contents. Mol Biochem Parasitol 2007; 154:95–97.
36. Dolečková K, Albrecht T, Mikeš L et al. Cathepsins B1 and B2 in the neuropathogenic schistosome *Trichobilharzia regenti*: distinct expression and presumptive role throughout the life cycle. Parasitol Res 2010; 107:751-755.
37. Jacobsen W, Cristians U, Benet L. In vitro evaluation of the disposition of a novel cysteine protease inhibitor. Drug Metab Dispos 2000; 28:1343-1351.
38. Xiao SH. Development of antischistosomal drugs in China, with particular consideration to praziquantel and the artemisinins. Acta Trop 2005; 96:153-167.
39. Renslo AR, McKerrow JH. Drug discovery and development for neglected parasitic diseases. Nat Chem Biol 2006; 12:701-710.
40. Abdulla MH, Lim KCh, Sajid M et al. Schistosomiasis mansoni: Novel chemotherapy using a cysteine protease inhibitor. PLoS Med 2007; 4:130-138.
41. Musil D, Zucic D, Turk D et al. The refined 2.15 A X-ray crystal structure of human liver cathepsin B: the structural basis for its specificity. EMBO J 1991; 10:2321-2330.
42. Turk B, Turk D, Turk V. Lysosomal cysteine proteases: more than scavengers. Biochim Biophys Acta 2000; 147:98-111.
43. Caffrey CR, Salter JP, Lucas KD et al. SmCB2, a novel tegumental cathepsin B from adult Schistosoma mansoni. Mol Biochem Parasitol 2002; 121:49-61.
44. Lecaille F, Kaleta J, Bröme D. Human and parasitic papain-like cysteine proteases: Their role in physiology and pathology and recent developments in inhibitor design. Chem Rev 2002; 102:4459-4488.
45. Rawlings ND, Barrett AJ. Introduction: the clans and families of cysteine peptidases. In Handbook of Proteolytic Enzymes, 2 edn (Barrett, A.J., Rawlings, N.D. and Woessner, J.F. eds), Elsevier, Amsterdam 2004, 1051-1071.
46. Merckelbach A, Hasse S, Dell R et al. cDNA sequences of Schistosoma japonicum coding for two cathepsin B-like proteins and Sj32. Ann Trop Med Parasitol 1994; 45:193-198.
47. Caffrey CR, Mathieu MA, Gaffney AM et al. Identification of a cDNA encoding an active asparaginyl endopeptidase of Schistosoma mansoni and its expression in Pichia pastoris. FEBS Lett 2000; 466:244-248.
48. Berriman M, Haas BJ, LoVerde PT et al. The genome of the blood fluke Schistosoma mansoni. Nature 2009; 460:352-358.
49. Pils B, Schultz J. 2004. Inactive enzyme-homologues find new function in regulatory processes. Mol Biochem Parasitol 2004; 340:399-404.

50. Atkinson HJ, Babbitt PC, Sajid M. The global cysteine peptidase landscape in parasites. Trends Parasitol 2009; 12:573-581.
51. Law RH, Smooker PM, Irving JA et al. Cloning and expression of the major secreted cathepsin B-like protein from juvenile Fasciola hepatica and analysis of immunogenicity following liver fluke infection. Infect Immun 2003; 71:6921-6932.
52. Meemon K, Grams R, Vichasri GS et al. Molecular cloning and analysis of stage and tissue-specific expression of cathepsin B encoding genes from Fasciola gigantica. Mol Biochem Parasitol 2004; 136;1-10.
53. Beckham S, Law RHP, Smooker PM et al. Production and processing of a recombinant Fasciola hepatica cathepsin B-like enzyme (FhcatB1) reveals potential processing mechanisms in the parasite. Biol Chem 2006; 387:1053-1061.
54. Illy Ch, Quraishi O, Wang J et al. Role of the occluding loop in cathepsin B activity. J Biol Chem 1997; 10:1197–1202.
55. Krupa JC, Hasnain S, Nägler DK et al. S2'substrate specificity and the role of His[110] and His[111] in activity of human cathepsin B. Biochem J 2002; 361:613-619.
56. Nägler DK, Storer AC, Portaro FC et al. Major increase in endopeptidase activity of human cathepsin B upon removal of occluding loop contacts. Biochemistry 1997; 41:12608-12615.
57. Quraishi O, Nägler DK, Fox T et al. The occluding loop in cathepsin B defines the pH dependence of inhibition by its propeptide. Biochemistry 1999; 38:5017-5023.
58. Berti PJ, Storer AC. Alignment/phylogeny of the papain superfamily of cysteine proteases. J Mol Biol 1995; 246:273-283.
59. Jolly ER, Chin ChS, Miller S et al. Gene expression patterns during adaptation of a helminth parasite to different environmental niches. Genome Biol 2007; 8:R65.
60. Liu F, Lu J, Hu W et al. New perspectives on host-parasite interplay by comparative transcriptomic and proteomic analyses of Schistosoma japonicum. PLoS Pathog 2006; 2:268-281.
61. Choe Y, Leonetti F, Greenbaum DC et al. Substrate profiling of cysteine proteases using a combinatorial peptide library identifies functionally unique specificities. J Biol Chem 2006; 281:12824-12832.
62. Greenbaum D, Medzihradszky KF, Burlingame A et al. Epoxide electrophiles as activity-dependent cysteine protease profiling and discovery tools. Chem Biol 2000; 7:569–581.
63. Dvořák J, Mashiyama ST, Braschi S et al. Differential use of protease families for invasion by schistosome cercariae. Biochimie 2008; 90:345-348.
64. Spiess E, Brüning A, Gack S et al. Cathepsin B activity in human lung tumor cell lines: ultrastructural localization, pH sensitivity and inhibitor status at the cellular level. The J Histochem Cytochem 1994; 42:917-929.
65. Khersonsky O, Roodveldt C, Tawfik DS. Enzyme promiscuity: evolutionary and mechanistic aspects. Curr Opin Chem Biol 2006; 10:498-508.
66. Rudolfová J, Hampl V, Bayssade-Dufour C et al. Validity reassessment of Trichobilharzia species using Lymnaea stagnalis as the intermediate host. Parasitol Res 2005; 95:79-89.
67. Salter JP, Lim KC, Hansell E et al. Schistosome invasion of human skin and degradation of dermal elastin are mediated by a single serine protease. J Biol Chem 2000; 275:38667-38673.
68. Salter JP, Choe Y, Albrecht H et al. Cercarial elastase is encoded by a functionally conserved gene family across multiple species of schistosomes J Biol Chem 2002; 277:24618-24624.
69. Baig S, Damian RT, Peterson DS. A novel cathepsin B active site motif is shared by helminth bloodfeeders Exp Parasitol 2002; 101:83-89.
70. Chanová M, Horák P. Terminal phase of bird schistosomiasis caused by Trichobilharzia regenti (Schistosomatidae) in ducks (Anas platyrhynchos f. domestica). Folia Parasitol 2007; 54:105-107.
71. Sojka D, Hajdušek O, Dvořák J et al. IrAE—An asparaginyl endopeptidase (legumain) in the gut of the hard tick Ixodes ricinus. Int J Parasitol 2007; 37:713–724.
72. Krautz-Peterson G, Skelly PJ. Schistosome asparaginyl endopeptidase (legumain) is not essential for cathepsin B1 activation in vivo. Mol Biochem Parasitol 2008; 159:54-58.
73. Dalton JP, Brindley PJ, Donnelly S et al. The enigmatic asparaginyl endopeptidase of helminth parasites. Trends Parasitol 2009; 25:59-61.
74. Mathieu MA, Bogyo M, Caffrey CR et al. Substrate specificity of schistosome versus human legumain determined by P1-P3 peptide libraries. Mol Biochem Parasitol 2002; 121:99-105.
75. Wulff C, Haeberlein S, Haas W. Cream formulations protecting against cercarial dermatitis by Trichobilharzia. Parasitol Res 2007; 101:98-111.
76. Stamatakis A. RAxML-VI-HPC: maximum likelihood-based phylogenetic analyses with thousands of taxa and mixed models. Bioinformatics 2005; 22:2688-2690.
77. Benson DA, Karsch-Mizrachi I, Lipman DJ et al. GenBank. Nucleic Acids Res 2009; 37:D26-31.
78. Sali A, Blundell TL. Comparative protein modelling by satisfaction of spatial restraints. J Mol Biol 1993; 246:273-283.

79. Podobnik M, Kuhelj R, Turk V et al. Crystal structure of the wild-type human procathepsin B at 2.5 A resolution reveals the native active site of a papain-like cysteine protease zymogen. J Mol Biol 1993; 271:774-788.
80. Stern I, Schaschke N, Moroder L et al. Crystal structure of NS-134 in complex with bovine cathepsin B: a two-headed epoxysuccinyl inhibitor extends along the elntire active-site cleft. Biochem J 2004; 381:511-517.
81. Schneidman-Duhovny D, Inbar Y, Nussinov R et al. PatchDock and SymmDock: servers for rigid and symmetric docking. Nucleic Acids Res 2005; 3:363-367.

CHAPTER 10

PROTEASES IN BLOOD-FEEDING NEMATODES AND THEIR POTENTIAL AS VACCINE CANDIDATES

David Knox

Moredun Research Institute, Pentlands Science Park, Bush Loan, Penicuik, Midlothian, Scotland, UK
Email: dave.knox@moredun.ac.uk

Abstract: Parasitic nematodes express and secrete a variety of proteases which they use for many purposes including the penetration of host tissues, digestion of host protein for nutrients, evasion of host immune responses and for internal processes such as tissue catabolism and apoptosis. For these broad reasons they have been examined as possible parasite control targets. Blood-feeding nematodes such as the barber-pole worm *Haemonchus contortus* that infect sheep and goats and the hookworms, *Ancylostoma spp.* and *Necator americanus*, affecting man, use an array of endo- and exopeptidases to digest the blood meal. Haemoglobin digestion occurs by an ordered and partly conserved proteolytic cascade. These proteases are accessible to host immune responses which can block enzyme function and lead to parasite expulsion and/or death. Thus they are receiving attention as components of vaccines against several parasitic nematodes of social and economic importance.

INTRODUCTION

Proteases are ubiquitous in biological systems, ranging from viruses to vertebrates. They have numerous functions including digestion of food, immune recognition of pathogens and are components of highly-regulated cascades such as the blood-clotting cascade, the complement system and apoptosis pathways. Proteases are divided into five major subgroups on the basis of their catalytic type namely aspartic, cysteine, metallo, serine and threonine proteases. It has been proposed that one of the main selection pressures to initiate the host-parasite relationship may have been nutrient acquisition for

Cysteine Proteases of Pathogenic Organisms, edited by Mark W. Robinson and John P. Dalton.

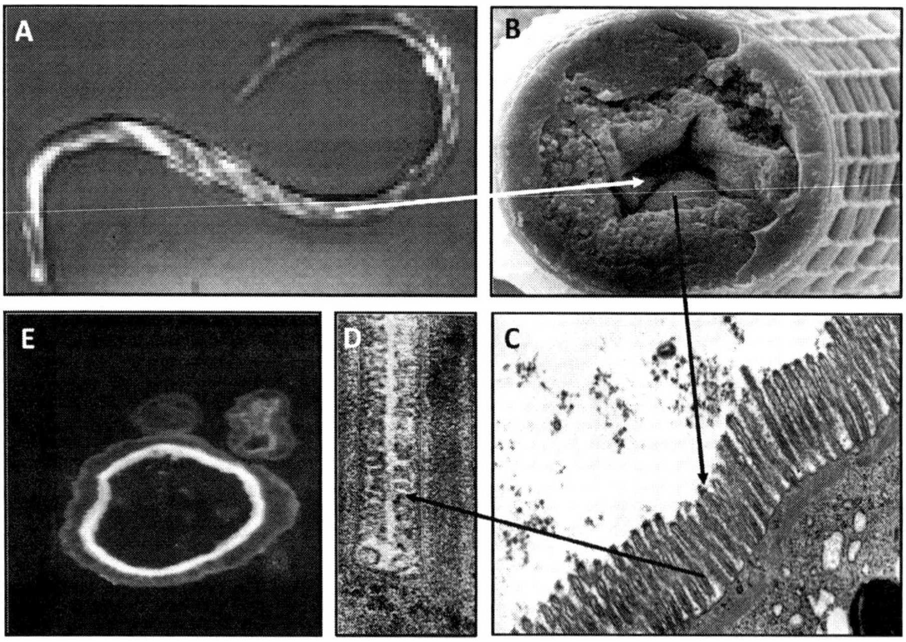

Figure 1. Adult *H. contortus* and the hookworms are blood-feeding. Panel A shows the intestine of adult *Haemonchus* filled with the blood meal (red against a white background). Panel B shows the intestine (centre) surrounded by body wall musculature within a thick outer cuticle. The intestine is lined with microvilli, the brushborder (panel C) and a coiled structure (contortin) is evident between the microvilli (panel D). The proteases targeted by vaccination are found on the surface of the microvilli (panel E) and antibody thus induced binds to the surface of the gut and, in most cases, inhibits the enzyme activity impairing worm nutrition resulting in stunted growth and worm expulsion.

the parasite.[1] However, parasite proteases have been implicated in the degradation of tissue barriers and evasion of host immune responses as well as metabolic roles such as protein turnover and processing within the parasite itself. The present author's interest in nematode proteases was stimulated by a series of papers published in the 1950s by Thorson. These papers followed a logical process of investigation starting with studies of the mechanism of immunity to *Nippostrongylus brasiliensis* (*muris*) in the rat.[2,3] Eye catching, from a vaccination point of view, were follow-up studies in the dog hookworm *Ancylostoma caninum*. Firstly, a study which demonstrated proteolytic activity in extracts of the oesophagus from the adult parasite that could be inhibited by immune serum[4] and another that showed the induction of acquired immunity in dogs by the injection of similar extracts.[5] Blood-feeding nematode pathogens digest haemoglobin as a source of nutrition[6] and the intestinal brush border is a rich source of proteases containing aspartyl, cysteinyl, metalloproteases as well as di- and aminopeptidases (Fig. 1). Haemoglobin digestion is conducted by a semi-ordered pathway which is remarkably similar in distantly related organisms such as *Plasmodium*, hookworms and the ovine abomasal nematode parasite *Haemonchus contortus*.[6] This similarity and evolutionary conservation will become evident in the descriptions below of the proteases used by *H. contortus* and hookworms to survive in the mammalian host.

HAEMONCHUS CONTORTUS

General Background

Proteases with acidic pH optima which hydrolysed haemoglobin were first described in a variety of adult parasitic helminths including the nematodes *Angiostrongylus cantonensis, Dirofilaria immitis, Trichuris muris* and *Ascaris suum*.[7] The highest levels were found in the blood-feeding genera. The highest specific activity was found in the intestines and activity could be inhibited by pepstatin, an aspartyl protease inhibitor. Finally, this activity resembled cathespin D rather than that of pepsin. Cysteine protease activity was also detected but not always implicated in haemoglobin digestion.[7] Evidence that adult *Haemonchus* contained elastase-like activity with a possible anticoagulant function was presented.[8] At the same time, Cox and coworkers[9-11] described a family of cysteine proteases with putative fibrinogen-degrading properties which were associated with moderate vaccine-induced protective immunity in lambs.[12] Moreover, Knox et al[13] showed that proteases (predominantly cysteine proteases) present in detergent extracts of adult *Haemonchus* were capable of degrading haemoglobin, albumin and fibrinogen. These proteases were inhibited by sera from lambs which were protected against *Haemonchus* challenge by immunisation with gut membrane proteins.[8] This entire body of work indicated that proteases were valid vaccine candidates and subsequent work has confirmed this opinion.

[^{3}H]leucine-labelled haemoglobin[14] was used to assess the role of secreted and intestinal cysteine proteases from adult *H. contortus* in worm metabolism. After 24 hour incubation, radioactivity was detectable within worm tissues and the process was inhibited by puromycin, a protein synthesis inhibitor. These observations indicated that haemoglobin was digested and the resulting amino acids incorporated into worm proteins. The latter was not affected by a specific cysteine protease inhibitor but the inhibitor did markedly reduce haemoglobin degradation in the culture medium supporting a role for cysteine proteases in extracorporeal digestion. The studies described above collectively indicated that proteases played a key role in parasite nutrient acquisition and, as such, would be good vaccine targets if accessible to the immune response. Concurrently, ground-breaking work emerged from Australia where Willadsen and coworkers[15] pioneered the use of "concealed antigens" as vaccine components in ticks—these antigens are not normally recognised by humoral immune responses induced by natural infection. Concealed antigens are usually expressed in the intestine and can be disrupted by vaccine-induced host humoral responses. Again coincidently, vaccination of lambs with a protein fraction enriched for a helical polymer (contortin) known to be present in the pharynx and intestine of L4 and adult worms, reduced worm burdens in vaccinated lambs by 78% compared to controls.[16] The protein polymer was insoluble and could be enriched from phosphate-buffered saline extracts of adult parasites by ultracentrifugation.[17] The function of contortin was undefined but it was suggested that it could be an immobilized anticoagulant.[16] These studies focussed attention on proteins (in particular proteases) expressed in the intestine of blood-feeding nematodes which may be essential for digestion (Table 1).

Proteases and Vaccination

The two most effective native vaccine antigens for *H. contortus* defined to date are both isolated from detergent-soluble extracts of the adult parasite, are both expressed in

Table 1. Protection induced by vaccination with proteases against *Haemonchus contortus* and against the hookworms *Ancylostoma caninum (Ac)* and *Ancylostoma ceylanicum (Ay)*

Antigen Acronym and Protease Class	Stage Expressed	Function	Reduction in Worms (host sp.)	Reduction in Epg[a] (host sp.)	Other Findings and Comments	References
H11 Gut; aminopeptidase	L4 and Adults	Haemoglobin digestion	>90% (sheep)	>90% (sheep)	4 isoforms Greatly reduced protection with recombinants	17
H-gal-GP Gut; Aspartyl- & metalloproteases	L4 and Adults	Haemoglobin digestion	72% (sheep)	93% (sheep)	Protease complex; contains other proteins Greatly reduced protection with recombinants	23, 24, 25, 26, 27, 28, 31, 32 93
TSBP Gut; Cysteine proteases	L4 and Adults	Haemoglobin digestion	47% (sheep)	77% (sheep)	Comprises 3 proteases *hmcp* 1, 4, 6 Recombinants have reduced efficacy	38, 39, 40, 42
Ac-CP-2 Gut; Cysteine protease	Adults	Haemoglobin digestion	18% (dogs)	61% (dogs)	Recombinant Worms stunted. Antibodies bound to gut of worms	74
Ac-APR-1 Gut; Aspartyl protease	Adults	Haemoglobin digestion	33% (dogs)	70% (dogs)	Reduced blood loss. Antibodies bound to gut of worms.	77
Ay-MTP-1 Oesophagus, L3 cuticle; metalloprotease	Activated L3	Penetration of tissue barriers	36% (hamster)	59% (hamster)	Fusion protein with ASP	76

the microvillar surface of the intestine and both show protease activity. The antigens have been termed H11, an aminopeptidase family,[18] and an integral membrane glycoprotein complex, termed *Haemonchus* galactose-containing glycoprotein complex (H-gal-GP).[19] Both independently induce high levels of protection (>90% reductions in egg output) in a range of sheep breeds when evaluated indoors under experimental conditions and are effective against anthelmintic-resistant strains of *H. contortus*. When tested as a combination they also protected newly weaned Merino lambs grazing contaminated pasture in New South Wales.[20] Enzyme activity associated with both antigens is inhibited by IgG antibodies from vaccine-protected sheep.

H11

H11 is a 110 kDa integral membrane glycoprotein complex expressed exclusively in the intestinal microvilli of the parasitic stages. It is a highly effective immunogen (>90% reductions in faecal egg output, >75% reduction in worm burden) against *H. contortus* challenge[18] and protection is closely correlated with specific systemic IgG titres to H11.[21] cDNA cloning and sequence analysis showed that H11 was an aminopeptidase[22] and has several isoforms[18] (H11-1 to H11-4). The native form of H11, purified by lectin-affinity chromatography, comprises a mixture of these isoforms and both aminopeptidase A and M-type activities.[18] The isoforms have a short N-terminal cytoplasmic tail, a transmembrane region and an extracellular region organised into four domains.[18] Enzyme activity is localised exclusively in the microvilli and is inhibited by H11 antisera in vitro, the level of inhibition observed being correlated to protection in vivo.[21] H11 is thought to be involved in the final breakdown of small peptides that are produced by digestion of dietary protein and the mechanism of protection may be antibody-induced disruption of nutrient uptake. Of note, the level of protection attained diminished markedly with dissociation and reduction insinuating that conformational epitopes contributed to the expression of protective immunity. The extracellular domains of H11-1, -2 and -3 were expressed as enzymatically active recombinant proteins in insect cells but were not protective in trials despite inducing antibodies which inhibited enzyme activity in the recombinant and native H11 proteins.[17] Vaccination with bacterially-expressed versions of these domains induced modest but consistent protective immunity in lambs. Several experiments[18] suggested that protection was not necessarily dependent on inhibition of enzyme activity. Glycan modification was considered central to the expression of protective immunity given that native H11 possesses highly immunogenic and novel N-linked glycans[22] which are quite distinct from glycans expressed by insect cells. Worms recovered from sheep immunized with native H11 have a coating of sheep antibody bound to the worm gut surface, whereas those recovered from sheep immunized with baculovirus H11 do not;[17] an observation which could be ascribed to the differing glycan components in the two vaccines.

H-gal-GP

H-gal-GP is found on the microvillar surface of the intestine and is purified from integral membrane protein extracts of the adult parasites using lectins with specificity for *N*-acetylgalactosamine.[24] As a vaccine, it reduces worm burdens and faecal egg output from challenged lambs by 72% and 93% respectively over a number of independent trials. Early biochemical analyses showed that preparations of the complex had haemoglobinase activity and contained aspartyl-, metallo- and, on occasion, cysteine proteases. The components of the complex have been precisely defined using N-terminal sequence analysis of the individual peptide components from the four major protein zones combined with cDNA library immunoscreening.[25-29] The level of protection correlates with antibody concentration indicating that protection is antibody-mediated. A major component of this highly-protective antigen complex is a family of four zinc metalloendopeptidases, designated MEPs 1-4 with MEP-1.[26,30] MEP1 was shown to be a putative zinc metallopeptidase which showed homology with the mammalian Type II integral membrane protein neutral endopeptidases.[21] The protein was localised to the luminal surface of the nematode gut whilst RT-PCR and immunoblot analyses showed that expression was restricted to the blood-feeding stages. Southern blotting indicated that MEP1 belonged to a multi-gene family. Further analyses showed that MEP1 and 3 both had a predicted Type II integral membrane protein structure.[30] In contrast MEPs 2 and 4 had signal peptides indicative of secreted proteins. Proteomic analysis of H-gal-GP showed that the extracellular domain of all 4 MEPs was cleaved close to the transmembrane region/signal peptide with additional cleavage sites mid-way along the polypeptide. A version of H-gal-GP lacking MEP3 was purified from adult worms using jacalin lectin and was an equally effective vaccine as entire H-gal-GP reducing egg output by 86% and worm burden by 52%.[31] A combination of all four MEPs, separated from the rest of the complex by gel filtration after solubilisation in 8 M urea, significantly reduced *H. contortus* egg counts by 45% and 50%. MEP3 alone or MEPs 1, 2 and 4 in combination, electro-eluted from the complex following SDS gel electrophoresis, each reduced egg counts by about 30%.[32] These data indicated that the MEPs all contributed to the protection induced by H-gal-GP. Protection was lost when sheep were immunized with fully reduced and denatured H-gal-GP or with bacterially-expressed recombinant forms of MEP 1 or the principal domains of MEP3, suggesting that conformational epitopes on the MEPs are required for immunity.[32] Another component of H-gal-GP was shown to be an aspartyl protease with homology to mammalian pepsinogen sequences (HcPEP1), again expressed on the luminal surface of the gut in the blood-feeding stages.[25] The enzyme was shown to be almost exclusively expressed by the blood-feeding parasite stages. Antibody produced to a bacterially-expressed version bound to the luminal surface of the gut in the adult parasite. At the same time, biochemical analyses had shown that H-gal-GP had haemoglobinase activity which was inhibited by the aspartyl protease inhibitor pepstatin (D.P. Knox, unpublished data). The proteinase may play a central role in digesting the blood meal and, for this reason, is considered a potential sub-unit vaccine candidate. A second aspartyl protease, designated HcPEP2, showed 50% sequence identity with HcPEP1.[33] A fraction of H-gal-GP prepared by gel filtration in 8 M urea, containing both HcPEP1 and 2, reduced faecal egg output by 48% and final worm burdens by 36% compared to controls. The HcPEP1 and 2 fraction did not protect if electro-eluted from SDS-dissociated H-gal-GP, nor did bacterially-expressed recombinant HcPEP1, suggesting that conformational epitopes are important for inducing immunity. Very recent structural analyses using electron microscopy suggests that the complex is the structural equivalent of a bacterial proteosome (J Trinnick and W.D. Smith, unpublished data). Antibody purified from vaccinated, protected lambs inhibits the aspartyl protease activity of the complex by at least 50% (Ekoja and Smith, Parasite Immunology, In press) suggesting that this may be the mechanism which mediates protection. The contribution of the metalloproteases to vaccine efficacy has yet to be clarified and antibody inhibition studies are ongoing. As discussed below, *Haemonchus* shares several protease classes implicated in blood meal digestion with hookworms and the data suggest evolutionary conservation of this process.

Does Glycan Contribute to Immunity-Induced by Vaccination with H11 and H-gal-GP?

A considerable portion of the antibody response induced in the host following immunisation with H11 and H-gal-GP and for that matter TSBP (see below), is directed at the glycan components. The hypothesis that the glycan moieties themselves may induce protective immunity was tested using the equivalent lectin-binding fraction from *C. elegans*.[33] *C. elegans* is closely related to *H. contortus*, both being Clade V nematodes and has similar glycan moieties with some unique to nematodes.[21] A subset of glycoproteins from *C. elegans*, isolated using ConA-lectin chromatography, were strongly recognised by the antiglycan component of the antibody response to H-gal-GP in sera from vaccinated and protected lambs. The efficacy of these *C. elegans* glycoproteins as immunogens against *H. contortus* challenge infection was tested in sheep, the point being to seek evidence that the antiglycan antibody response could protect lambs against infection. Despite the generation of a high titre systemic IgG antibody response to the *C. elegans* glycoproteins which strongly recognised all the components of H-gal-GP and bound to the microvillar surface of the gut of *H. contortus*, no protection against challenge infection was observed.[34] These data suggested that glycan alone was not protective but this does not rule out the possibility that it plays a role in inducing the correct antibody response with the required avidity.

Cysteine Proteases

Cathepsin B-like cysteine proteases generally constitute large multi-gene families in both parasitic and nonparasitic helminths, including blood-feeders such as *H. contortus*, hookworms and the schistosomes. In addition, they are found in free-living *Caenorhabditis* sp. In each case they are expressed in the nematode gut and/or found in excretions/secretions and, on this basis, have been implicated in digestion of proteins for nutrient acquisition. They are now under close scrutiny as vaccine candidates in a variety of helminths. The cathepsin B gene family has undergone enormous expansion in blood-feeding helminths representing approximately 16% of all intestinal transcripts in *H. contortus*[35] and are the most abundant and diverse protein family in the gut of this parasite. Cathepsin Bs are also highly represented in the hookworm intestinal transcriptome.[36] This diversity may lead to redundancy of function and differing antigenicity with implications for recombinant protein vaccine efficacy making them unsuitable targets for monovalent antihelminth vaccines. It noteworthy that antibodies raised in mice to recombinant protein forms of four distinct cathepsin Bs expressed in the intestine of the hookworm *Necator americanus* showed minimal cross-reactivity with each other while each antiserum did bind to the intestine of the adult parasite.[36] Despite these factors, cathepsin Bs are showing promise as components of vaccines against blood-feeding nematodes.

Twelve-month old lambs were immunised with a fraction enriched for fibrinogen-degrading proteins derived from the intestine of adult *H. contortus*.[12] Modest protective immunity, evidenced by significant reductions in mean faecal worm egg counts, adult worm burdens and modified female fecundity ratios at necropsy, were detected in vaccinated sheep compared to control animals. A major component of this protective extract was a cathepsin B-like protease with a predicted protein sequence of 342 amino acids with an N-terminal signal sequence indicative of secretion.[9] The genomic sequence was determined and designated *AC-2*.[10] The *AC-2* gene spanned 8 kb, had 11 introns one of which interrupted the proposed active site region. Importantly, Southern blots indicated that the protease was encoded by a small gene family. Northern blots indicated that mRNA transcripts for the gene family were present at low levels in a mixed population of L3 and L4 larvae but were highly abundant in adult worms. These data correlated with the onset of blood-feeding, indicative of a role in blood digestion. Further members of this gene family were identified with one, AC-3, being linked in tandem to the *AC-2* gene.[11] The authors described other family members, designated AC-4 and AC-5 and showed that the predicted amino acid sequences of AC-3, AC-4 and AC-5 shared 64-77% identity with one another and with the previously reported AC-1 and AC-2 sequences. The amino acids surrounding the active site cysteine were highly conserved, as were the positions of other cysteine residues in the mature protein sequences.

Fractions, greatly enriched for cysteine proteases (TSBP), were prepared from PBS soluble and detergent soluble (Triton X-100) membrane proteins by affinity chromatography using Thiol-Sepharose, a matrix which selectively binds proteins with free sulphydryl (SH-) groups such as cysteine proteases. Lambs vaccinated with the membrane fraction showed significant levels of protection (50% reduction in worm burdens) against homologous challenge in sheep.[38] Immuno-localisation studies showed that these proteases were present in the microvillar surface of the intestine. Further fractionate/vaccinate studies were undertaken, each with the purpose of refining the antigen subset. TSBP cysteine proteinases were purified by ion-exchange or by affinity-chromatography using recombinant *H. contortus* cystatin[39] and were just as efficacious as unfractionated TSBP in sheep vaccination trials providing further confirmation that TSBP-induced protection could be ascribed to the cysteine proteases. These proteases were further defined by cDNA library screening with sera from vaccinated lambs[40] which identified three immuno-positive cDNAs encoding cathepsin B-like cysteine proteinases, hmcp1, 4 and 6 (hmcp 2, 3 and 5 being PCR homologues not implicated in protection to date). A combination of bacterially-expressed recombinant glutathione *S*-transferase (GST) fusion forms of hmcp 1, 4 and 6 were tested as a vaccine with encouraging results reducing worm burdens by a statistically significant 38% but not faecal egg output compared to challenge controls.[41] A major proportion of the IgG response detected was directed against the GST component and the trial was repeated using nonfusion recombinant proteins with a similar impact on the outcome of a challenge infection (Fig. 2)[42].

Cysteine Protease Genetic Diversity

The *hmcp* genes (cloned from a UK isolate of the parasite) showed about 70% amino acid identity to the AC family members (USA isolate) and attempts to amplify AC-1 from the UK isolate consistently failed (Redmond and Knox unpublished data). Geographical strain diversity could have considerable implications for the production of a globally-effective recombinant vaccine. Around this time, Karanu and colleagues described differences in secretory protease profiles between two *Haemonchus* isolates derived from the same goat herd in the same geographic location in Kenya and with a USA isolate.[43] Moreover, it was noted that experiments at this time had not identified a single secretory protease from *H. contortus* that was conserved among the restricted number of isolates tested.[44] However, AC sequences were identified in the *Haemonchus*

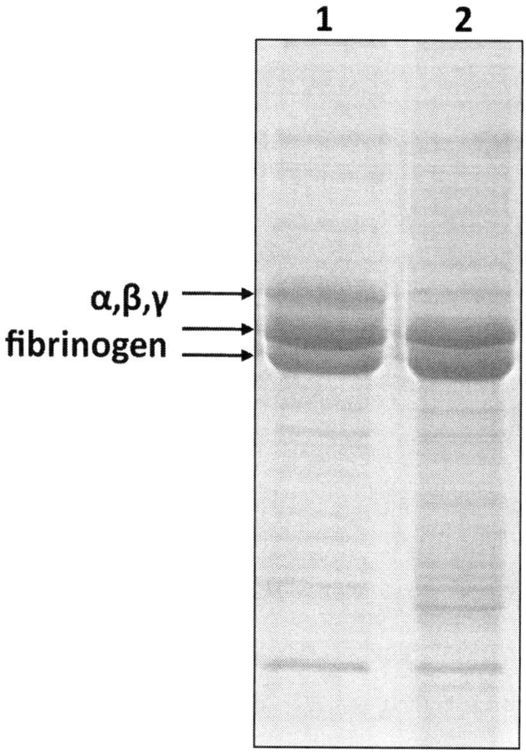

Figure 2. Fibrinogen degradation by contortin. The upper a-fibrinogen peptide is cleaved by the dipeptidyl-peptidase IV activity of contortin, an action which impairs cross-linking in the fibrin clot.

expressed sequence tag (EST) database comprising ~5,000 sequences established from a UK worm isolate and held at NEMBASE. In addition, a more recent analysis of cathepsin B-like cysteine proteases expressed in the adult *H. contortus* intestine conducted using material from the USDA Beltsville isolate identified ESTs encoding the *hmcp* sequences.[45] Moreover, a study of genetic variability in cysteine proteases was undertaken which analysed polymorphism of 5 previously reported genes (AC-1, AC-3, AC-4, AC-5 and GCP-7) by PCR-SSCP and sequencing procedures in two different strains of *H. contortus* derived from North American and Spain.[46] The SSCP data identified 20 different alleles for the 5 loci assessed with loci for AC-1, AC-3, AC-4 and GCP-7 having 5, 8, 2 and 4 alleles, respectively. All these loci were polymorphic with allele frequencies ranging from 0.0070 to 0.8560 and these were significantly different between strains. Of note, AC-5 did not display significant polymorphism between strains. Nucleotide analyses indicated that variations between strains were translated in some cases into changes in the amino acid sequence. This work supported previous observations based on sequence comparisons.[47] These authors compared several domains which define cathepsin B-like function including the occluding loop, S2 and S2 subsites and the pro region. They subdivided the sequences into four groups and identified putative orthologues in *Ascaris suum* and *C. elegans* and made several noteworthy conclusions. First, that variability in the domains analyzed suggested substantial functional diversity in enzymatic properties of nematode cathepsin Bs. Second, that nutrient digestion was a potential factor promoting cysteine protease diversification in these organisms (which could affect antigenicity). A follow-up analysis quantified the extent of diversity of *H. contortus* cathepsin B-like genes from available EST datasets.[48] Contig analysis of 686 *cbl* ESTs from a USA isolate resolved 123 clusters. Sixty percent of these clusters were present in a UK isolate. These analyses also indicated the likelihood of antigenic diversity which is relevant to vaccine strategies. The authors noted that the extreme abundance and diversity of intestinal cathepsin B-like transcripts appear to be relative specializations for *H. contortus*. Also of relevance in this context is a study from Utrecht[49] which identified seven cathepsin B-like cysteine proteases in the excretory-secretory (ES) products of *H. contortus*. This used a combination of two-dimensional (2-D) zymography and biotinylated inhibitors to localize active proteases in 2-D gels. Mass spectrometry identified these as AC-4, HMCP1, HMCP2 and GCP7 plus three novel CBLs encoded by clustered ESTs. *H. contortus* ES products were fractionated using Thiol-Sepharose affinity chromatography and the bound proteins eluted in two steps, 25 mM cysteine followed by a harsher reducing agent, 25 mM Dl-dithiothreitol (DTT).[50] Sheep were vaccinated with these eluate proteins and significant protection against a single challenge infection was stimulated by proteins in the DTT-eluted fraction in which egg output and worm burden were reduced by 52% and 50%, respectively compared to controls. The DTT-eluted fraction consisted predominantly of cysteine proteases but also contained metalloprotease activity. Subsequently, the same group[51] used a recombinant *H. contortus* cystatin affinity column to obtain a sub-fraction of ES enriched for cysteine protease activity. Only a single cysteine protease band was evident in substrate gels after this procedure which was identified as AC-5 using mass spectrometry analysis . Lambs vaccinated with cystatin-binding proteins had reduced worm burdens and faecal egg output (36% and 32% respectively compared to controls and after challenge higher local and systemic IgA and IgG responses were evident in the cystatin-binding vaccinate group.

A sequence comparison of cysteine proteases from blood-feeding helminths reported to play a role in the degradation of host hemoglobin was undertaken.[52] Species included *Haemonchus, Schistosoma spp.* and the hookworm, *A. caninum*. For comparative purposes, they included several cysteine proteases from *C. elegans* and from other nonhelminth species. The sequences were aligned using a specialized alignment programme which can identify short highly-conserved sequence blocks. The study highlighted an amino acid motif from the active site region that was present only in the blood-feeding helminths. They then used the pattern search tool Prosite with the motif from the Asn active site region (Y-W-[IL]-[IV]–A-N-S-W-X–X–D-W-G-E) as well as a BLAST search tool. They identified an additional 8 cathepsin B-like cysteine proteases containing the helminth blood-feeder motif. In each case, the possibility that a cysteine protease played a role in blood–feeding was supported by existing literature.

The ES products of adult *H. contortus* contain several proteases[8,43] and their substrate specificity is somewhat contradictory. They show closest homology to the cathepsin Bs yet in general terms, have substrate specificity against small peptide substrates indicative of cathepsin Ls; the activity in the intestine and ES of adult *Haemonchus* showing greater preference for Phe-Arg than Arg-Arg dipeptide substrates indicative of cathepsin L and B-like proteases respectively. Cathepsin B-like substrate preference has been attributed to a Glu_{245} residue (mammalian cathepsin B numbering) in the P2 position in the active site, this residue in the equivalent position being absent in the majority of the cathepsin B-like sequences isolated from *Haemonchus* at this time.[45] As an intriguing aside, a family of cysteine proteases was identified in *C. elegans*[53] (all of the intestinal proteases identified in *H. contortus* had homologues in *C. elegans*[54]). RNA interference (RNAi) was used in *C. elegans* to knockdown expression of several family members simultaneously and treated worms showed no phenotype and appeared to feed normally.[55]

Asparaginyl endopeptidases (or legumains) may play a role in the activation of cathepsin B and L-like proteinases by cleavage of the peptide backbone between the prosegment and mature enzyme domains although this function is still open to some debate.[56] An EST sequence encoding a full-length legumain from *H. contortus* was characterized.[57] RT-PCR showed that the legumain transcript was present from the L4 life-cycle stage onwards, coincident with the onset of blood feeding and the expression of intestinal cysteine protease genes. Immuno-localisation showed that the enzyme was expressed near the microvillar surface of the intestinal cells.

Contortin

The suggestion that contortin may be an immobilised anticoagulant received recent confirmation[58] in an analysis of a standard contortin preparation using mass spectrometry. Contortin comprised two major proteins, Hc-PCP1 and HcPCP2, with homology to prolyl carboxypeptidases. The proteins shared 64% amino acid sequence identity and contained two prolyl carboxypeptidase S28 type domains organized in a tandem repeat. The transcripts of both genes are present from the L4 stage onwards,

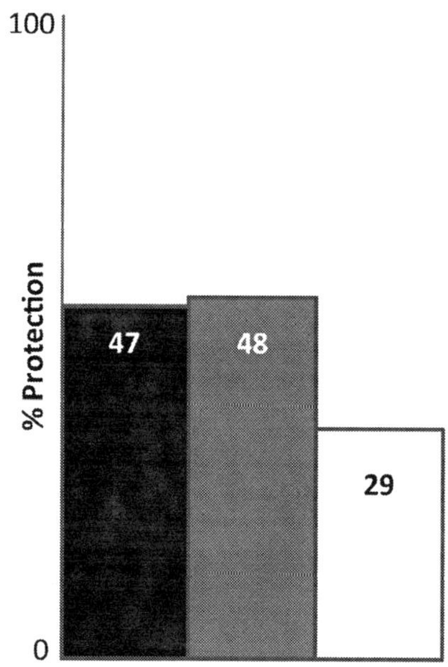

Figure 3. Protection induced in lambs, as judged by final worm burdens, against *Haemonchus* by vaccination with a native cysteine protease enriched fraction termed TSBP (black), a sub-fraction containing cysteine proteases alone (grey) and a bacterially-expressed recombinant cocktail vaccine (white) comprising the 3 known cysteine protease components of TSBP. Protection induced by the recombinants is lower than that by the native proteins inferring that correct protein folding contributes to the complete induction of protection.

coinciding with the onset of blood feeding. Contortin has dipeptidyl IV activity and this rapidly (within seconds) inhibited blood coagulation. Mass spectrometry showed that contortin degraded the C-terminal end of the alpha-fibrinogen chain which is essential for clot formation (Fig. 3). This cleavage was notably rapid, degradation occurring within seconds. Some preliminary work requiring confirmation (Knox, unpublished data) suggests that antibody to contortin inhibits DPP IV activity but it remains to be seen if this correlates with protection.

Ostertagia Ostertagi

Targeting intestinal proteases as an approach to vaccination may extend beyond obligate blood-feeding nematode genera. Homologues of *H. contortus* intestinal cysteine proteases were identified in the bovine abomasal parasite *Ostertagia ostertagi* by screening a genomic DNA library with a *H. contortus* cathepsin B-like cysteine protease cDNA as a hybridization probe.[59] Three cathepsin B-like genes were identified with 66% homology to each other and similar levels of homology to some *Haemonchus* AC genes, particularly in the region encoding the mature protein. In addition, the exon/intron structure of homologous genes was conserved between the species. The *O. ostertagi* CP-3 protein was most closely related to the *H. contortus* AC-5 protein. Geldhof et al[60] prepared cysteine protease-enriched fractions from both a detergent-soluble extract of adult *O. ostertagi* and ES products by thiol-sepharose chromatography. They tested their protective efficacy against *O. ostertagi* in cattle and only the ES-derived proteins were protective. In a follow-up trial[61], ES-Thiol was sub-fractionated using MonoQ-Sepharose anion exchange chromatography to determine whether the cysteine protease component or a dominant protein, an ASP homologue, were responsible for the induced protection. Both the cysteine protease and ASP fractions were protective in calves (reductions in cumulative faecal egg output of 74% and 80% respectively).

HOOKWORMS—*NECATOR AMERICANUS* AND *ANCYLOSTOMA SPP*

General Background

There has been long-standing interest in the anticoagulant properties of the ES from the dog hookworm *A. caninum.*[62-65] The latter[65] showed that this anticoagulant activity could be attributed to an elastase-like 36 kDa protease which could degrade fibrinogen into five smaller polypeptides each with anticoagulant properties and convert plasminogen to a mini-plasminogen-like molecule. The protease was purified from adult parasite extracts by several chromatographic steps, had optimal activity at pH 9-11 and was inhibited by EDTA, indicative of a metalloprotease.[66] The enzyme could degrade elastin and trypsin-labile glycoproteins in a rat vascular smooth muscle extracellular matrix. Moreover, the enzyme could be detected in extracts from the infective larval stage suggesting that the enzyme has a function at an earlier point in the parasite life-cycle. *A. caninum* L3 infective larvae were stimulated to feed in vitro and released two metalloproteases (50 kDa and 90 kDa) which were associated with feeding in vitro by the use of inhibitors.[67] It was suggested that the adult hookworm utilizes the protease for digesting the bolus of intestinal mucosa lodged in its buccal capsule and plays a haemorrhagic role.[66] Moreover, metalloproteases released from the Ancylostoma spp. degraded human fibronectin to a 60 kDa polypeptide intermediate, but had no effect on solubilized bovine elastin or human laminin.[68] Gelatinolytic protease activity at pH 8.5 was detected in exsheathing fluid (EF) and ES products of infective L3 from *N. americanus*.

EF contained a single cysteine protease (116 kDa) and proteases in the size range 62 to 219 kDa in the L3 ES products. Evidence presented suggested that ES comprised a mixture of cysteine, metallo and serine proteases. ES degraded IgG, IgA and IgM (but not IgD or IgE) at the Fc region, this being inhibited by PMSF indicative of the action of a serine protease.[69] Taken together these studies suggest that proteases are used by hookworm larvae to establish infection, by the adults for feeding purposes and by both stages to evade the host immune response. Further evidence for the latter role was provided when it was shown that adult *N. americanus* ES products inhibited eosinophil recruitment in vivo by proteolytic degradation of eotaxin.[70] EDTA and phenanthroline inhibited this activity, indicative of the action of metalloproteases. These authors suggested that the production of enzymes inactivating eotaxin may be a strategy employed by helminths to prevent recruitment and activation of eosinophils at the site of infection.

Proteases and Vaccination

Cysteine Proteases

Adult and larval extracts as well as ES from adult *A. caninum* contain cysteine proteases.[71] One gene, AcCP-1, encoded a cathepsin B-like enzyme with closest identity to *H. contortus* cysteine proteases (61%).[72] This gene encoded a protease of 343 amino acids with a mature form of 255 amino acids. A second gene, AcCP-2, encoded a mature cysteine protease of 254 amino acids that showed 86% identity to AcCP-1. Antibody to recombinant AcCP-1 reacted with the oesophageal, amphidial and excretory glands in sections of male and female adult hookworms and with an antigen of approximately 40 kDa in adult ES products. Homology modelling of AcCP-1 (using the crystal structure of human cathepsin B as a template) predicted that substrate binding and specificity differed between AcCP-1 and cathepsin B and showed that AcCP-1 would preferentially cleave the cathepsin L substrate Phe-Arg over Arg-Arg, the bond favoured by cathepsin B.[73] This offers a structural explanation why this cathepsin B-like protease from hookworm shows cathepsin L-like substrate specificity. Ac-CP-2 was expressed as a catalytically-active recombinant protein and used to vaccinate dogs prior to challenge infection.[74] Compared to controls, the vaccinates showed a decrease in faecal egg output and the number of female worms while adult parasites surviving in the intestine at post mortem were smaller than those from control dogs. Antibodies stimulated by vaccination with Ac-CP-2 inhibited proteolysis of a peptide substrate and bound to the intestinal brush border of the parasite. These data suggested that the parasite imbibed antibody with the blood meal and that these antibodies impaired digestion of the blood meal and resultant nutrient acquisition.

The human hookworm, *N. americanus* also expresses a number of cysteine protease genes.[75] Four genes forming a robust clade with other hookworm genes were described and were most similar to *Haemonchus* intestinal counterparts. All were up-regulated in the blood-feeding adults compared to the L3 stage and were expressed in the gut tissue. Of these, (*Na-cp-2*, *-3*, *-4* and *-5*) *Na*-CP-3 was expressed in a catalytically-inactive form in yeast and it was shown to undergo autoactivation at low pH. This recombinant digested gelatin, the fluorogenic peptide substrate Z-Phe-Arg-AMC but not Z-Arg-Arg-AMC and did not degrade haemoglobin unless this had been partially broken down by prior incubation with aspartyl proteases from *N. americanus*. The remaining proteases (*Na*-CP-2, -4 and -5) were expressed as insoluble inclusion bodies in *Escherichia coli*. Mouse-derived antibodies to all 4 proteases bound to the intestine of adult worms but, intriguingly, showed minimal cross-reactivity with the other gene products.

Aspartyl Proteases

The first indication that hookworm aspartyl proteases may be valid vaccine targets came in a study where dogs were vaccinated with 3 different recombinant fusion proteins, one of which was an aspartyl protease, Ac-APR-1.[76] The dogs mounted a strong antibody response to the protein and the numbers of adult worms retrieved from small intestine was reduced compared to controls. Moreover, worms appeared to be displaced from their predilection site into the colon, the effect being particularly pronounced for female worms and correlated with antibody titre. These data were consolidated by observations that dogs vaccinated with an enzymatically active yeast expressed recombinant version of Ac-APR-1 showed significantly reduced worm burdens and faecal egg outputs induced compared to control dogs following challenge infection with *A. caninum* larvae.[77] The vaccine reduced blood loss and the development of anaemia. IgG from vaccinated animals inhibited activity of the recombinant enzyme and bound to the intestines of worms retrieved from vaccinated dogs. Subsequently, the same group[78] expressed the equivalent protease (Na-APR-1) from the human hookworm *N. americanus* in a catalytically active form and as an inactive mutant which retained the ability to bind substrates. Dogs vaccinated with the mutant enzyme were significantly protected against infection with *A. caninum* and data suggested that the protective effect was due to conformational changes induced by neutralising antibody responses to vaccination. Antibodies binding to native Na-APR-1 not only neutralized enzymatic activity of Na-APR-1 but also neutralized activity of APR-1 orthologues from three other hookworm species that infect humans. Moreover, anti-Na-APR-1 IgG1 was the most prominently detected antibody in sera from people resident in high-transmission areas for *N. americanus*, indicating that natural boosting may occur in exposed humans. A panel of monoclonal antibodies with neutralizing activity was established and used to define a particular epitope, AGPKAQVEAIQKY (A(291)Y) responsible for this activity.[79] The epitope was fused to other helminth vaccine antigens, including *S. mansoni* Sm-TSP-2 and *N. americanus* Na-GST-1 and antibodies against both chimeras neutralized the enzymatic activity of Na-APR-1. This observation indicates that multivalent helminth vaccines are a possibility. The driver for this ambition is discussed in reference 80.

Metalloproteases

A zinc metalloendopeptidase cDNA (Ac-mep-1) was cloned from *A. caninum* adult hookworms[81] with a predicted molecular weight of 98.7 kDa and contained predicted zinc-binding domains (HExxH and ENxADxGG), signature sequences of the Neutral

Endopeptidase 24.11 (neprilysin) family of enzymes.[81] The protease had sequence homology to the MEPs from *H. contortus* described above and was only expressed in adult worms. Moreover, the protease was localised to the microvilli of the worm intestine indicative of a role in digestion of the parasite blood meal.

Haemoglobin Digestion and Protease Substrate Specificity

Hookworms frequently reach the gut of nonpermissive hosts but fail to successfully feed, develop and reproduce.[82] As it has been suggested that host-parasite co-evolution compromised the ability of hookworms to feed in nonpermissive hosts, the digestion of human and dog haemoglobin by baculovirus-expressed cathepsin D-like protease from the canine hookworm *A. caninum* (Ac-APR-1) and its counterpart from the human hookworm *N. americanus* (Na-APR-1) was compared. Both cleaved human and dog haemoglobin in vitro but each were more efficient at degrading the protein from the permissive host and cleavage occurred at specific sites. Intriguingly, both proteases have identical residues lining their active site clefts but showed differing substrate specificities. Moreover, Williamson et al[83] described the cloning and functional expression of an aspartic protease, Na-APR-2, from the human hookworm *N. americanus* which they noted was more similar to a family of nematode-specific, aspartic proteases (nemepsins). The recombinant protein cleaved human hemoglobin and serum proteins almost twice as efficiently as the same substrates from the dog. Combined, these studies suggest that hookworm aspartyl proteases contribute to host specificity. An extension of these studies,[84] showed that Ac-APR-1 and Na-APR-1 were expressed in the L3 infective stage as well as adult worms. Both proteases degraded skin macromolecules and serum proteins with an efficiency reflecting host-specificity. This study also showed that despite having active site clefts comprised of identical primary sequences, residues in the S3 pocket adopted different conformations which may account for the differing substrate preferences. Antisera against both proteases partially inhibited (16-26%) migration of hookworm L3 through hamster skin in vitro, further implying a connective tissue invasive role for these enzymes in addition to digestion of serum and erythrocyte proteins for nutrition.

Haemoglobin digestion in *Plasmodium* is undertaken by at least three different mechanistic classes of enzyme in an ordered manner.[85] Initial haemoglobin degradation is undertaken by aspartic proteases followed by degradation into smaller peptides by cysteine proteases and then metalloproteases in sequence. Exopeptidases complete the digestion to constituent amino acids. The preceeding descriptions of blood meal digestion in *Haemonchus* reinforces the view that this degradation pathway is conserved in blood-feeding nematodes.[6] Catalytically active forms of Na-APR-1, a cysteine protease Na-CP-3 and the metalloprotease Na-MEP-1 were used to assess their roles in haemoglobin digestion by adult *N. americanus*.[86] Recombinant Na-APR-1 cleaved intact human haemoglobin. In contrast, Na-CP-3 and Na-MEP-1 could not cleave intact haemoglobin but did cleave globin fragments released by hydrolysis with Na-APR-1, data the authors considered implying an ordered process of hemoglobinolysis. Multiple (74) cleavage sites within haemoglobin alpha- and beta-chains were identified after digestion with all 3 proteases and the proteases demonstrated promiscuous subsite specificities within haemoglobin. Aromatic and hydrophobic P1 residues and hydrophobic P1' residues were preferred by Na-APR-1 and hydrophobic P1 residues by Na-MEP-1.

Proteases and Establishment in the Host

Metalloproteases have been implicated in host penetration by hookworm infective L3s including skin penetration and tissue migration.[68,69] Metalloprotease secretion was initiated when third-stage infective larvae were stimulated to feed in vitro, two distinct zones of proteolysis at 50 kDa and 90 kDa being detected by substrate gel electrophoresis.[68] These authors suggested that the specific release of proteases associated with the initiation of feeding serves an integral function in the transition of the free-living stage to parasitism. A protease associated with this activity, Ac-MTP-1, was cloned and showed sequence homology to members of a family of zinc metalloproteases called the astacins.[87] These proteases have a short N-terminal signal peptide indicative of secretion, a propeptide and a catalytic domain containing the characteristic zinc-binding region and "Met turn".[88] *Ac*-MTP-1 also contains a C-terminal epidermal growth factor-like domain and a CUB like domain. The precise function of *Ac*-MTP-1 is unknown but it may have a variety of functions by reference to the broader literature on astacins.[87] These include digestion and hatching, growth factor processing and pattern formation in embryos. They suggested that *Ac*-MTP-1 synthesis and its release in ES products may have a regulatory role during the developmental changes associated with the activation to parasitism and the subsequent molt to the L4 during infection. They also suggested that the protease may activate host TGF-ß molecules during infection which could stimulate parasite development directly, or determine tissue predilection site. Alternatively, Ac-MTP-1 may have a role in immune evasion such as the inhibition of neutrophil infiltration.[87] Whatever its precise function, the protease has been evaluated as a possible component of a hookworm vaccine. Hotez et al[89] vaccinated laboratory dogs with a recombinant fusion protein encoding Ac-MTP-1 and vaccinated and control dogs were then challenged by s.c. injection of 500 L3 of the canine hookworm *A. caninum*. The vaccinated dogs had IgG2 and IgE isotype responses specific to anti-Ac-MTP-1-fusion protein recognising a protein of the estimated apparent molecular weight of Ac-MTP-1 in activated L3 secretory products. The IgG2 titre had a statistically significant inverse association with intestinal adult hookworm burden and quantitative egg counts. These promising data have been reinforced by a vaccination study where Syrian Golden hamsters were vaccinated with the recombinant fusion proteins Ay-ASP-2 and Ay-MTP-1 from the infective larvae of the hookworm *Ancylostoma ceylanicum*.[90] This cocktail reduced worm burden and faecal egg counts by 36% and 59% respectively ($p<0.001$) and was beneficial for parameters of health including blood hemoglobin levels and body weight compared to each antigen or adjuvant alone. Taken together, these data suggest that combination of two or more antigens may present an effective vaccine development strategy to improve protection and/or disease symptoms in affected individuals. Adult *A. caninum* also secrete an astacin-like protease (Ac-MTP-2)[91]. The full-length cDNA encoded 233 amino acids and included a conserved Met-turn sequence (SXMHY) and zinc-binding signature sequence. However, this (GXXXEHXRXER) is truncated compared to other astacins (HHXXGXXHEXXRXDR). Enzymically active recombinant Ac-MTP-2 protein was inhibited by 1,10 phenanthroline and by Ac-TMP, a putative tissue inhibitor of metalloprotease that is abundant in adult ES products. Ac-MTP-2 was localized to the oesophageal glands uterus. The authors suggested that Ac-MTP-2 functions in the extracorporeal digestion of the intestinal mucosal plug lodged in the buccal capsule of the adult parasite.[91]

CONCLUSION

The above descriptions emphasise the potential of proteases as targets for vaccines or drugs to control blood-feeding helminths. Where comparisons are possible, there is a strong suggestion that native parasite proteases stimulate greater protective immunity than their recombinant counterparts. This has been particularly evident with H11 and H-gal-GP where the best recombinant protein vaccine efficacy obtained to date does not approach the levels obtained with native counterparts, even when cocktail vaccines are tested.[92] Likely reasons are the recombinants lack important conformational epitopes or they do not have appropriate posttranslational processing. Analyses have shown that the major N-linked glycans found in H11 had up to three fucose residues attached to chitobiose cores, structures which are predicted to be highly immunogenic in mammals.[22] These carbohydrates may contribute to the protective immune responses stimulated by vaccination with gut-expressed glycoproteins. Indeed, as noted earlier, a considerable component of the IgG response in lambs immunised with H-gal-GP is directed against the glycan component. Bacterial expression is likely to be inappropriate because the recombinant product will not be processed in the correct manner. Insect cell and yeast based expression systems have been used as alternatives and there are numerous examples of successful expression of functionally active, glycosylated helminth proteases in these systems. However, even this capability is not necessarily the solution. Functionally active recombinant versions of the H11 isoforms were expressed in baculovirus but did not stimulate protective immunity against *Haemonchus* in lambs (S.E. Newton, personal communication). Qualitative differences were evident in the antibodies induced by native H11 compared to the insect cell product, antibodies to the native protein being found coating the surface of the gut of worms recovered from vaccinated sheep but not in worms from lambs given the recombinant proteins. This may be due to inappropriate glycosylation or the glycan may mask protective epitopes. Further work is required to define the precise nature of the epitope(s) which are critical for the expression of protective immunity. Whatever the reason, the reduced efficacy of recombinant antigens compared to their native counterparts has implications for interpreting the outcome of recombinant vaccine trials. Antigens showing moderate efficacy as recombinant proteins may actually be highly effective in native form and the degree of protection observed in a trial has major implications for the further progression of an antigen in any vaccine production pipeline. One approach to the issue of antigen expression is to express nematode genes nematode-derived cell lines[92] or in *C. elegans.*[94] This has been achieved by microinjection of a plasmid construct carrying the gene of interest downstream of a suitable *C. elegans* promoter.[94] A cathepsin L from *Haemonchus* was expressed in an enzymically active, glycosylated form in under control of the promoter of the orthologous *C. elegans* cathepsin L gene. The recombinant protein carried a His tag which was exploited to purify the protein from liquid cultures by nickel affinity chromatography in sufficient quantity for a vaccine trial.[94] In more recent work, an isoform of H11 has been expressed using this system and is undergoing functional analysis including comparing the glycan structures with the native protein (Britton, unpublished data). Another approach is the use of Phage Display Libraries to identify and produce peptide phage "mimotopes" which mimic the shape of structural epitopes, including carbohydrate. The approach has been used to screen cathepsin L mimotopes of *Fasciola hepatica* in sheep. Seven positive clones which were recognized by rabbit antisera to cathepsin L1/L2 from *Fasciola,* each with sequences which could be mapped to the native protein, were isolated. Lambs vaccinated with some individual clones and some mixtures had reduced worm burdens and egg outputs compared to controls indicating that the approach may provide an alternative method for vaccine production.[95]

Perhaps the biggest technological shift impacting on helminth protease discovery and functional definition is the rapid advance in genome sequencing technologies. For the blood-feeding nematodes, sequencing initiatives have already revealed numerous, previously uncharacterized proteases. Bioinformatics analyses can be used to predict function and whether or not a protease is likely to be secreted but a key issue, in terms of identifying control targets, is whether the protease is essential for parasite survival. The abundance of potential protease targets being revealed by mass sequencing necessitates more rapid and specific methods to identify which gene products are practically useful as targets for control. One option is RNAi, a process which down-regulates the expression of a gene by transcript removal following the addition of gene-specific double-stranded RNA (dsRNA). The technique has been used widely in *C. elegans* to define gene function on the basis of phenotype.[96] RNAi appears to be very inefficient (<30% of genes susceptible) in *H. contortus*[97] but some genes are sensitive including, in current work, the H11-1 isoform (Samarasinghe, Knox and Britton, unpublished data). Knockdown of this gene transcript in L3 larvae prior to infection of lambs resulted in >50% reductions in worm burdens, egg output and H11 enzyme activity compared to controls. These data are encouraging at two levels. First, larvae exposed to dsRNA treatment retain their infectivity which enables follow-up in vivo testing and, second, that the phenotypic outcome of RNAi in vivo does correlate with a vaccination approach where the function of the target protein product is impaired by antibody or cell-mediated host immune responses. RNAi is in its infancy as a tool for use in the parasitic nematodes of man and livestock but it has the potential to allow precise screening of the gene datasets for control targets for vaccination or chemotherapy as well as defining gene function.

REFERENCES

1. Halton D. Nutritional adaptations to parasitism within the platyhelminthes. Int J Parasitol 1997; 27:693-704.
2. Thorson RE. Studies on the mechanism of immunity in the rat to the nematode, Nippostrongylus muris. Am J Hyg 1953; 58:1-15.
3. Thorson RE. Effect of immune serum from rats on infective larvae of Nippostrongylus muris. Exp Parasitol 1954; 3:9-15.
4. Thorson RE. Proteolytic activity in extracts of the esophagus of adults of Ancylostoma caninum and the effect of immune serum on this activity. J Parasitol 1956; 42:21-25.
5. Thorson RE. The stimulation of acquired immunity in dogs by injections of extracts of the esophagus of adult hookworms. J Parasitol 1956b; 42:501-504.
6. Williamson A, Lecchi P, Turk BE et al. A multi-enzyme cascade of hemoglobin proteolysis in the intestine of blood-feeding hookworms. J Biol Chem 2003; 279:35950-35957.
7. Timms AR, Bueding E. Studies of a proteolytic enzyme from Schistosoma mansoni. Br J Pharmacol Chemother 1959; 14:68-73.
8. Knox DP, Jones, DG. Studies on the presence and release of proteolytic enzymes (proteinases) in gastro-intestinal nematodes of ruminants. Int J Parasitol 1990; 20:243-249.
9. Cox GN, Pratt D, Hageman R et al. Molecular cloning and primary sequence of a cysteine protease expressed by Haemonchus contortus adult worms. Mol Biochem Parasitol 1990; 41:25-34.
10. Pratt D, Cox GN, Milhausen MJ et al. A developmentally regulated cysteine protease gene family in Haemonchus contortus. Mol Biochem Parasitol 1990; 43:181-191.
11. Pratt D, Armes LG, Hageman R et al. Cloning and sequence comparisons of four distinct cysteine proteases expressed by Haemonchus contortus adult worms. Mol Biochem Parasitol 1992; 51:209-218.
12. Boisvenue RJ, Stiff MI, Tonkinson LV et al. Fibrinogen-degrading proteins from Haemonchus contortus used to vaccinate sheep. Am J Vet Res 1992; 53:1263-1265.

13. Knox DP, Redmond DL, Jones DG. Characterization of proteinases in extracts of adult Haemonchus contortus, the ovine abomasal nematode. Parasitology 1993; 106:395-404.
14. Fetterer RH, Rhoads ML. The in vitro uptake and incorporation of hemoglobin by adult Haemonchus-contortus. Vet Parasitol 1997; 69:77-87.
15. Willadsen P, Kemp DH. Vaccination with 'concealed' antigens for tick control. Parasitol Today 1988 J; 4:196-198.
16. Munn EA, Greenwood CA, Coadwell WJ. Vaccination of young lambs by means of a protein fraction extracted from adult Haemonchus contortus. Parasitology 1987; 94:385-97.
17. Munn EA. A helical, polymeric extracellular protein associated with the luminal surface of Haemonchus contortus intestinal cells. Tissue Cell 1977; 9:23-34.
18. Newton SE, Munn EA. The development of vaccines against gastrointestinal nematode parasites, particularly Haemonchus contortus. Parasitol Today 1999; 15:116-122.
19. Smith SK, Smith WD. Immunisation of sheep with an integral membrane glycoprotein complex of Haemonchus contortus and with its major polypeptide components. Res Vet Sci 1996; 60:1-6.
20. LeJambre LF, Windon RG, Smith WD. Vaccination against Haemonchus contortus: performance of native parasite gut membrane glycoproteins in Merino lambs grazing contaminated pasture. Vet Parasitol 2008; 153:302-312.
21. Munn EA, Smith TS, Smith H et al. Vaccination against Haemonchus contortus with denatured forms of the protective antigen H11. Parasite Immunol 1997; 19:243-248.
22. Smith TS, Graham M, Munn EA et al. Cloning and characterization of a microsomal aminopeptidase from the intestine of the nematode Haemonchus contortus. Biochim Biophys Acta 1997; 1338:295-306.
23. Haslam SM, Coles GC, Munn EA et al. Haemonchus contortus glycoproteins contain N-linked oligosaccharides with novel highly fucosylated core structures. J Biol Chem 1996; 271:30561-30570.
24. Smith WD, Smith SK, Murray JM. Protection studies with integral membrane fractions of Haemonchus contortus. Parasite Immunol 1994; 16:231-41.
25. Longbottom D, Redmond DL, Russell M et al. Molecular cloning and characterisation of a putative aspartyl proteinase associated with a gut membrane protein complex from adult Haemonchus contortus. Mol Biochem Parasitol 1997; 88:63-72.
26. Redmond DL, Knox DP, Newlands G et al. Molecular cloning and characterisation of a developmentally regulated putative metallopeptidase present in a host protective extract of Haemonchus contortus. Mol Biochem Parasitol 1997; 85:77-87.
27. Smith SK, Pettit D, Newlands GF et al. Further immunization and biochemical studies with a protective antigen complex from the microvillar membrane of the intestine of Haemonchus contortus. Parasite Immunol 1999; 21:187-199.
28. Newlands GF, Skuce PJ, Knox DP et al. Cloning and characterization of a beta-galactoside-binding protein (galectin) from the gut of the gastrointestinal nematode parasite Haemonchus contortus. Parasitology 1999; 119:483-490.
29. Skuce PJ, Newlands GF, Stewart EM et al. Cloning and characterisation of thrombospondin, a novel multidomain glycoprotein found in association with a host protective gut extract from Haemonchus contortus. Mol Biochem Parasitol 2001; 117:241-244.
30. Newlands GF, Skuce PJ, Nisbet AJ et al. Molecular characterization of a family of metalloendopeptidases from the intestinal brush border of Haemonchus contortus. Parasitology 2006; 133:357-368.
31. Smith WD, Smith SK, Pettit D et al. Relative protective properties of three membrane glycoprotein fractions from Haemonchus contortus. Parasite Immunol 2000; 22:63-71.
32. Smith WD, Newlands GF, Smith SK et al. Metalloendopeptidases from the intestinal brush border of Haemonchus contortus as protective antigens for sheep. Parasite Immunol 2003; 25:313-323.
33. Smith WD, Skuce PJ, Newlands GF et al. Aspartyl proteases from the intestinal brush border of Haemonchus contortus as protective antigens for sheep. Parasite Immunol 2003; 25:521-530.
34. Redmond DL, Geldhof P, Knox DP. Evaluation of Caenorhabditis elegans glycoproteins as protective immunogens against Haemonchus contortus challenge in sheep. Int J Parasitol 2004; 34:1347-1353.
35. Jasmer DP, Mitreva MD, McCarter JP. mRNA sequences for Haemonchus contortus intestinal cathepsin B-like cysteine proteases display an extreme in abundance and diversity compared with other adult mammalian parasitic nematodes. Mol Biochem Parasitol 2004; 137:297-305.
36. Ranjit N, Jones MK, Stenzel DJ et al. A survey of the intestinal transcriptomes of the hookworms, Necator americanus and Ancylostoma caninum, using tissues isolated by laser microdissection microscopy. Int J Parasitol 2006; 36:701-710.
37. Ranjit N, Zhan B, Stenzel DJ et al. A family of cathepsin B cysteine proteases expressed in the gut of the human hookworm, Necator americanus. Mol Biocheml Parasitol 2008; 160:90-99.
38. Knox DP, Smith SK, Smith WD. Immunization with an affinity purified protein extract from the adult parasite protects lambs against infection with Haemonchus contortus. Parasite Immunol 1999; 21:201-210.

39. Redmond DL, Knox DP. Protection studies in sheep using affinity-purified and recombinant cysteine proteinases of adult Haemonchus contortus. Vaccine 2004; 22:4252-4261.
40. Skuce PJ, Redmond DL, Liddell S et al. Molecular cloning and characterization of gut-derived cysteine proteinases associated with a host protective extract from Haemonchus contortus.Parasitology 1999; 119:405-412.
41. Redmond DL, Knox DP. Protection studies in sheep using affinity-purified and recombinant cysteine proteinases of adult Haemonchus contortus. Vaccine 2004; 22:4252-4261.
42. Redmond DL, Knox DP. Further protection studies using recombinant forms of Haemonchus contortus cysteine proteinases. Parasite Immunol 2006; 28:213-219.
43. Karanu FN, Rurangirwa FR, McGuire TC et al. Haemonchus contortus: identification of proteases with diverse characteristics in adult worm excretory-secretory products. Exp Parasitol 1993; 77:362-371.
44. Karanu FN, Rurangirwa FR, McGuire TC et al. Haemonchus contortus: inter- and intrageographic isolate heterogeneity of proteases in adult worm excretory-secretory products. Exp Parasitol 1997; 86:89-91.
45. Shompole S, Jasmer DP. Cathepsin B-like cysteine proteases confer intestinal cysteine protease activity in Haemonchus contortus. J Biol Chem 2001:276:2928-2934.
46. Ruiz A, Molina JM, Njue A et al. Genetic variability in cysteine protease genes of Haemonchus contortus. Parasitology 2004; 128:549-559.
47. Rehman A, Jasmer DP. Defined characteristics of cathepsin B-like proteins from nematodes: inferred functional diversity and phylogenetic relationships. Mol Biochem Parasitol 1999; 102:297-310.
48. Jasmer DP, Mitreva MD, McCarter JP. mRNA sequences for Haemonchus contortus intestinal cathepsin B-like cysteine proteases display an extreme in abundance and diversity compared with other adult mammalian parasitic nematodes. Mol Biochem Parasitol 2004; 137:297-305.
49. Yatsuda AP, Bakker N, Krijgsveld J et al. Identification of secreted cysteine proteases from the parasitic nematode Haemonchus contortus detected by biotinylated inhibitors. Infect Immun 2006; 74:1989-1993.
50. Bakker N, Vervelde L, Kanobana K et al. Vaccination against the nematode Haemonchus contortus with a thiol-binding fraction from the excretory/secretory products (ES). Vaccine 2004; 22:618-628.
51. De Vries E, Bakker N, Krijgsveld J et al. An AC-5 cathepsin B-like protease purified from Haemonchus contortus excretory secretory products shows protective antigen potential for lambs. Vet Res 2009; 40:41.
52. Baig S, Damian RT, Peterson DS. A novel cathepsin B active site motif is shared by helminth bloodfeeders. Exp Parasitol 2002; 101:83-89.
53. Larminie CG, Johnstone IL. Isolation and characterization of four developmentally regulated cathepsin B-like cysteine protease genes from the nematode Caenorhabditis elegans. DNA Cell Biol 1996; 15:75-82.
54. Jasmer DP, Roth J, Myler PJ. Cathepsin B-like cysteine proteases and Caenorhabditis elegans homologues dominate gene products expressed in adult Haemonchus contortus intestine. Mol Biochem Parasitol 2001; 116:159-169.
55. Geldhof P, Molloy C, Knox DP. Combinatorial RNAi on intestinal cathepsin B-like proteinases in Caenorhabditis elegans questions the perception of their role in nematode biology. Mol Biochem Parasitol 2006; 145:128-132.
56. Dalton JP, Brindley PJ, Donnelly S et al. The enigmatic asparaginyl endopeptidase of helminth parasites. Trends Parasitol 2009; 25:59-61.
57. Oliver EM, Skuce PJ, McNair CM et al. Identification and characterization of an asparaginyl proteinase (legumain) from the parasitic nematode, Haemonchus contortus. Parasitology 2006; 133:237-244.
58. Geldhof P, Knox D. The intestinal contortin structure in Haemonchus contortus: An immobilised anticoagulant? Int J Parasitol 2008; 38:1579-1588
59. Pratt D, Boisvenue RJ, Cox GN. Isolation of putative cysteine protease genes of Ostertagia ostertagi. Mol Biochem Parasitol 1992; 56:39-48.
60. Geldhof P, Claerebout E, Knox D et al. Vaccination of calves against Ostertagia ostertagi with cysteine proteinase enriched protein fractions. Parasite Immunol 2002; 24:263-270.
61. Meyvis Y, Geldhof P, Gevaert K et al. Vaccination against Ostertagia ostertagi with subfractions of the protective ES-thiol fraction. Vet Parasitol 2007; 149:239-245.
62. Thorson RE. Proteolytic activity in extracts of the esophagus of adults of Ancylostoma caninum and the effect of immune serum on this activity. J Parasitol 1956:42:21-25.
63. Eiff JA. Nature of an anticoagulant from the cephalic glands of Ancylostoma caninum. J Parasitol 1966; 52:833-843.
64. Spellman GG Jr, Nossel HL Anticoagulant activity of dog hookworm. Am J Physiol 1971; 220:922-927.
65. Hotez PJ, Cerami A. Secretion of a proteolytic anticoagulant by Ancylostoma hookworms. J Exp Med 1983; 157:1594-1603.
66. Hotez PJ, Trang NL, McKerrow JH et al. Isolation and characterization of a proteolytic enzyme from the adult hookworm Ancylostoma caninum. J Biol Chem 1985; 260:7343-7348.
67. Hawdon JM, Jones BF, Perregaux MA et al. Ancylostoma caninum: metalloprotease release coincides with activation of infective larvae in vitro. Exp Parasitol 1995; 80:205-211.

68. Hotez P, Haggerty J, Hawdon J et al. Metalloproteases of infective Ancylostoma hookworm larvae and their possible functions in tissue invasion and ecdysis. Infect Immun 1990; 58:3883-3892.
69. Kumar S, Pritchard DI Secretion of metalloproteases by living infective larvae of Necator americanus. J Parasitol 1992; 78:917-919.
70. Culley FJ, Brown A, Conroy DM et al. Eotaxin is specifically cleaved by hookworm metalloproteases preventing its action in vitro and in vivo. J Immunol 2000; 165:6447-6453.
71. Dowd AJ, Dalton JP, Loukas AC et al. Secretion of cysteine proteinase activity by the zoonotic hookworm Ancylostoma caninum. Am J Trop Med Hyg 1994; 51:341-347.
72. Harrop SA, Sawangjaroen N, Prociv P et al. Characterization and localization of cathepsin B proteinases expressed by adult Ancylostoma caninum hookworms. Mol Biochem Parasitol 1995; 71:163-171.
73. Brinkworth RI, Brindley PJ, Harrop SA. Structural analysis of the catalytic site of AcCP-1, a cysteine proteinase secreted by the hookworm Ancylostoma caninum. Biochim Biophys Acta 1996:1298:4-8.
74. Loukas A, Bethony JM, Williamson AL et al. Vaccination of dogs with a recombinant cysteine protease from the intestine of canine hookworms diminishes the fecundity and growth of worms. J Infect Dis 2004; 189:1952-1961.
75. Ranjit N, Zhan B, Stenzel DJ et al. A family of cathepsin B cysteine proteases expressed in the gut of the human hookworm, Necator americanus. Mol Biochem Parasitol 2008; 160:90-99.
76. Hotez PJ, Ashcom J, Zhan B et al. Effect of vaccination with a recombinant fusion protein encoding an astacinlike metalloprotease (MTP-1) secreted by host-stimulated Ancylostoma caninum third-stage infective larvae. J Parasitol 2003; 89:853-855.
77. Loukas A, Bethony JM, Mendez S et al. Vaccination with recombinant aspartic hemoglobinase reduces parasite load and blood loss after hookworm infection in dogs. PLoS Med 2005; 2:e295.
78. Pearson MS, Bethony JM, Pickering DA et al. An enzymatically inactivated hemoglobinase from Necator americanus induces neutralizing antibodies against multiple hookworm species and protects dogs against heterologous hookworm infection. FASEB J 2009; 23:3007-3019.
79. Pearson MS, Pickering DA, Tribolet L et al. Neutralizing antibodies to the hookworm hemoglobinase Na-APR-1: implications for a multivalent vaccine against hookworm infection and schistosomiasis. J Infect Dis 2010; 201:1561-1569.
80. Hotez PJ, Bethony JM, Oliveira SC et al. A. Multivalent anthelminthic vaccine to prevent hookworm and schistosomiasis. Exp Rev Vaccines 2008; 7:745-752.
81. Jones BF, Hotez PJ. Molecular cloning and characterization of Ac-mep-1, a developmentally regulated gut luminal metalloendopeptidase from adult Ancylostoma caninum hookworms. Mol Biochem Parasitol 2002; 119:107-116.
82. Williamson AL, Brindley PJ, Abbenante G et al. Cleavage of hemoglobin by hookworm cathepsin D aspartic proteases and its potential contribution to host specificity FASEB J 2002; 16 :1458-1460.
83. Williamson AL, Brindley PJ, Abbenante G et al. Hookworm aspartic protease, Na-APR-2, cleaves human hemoglobin and serum proteins in a host-specific fashion. J Infect Dis 2003; 187:484-494.
84. Williamson AL, Brindley PJ, Loukas A. Hookworm cathepsin D aspartic proteases: contributing roles in the host-specific degradation of serum proteins and skin macromolecules. Parasitology 2003; 126:179-185.
85. Gluzman IY, Francis SE, Oksman A et al. Order and specificity of the Plasmodium falciparum hemoglobin degradation pathway. J Clin Invest 1994; 93:1602-1608.
86. Ranjit N, Zhan B, Hamilton B et al. Proteolytic degradation of hemoglobin in the intestine of the human hookworm Necator americanus. J Infect Dis 2009; 199:904-912.
87. Zhan B, Hotez PJ, Wang Y et al. A developmentally regulated metalloprotease secreted by host-stimulated Ancylostoma caninum third-stage infective larvae is a member of the astacin family of proteases. Mol Biochem Parasitol 2002; 120:291-296.
88. Bond JS, Beynon RJ. The astacin family of metalloendopeptidases. Protein Sci 1995; 4:1247-1261.
89. Hotez PJ, Ashcom J, Zhan B et al. Effect of vaccination with a recombinant fusion protein encoding an astacin-like metalloprotease (MTP-1) secreted by host-stimulated Ancylostoma caninum third-stage infective larvae. J Parasitol 2003; 89:853-855.
90. Mendez S, Zhan B, Goud G et al. Effect of combining the larval antigens Ancylostoma secreted protein 2 (ASP-2) and metalloprotease 1 (MTP-1) in protecting hamsters against hookworm infection and disease caused by Ancylostoma ceylanicum. Vaccine 2005; 23:3123-3130.
91. Feng J, Zhan B, Liu Y et al. Molecular cloning and characterization of Ac-MTP-2, an astacin-like metalloprotease released by adult Ancylostoma caninum. Mol Biochem Parasitol 2007; 152:132-138.
92. Coyne CP, Brake D. Characterisation of Haemonchus contortus-derived cell populations propagated in vitro in a tissue culture environment and their potential to induce protective immunity in sheep. Int J Parasitol 2001; 31:359-376.
93. Cachat E, Newlands GF, Ekoja SE et al. Attempts to immunize sheep against Haemonchus contortus using a cocktail of recombinant proteases derived from the protective antigen, H-gal-GP. Parasite Immunol 2010; 32:414-419.

94. Murray L, Geldhof P, Clark D et al. Expression and purification of an active cysteine protease of Haemonchus contortus using Caenorhabditis elegans. Int J Parasitol 2007; 37:1117-1125.
95. Villa-Mancera A, Quiroz-Romero H, Correa D et al. Induction of immunity in sheep to Fasciola hepatica with mimotopes of cathepsin L selected from a phage display library. Parasitology 2008; 135:1437-1445.
96. Kamath RS, Ahringer J. Genome-wide RNAi screening in Caenorhabditis elegans. Methods 2003; 30:313-321.
97. Geldhof P, Murray L, Couthier A et al. Testing the efficacy of RNA interference in Haemonchus contortus. Int J Parasitol 2006; 36:801-810.

CHAPTER 11

CYSTEINE PROTEASES FROM BLOODFEEDING ARTHROPOD ECTOPARASITES

Daniel Sojka,[1] Ivo M.B. Francischetti,[2] Eric Calvo[2] and Michalis Kotsyfakis*[,3]

[1]Laboratory of Vector Immunology, Institute of Parasitology, Biology Centre of the Academy of Sciences of the Czech Republic, Ceske Budejovice, Czech Republic; [2]Laboratory of Malaria and Vector Research, National Institute of Allergy and Infectious Diseases, National Institutes of Health, Rockville, Maryland, USA; [3]Laboratory of Genomics and Proteomics of Disease Vectors, Institute of Parasitology, Biology Centre of the Academy of Sciences of the Czech Republic, Czech Republic

**Corresponding Author: Michalis Kotsyfakis—Email: kotsyfakis@paru.cas.cz*

Abstract: Cysteine proteases have been discovered in various bloodfeeding ectoparasites. Here, we assemble the available information about the function of these peptidases and reveal their role in hematophagy and parasite development. While most of the data shed light on key proteolytic events that play a role in arthropod physiology, we also report on the association of cysteine proteases with arthropod vectorial capacity. With emphasis on ticks, specifically *Ixodes ricinus*, we finally propose a model about the contribution of cysteine peptidases to blood digestion and how their concerted action with other tick midgut proteases leads to the absorbance of nutrients by the midgut epithelial cells.

INTRODUCTION

Cysteine proteases are involved in different physiological procedures associated with the development and tissue homeostasis of hematophagous arthropod ectoparasites. Most of these proteases belong to the papain-like superfamily of cysteine proteases, while legumains/asparaginyl endopeptidases and caspases that play a role in mosquito biology have also been reported. Here we review the available information about cysteine protease activities in bloodfeeding arthropod ectoparasites with particular emphasis on their varied roles at the molecular level. The chapter is divided into three parts; cysteine protease activities in (i) ticks, (ii) mosquitoes and (iii) other bloodfeeding arthropods.

Cysteine Proteases of Pathogenic Organisms, edited by Mark W. Robinson and John P. Dalton.

CYSTEINE PROTEASES OF TICKS

Ticks are mites (Acari) classified to the suborder Ixodida. Approximately 850 species described worldwide are represented by two major tick families, Ixodidae (hard ticks) and Argasidae (soft ticks).[1] Both groups are obligate bloodfeeding ectoparasites of terrestrial vertebrates and are important vectors of human and animal diseases, transmitting viruses, bacteria (notably *Rickettsia*), protozoa and fungi.[2] Ticks express cysteine peptidases with important roles in two major physiological events that are crucial to the ectoparasitic lifestyle: digestion of host blood and embryogenesis.

Blood feeding and digestion are essential events for ticks. Vertebrate blood is a rich source of nutrients for energy metabolism to support their demanding activities. There are remarkable differences in bloodfeeding strategies between hard and soft ticks.[3] In soft ticks, nymphs and adults of both sexes feed rapidly and drop off the host within half an hour to hours. Feeding and oviposition of female soft ticks are repeated and feeding performance is not related to mating status. Virgin females may conserve nutrients upon fertilization that can occur in up to 150-200 days in *Ornithodoros moubata.*[4] Hard ticks feed only once in each life-cycle stage. Adult females take in a single large blood meal, lay down a single large clutch of eggs and then die.[5] The process of female feeding lasts for several days and consists of the slow feeding period (6-9 days) followed by rapid engorgement (12-24 hrs prior to detachment)[6] in mated females.[7,8]

Despite being bloodfeeding arthropods, ticks are believed to digest blood proteins intracellularly, within the endosomes of their gut cells. Thus, the gut lumen serves as their nutrient storage organ.[9] Digestive gut cells use both receptor-mediated and fluid-phase endocytosis to take up the liquid blood meal from the gut lumen.[5] Proteolytic degradation occurs in the endosomal pH environment, well below the pH 6.3-6.5 of the gut contents.[7] The acidic optimum of proteolysis in tick gut cells is around pH 3.0[10] and is consistent with the activity range of specific cysteine and aspartic protease activities in the midgut, as shown with specific substrates and inhibitors in *Boophilus (Riphicephalus) microplus*.[11]

Tick Cathepsin-Like Cysteine Proteases

Purified enzymes/activities and genes encoding for cysteine and aspartic proteases, both putatively involved in blood digestion, have been reported individually from several tick species in the last decade. Cathepsin L from *B. microplus* (BmCL1) was identified in 2000.[12] Later, RT-PCR and western blot profiling using antibodies against the *Escherichia coli*-expressed recombinant BmCL1 confirmed that the enzyme is solely expressed in the gut of tick feeding stages. BmCL1 is expressed as a 42 kDa precursor protein with a 23 kDa mature peptidase domain. Immunogold electron microscopy localized the antigens to secretory vesicles of gut epithelial cells; however, localization has been proposed from the gut ultrastructure without any additional histochemical markers.[13] Predominant gut cysteine peptidases of 40 kDa and 48 kDa were identified in another hard tick species, *Haemaphysalis longicornis*, by performing inhibition studies against the gelatin substrate.[14] Two mRNAs encoding proteins with homology to cathepsin L were also identified by RT-PCR using cDNA isolated from *H. longicornis* gut.[15] Recently, a gut-associated *H. longicornis* cathepsin L (HlCPL-A) was prepared as recombinant enzyme in *E. coli*, purified as both zymogen and mature enzyme and functionally characterized.[16] Expression of the HlCPL-A gene was found to be up-regulated during feeding on vertebrate hosts. Recombinant HlCPL-A showed an optimum activity at pH 3.6 against both fluorescent peptidyl and bovine haemoglobin substrates.[16] Moreover, an individual enzyme classed as a cathepsin B, termed "longipain", was described from *H. longicornis* and proposed to have a critical role in the transmission of *Babesia spp.*[17] According to the authors, the longipain message is solely expressed in the gut tissue and up-regulated during tick feeding. Antibodies recognize the 39 kDa mature domain of the enzyme in immunoblots. Longipain appears to be expressed in the lysosomal vacuoles and gut cell surface and possesses the "occluding loop" that is typical of cathepsin B enzymes and necessary for its exopeptidase activity. Longipain expressed in *Pichia pastoris* showed unusually different pH optima for ZFR-AMC and ZRR-AMC peptidyl substrates. RNAi knockdown of longipain demonstrated its role in blood digestion and transmission of *Babesia sp.*[17]

Tick Asparaginyl Endopeptidases

The CD clan member *Ixodes ricinus* asparaginyl endopeptidase/legumain (IrAE) is the first member of its class described in arthropods.[18] It was cloned from a gut cDNA library of partially engorged females. Semi-quantitative RT-PCR profiling of IrAE mRNA levels in different tissues from feeding female ticks revealed that the enzyme message was expressed specifically in the gut. Immunofluorescence microscopy with specific serum raised against the *E. coli*-expressed recombinant zymogen localized IrAE in the digestive vesicles of gut cells and to the peritrophic matrix. Expression of the IrAE protein on the gut cell surface was further confirmed by immunogold electron microscopy. Recombinant IrAE produced in *P. pastoris* resembled its ortholog from the helminth parasite *Schistosoma mansoni* (SmAE) in its ability to auto-catalytically activate and to process schistosomal cathepsin B to its mature form. It has a strict specificity for asparagine at P1 and both recombinant and native enzymes show a drop of activity dependent on pH values (the enzyme loses its activity at pH values greater than 6.0).[18] Two comparative studies on the specific inhibition of IrAE and SmAE have been published. A set of novel selective legumain-specific inhibitors, the aza peptide Michael acceptors and epoxides, show very similar preferences for both the helminth and tick enzymes, suggesting their similar substrate preferences/enzymatic roles.[19,20] Two different legumains/AEs have been also described in *H. longicornis*; HlLgm1[21] and HlLgm2.[22] Based on the collective findings of the authors, the HlLgm1 and HlLgm2 expression profiles[23] are associated with the tick gut (where they have also been localized by immunohistochemistry) suggesting a role in blood feeding. The authors subsequently expressed an active recombinant enzyme in *E. coli*. However, some of the biochemical properties of the recombinant protein, including a pH optimum of about pH 8.0, were quite unusual for legumains and most likely will require further investigation. In a more recent analysis,[24] RNAi knockdown of HlLgm1 and HlLgm2 gene expression resulted in tick rejection from the host before the ticks reached repletion; a significant reduction in engorged tick body weight was also observed. Silencing legumain gene expression by RNAi also had a major impact on embryogenesis by delaying oviposition (characterised by a reduced number of eggs and structurally deformed eggs that failed to hatch). According to Alim et al,[24] HlLgm1 and HlLgm2 play a role in gut cell proliferation and morphology as demonstrated by the disruption of developmental events within the midgut cells and certain damage of midgut tissue in legumain gene-silenced ticks.

Mechanism of Haemoglobin Digestion by Tick Proteases

Based on the numerous individual reports about tick digestive cysteine peptidases and keeping in mind that gut-related aspartic[25,26] and leucine[27] aminopeptidases have been reported from ticks, it is evident that degradation of host proteins in tick gut cells is a complex process that is not performed by a sole "hemoglobinase" enzyme activity. Rather, the digestive process relies on a network of gut-associated cysteine and aspartic proteases analogous to what has been well studied in *S. mansoni*[28,29] and other flatworm parasites.[30] The digestive network consists of key cysteine endopeptidases of the papain family (clan CA) cathepsins B, C and L, asparaginyl endopeptidase (clan CD) and the aspartic peptidase cathepsin D (clan AA). Furthermore, the presence of a multienzyme complex of digestive cysteine and aspartic peptidases was supported by analysis of the first tick gut expressed sequence tag (EST) project by Anderson et al.[31] The analysis of 1,679 ESTs from midgut cDNA libraries of *Dermacentor variabilis* female ticks (the mialome, as it was named) at varying stages of feeding was performed. Fourteen different cysteine peptidase transcripts were found; of the 24 total cysteine peptidase ESTs, 19 ESTs were found in the 6-day-fed female guts, whereas only 5 ESTs were found in the 2-day-fed library, further confirming that expression of midgut cysteine peptidases is induced relatively late during hard tick feeding, in agreement with previous reports on individual cysteine peptidases. Phylogenetic analysis of these sequences identified three major groups of cysteine peptidases: legumain-like, cathepsin B-like and cathepsin L-like.[31]

To fully address the hypothesis that a conserved multienzyme network, comprised of cysteine and aspartic peptidases, functions to digest host blood in ticks and helminthes, a wide screening for the (i) expression and (ii) activity of schistosomal enzyme homologues was performed. A single tick species at a certain bloodfeeding stage was selected; the partially engorged females of *I. ricinus,* the major Lyme disease vector in Europe. (i) Reverse genetics/PCR-based screening with gut cDNA template and primers that amplify conserved protease domains revealed expression of mRNA sequences encoding cathepsins B, C and L (IrCB1, IrCC1, IrCL1), the aspartic peptidase (IrCD1) and the asparaginyl endopeptidase (IrAE1).[18,26] RT-PCR profiling revealed their association to gut tissue and, importantly, their simultaneous expression in tick gut cells upon feeding. (ii) Recent biochemical screening uncovered the entire enzymatic core of hemoglobinolysis in the gut cells of *I. ricinus* feeding females.[32] The activity-based profiling employed enzyme-specific substrates and inhibitors to define endogenous peptidase activities in tick gut tissue extracts (GTE). The overall hemoglobinolytic activity measured in GTE during feeding is highly elevated from the 6th day after attachment to the host, which corresponds to partially engorged ticks. Authentic enzyme activities associated to coding sequences for IrCB1, IrCL1, IrCC1, IrAE1 and IrCD1[26] were clearly detected and their optimal pH working conditions were estimated. Their exclusive role in tick gut hemoglobinolysis was confirmed by zero hemoglobin degradation in assays when a set of inhibitors targeting the full spectrum of the identified peptidases was applied. Moreover, all the endogenous forms of the described enzymes were visualized by activity-based probes on SDS-PAGE gels. Their dominant abundance in GTE was also confirmed by mass spectrometry. For hemoglobinolytic studies, bovine haemoglobin was incubated with GTE in the presence of various protease-specific inhibitors. Haemoglobin fragments were purified and analyzed by mass spectrometry. A specific haemoglobin cleavage map with marked cleavage sites for each studied protease type was created.[32] Collectively, the large amount of data obtained from the reverse genetics approach and the biochemical

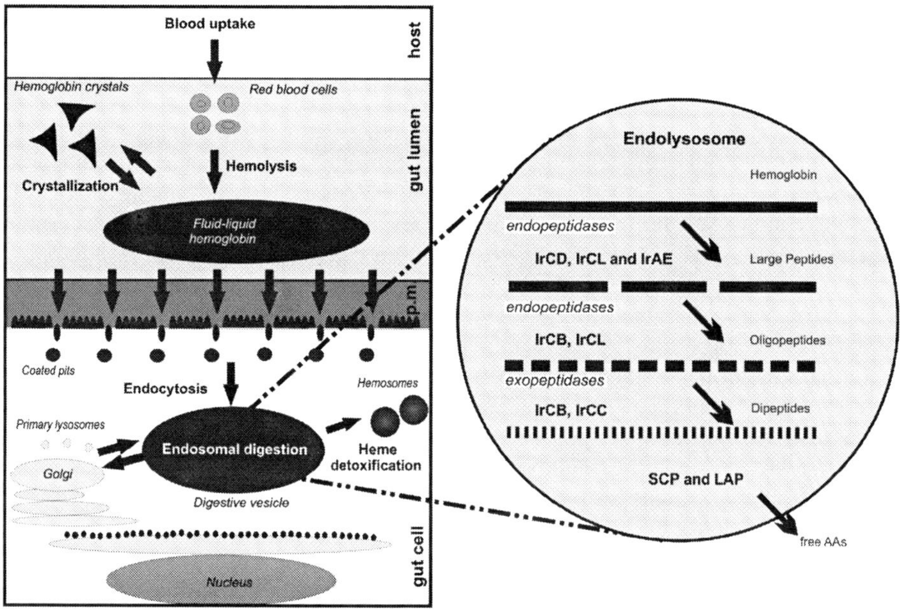

Figure 1. Schematic image of the uptake and digestion of host hemoglobin by *Ixodes ricinus* female ticks. In focus: the model of the proteolytic degradation cascade of host hemoglobin in endosomes of the gut cells as proposed by Horn et al[32] with marked positions of the cysteine and aspartic peptidases identified from the gut cDNA.[26]

mapping of tick gut hemoglobinolysis has resulted in the proposal of a mechanistic model for hemoglobinolysis in *I. ricinus* midgut cells (Fig. 1). Upon fusion of endosomes with lysosomes that carry the proteolytic machinery, IrCD endopeptidases, supported by IrCL and IrAE are responsible for the primary globin cleavage. Production of small secondary fragments is performed primarily by the IrCB. The endopeptidase-cleaved peptides are further processed by the exopeptidase (carboxy-dipeptidase) activity of IrCB and (amino-dipeptidase) IrCC. Single amino acids are then released from the resulting dipeptides by leucine or serine monopeptidases.[32]

Most of the research reviewed here contributes to the understanding of the function of the digestive cysteine proteases in hard ticks. Significantly less has been reported about the soft ticks, although the presence of similar enzyme activities, namely cathepsin B, C and legumain/AE, has been reported in *O. moubata.*[33]

The Role of Cysteine Proteases in Tick Embryogenesis

The second major physiological event in the tick body where cysteine proteases have been reported to play a role is embryogenesis, specifically the degradation of yolk proteins. During embryonic development, ticks rely on yolk reserve substances, mainly vitellin. Degradation is limited to the acid environment of the yolk spheres (specialized organelles that are considered as lysosomal-like organelles) since they contain both substrate (vitellin) and the degradative machinery. There is limited data on the particular enzymes involved in vitellin processing. Except for the reported yolk-degrading aspartic peptidases, the yolk proteolytic activity has been essentially attributed to a cathepsin L-like enzyme through substrate and inhibitor specificity studies. In the soft tick *O. moubata*, the activity has been demonstrated as two bands of 37 kDa and 39 kDa on gelatin gels with maximum degradation efficiency at pH 3-4. At neutral pH, the protein binds vitellin.[34] In a later study it was shown that the enzyme is stored as an inactive zymogen that can be activated by a low pH environment.[35] However, molecular or biochemical identification of the specific protease activity, e.g., based on cDNA or mass spectrometry data, is still missing. This mechanism seems not to be soft tick specific, as a vitellin-degrading cysteine endopeptidase (VTDCE) has been identified from eggs of the hard tick *B. microplus*, as well.[36] Purified endogenous VTDCE is tightly associated to vitellin, is inhibited by E64 and appears as a dimer of 17 kDa and 22 kDa. Vaccination of bovines with purified VTDCE induces a partial protective immune response against *B. microplus* infestation.[37] A further vitellin-degrading enzyme (RmLCE) was purified and identified from larval extracts of *B. microplus* larvae.[38] It has been proposed that VTDCE is a maternally-derived protease with restricted specificity and that RmLCE is synthesized in larvae to complete hydrolysis of the remaining vitellin peptides. Such an enzyme might help tick survival by providing amino acids for protein catabolism enabling larval survival until the first blood meal is achieved.

Unlike the digestive machinery, the yolk-processing cascades have remained conserved among bloodfeeding insects and ticks. Based on the similarities with gut cells, i.e., receptor-mediated endocytosis followed by subsequent endosomal degradation of vitellin, we speculate on the existence of a similar degradation enzyme matrix of cysteine and aspartic peptidases in guts and eggs/larvae. Different enzyme isoforms, possibly mutated duplicates of ancestral single-copy genes, encoding peptidases with varying substrate specificity for vitellin should be identified and their specific role confirmed by expression profiling, RNAi knock-down and proteomic analyses.

CYSTEINE PROTEASES IN MOSQUITOES

Cysteine Proteases and Mosquito Embryogenesis

An unusual cathepsin-B-like cysteine protease has been cloned and characterized from the yellow fever- and malaria-transmitting mosquito *Aedes aegypti*.[39] It is named vitellogenic cathepsin B (VCB) and it is produced in the fat body of vitellogenic female mosquitoes after blood feeding. Exclusively secreted by their fat body as a latent proenzyme, it is stored in yolk bodies of the developing oocytes in the inactive proenzyme form of 44 kDa.[39] The kinetics of its secretion by the vitellogenic fat body are similar to those of the yolk protein precursors and its synthesis does not take place in the ovaries. Instead, the protein is taken up by the ovaries through endocytosis, again similar to the yolk protein precursors. Upon embryogenesis it is processed to a mature 33 kDa VCB that is active against Z-Arg-Arg-pNA, a cathepsin B-specific substrate. The fat body-secreted, hemolymph form of VCB cannot be activated by acidic pH alone through autocatalysis, but acidic pH was sufficient for the activation of VCB in ovarian extracts, suggesting a physiological mechanism that ensures uptake of VCB by the oocytes in the latent, inactive proenzyme form and ensures its activation only in the mosquito embryo.[39] There, VCB degrades vitellogenin, the major yolk protein precursor. Vitellogenin degradation is inhibited by the thiol protease inhibitor E-64, further confirming the involvement of cysteine protease activity. Finally, the addition of an anti-VCB antibody to the embryonic extract prevented cleavage of vitellogenin, strongly indicating that VCB is involved in embryonic degradation of vitellin in vivo. In conclusion, VCB, similar to other key proteins in mosquito embryogenesis, is produced by an extraovarian tissue (fat body) and is endocytosed by the follicular cells, where it accelerates yolk protein degradation, a key event for the production of nutrients for the developing embryo.[39] An orthologous or homologous cathepsin in the mosquito *Anopheles gambiae* shows a blood-inducible pattern of expression in the fat body by quantitative real-time PCR (qPCR) analysis with a maximum accumulation of 35- to 74-fold, 24 hrs. after every blood meal.[40]

Finally, a study on the role of cathepsins in the development and maturation of the ovaries of the mosquito *Culex pipiens pallens* showed that cathepsin B- and L-like proteinases gradually accumulate in the developing ovaries after a blood meal and that they require more than 10 min of preincubation under acidic conditions to reach their maximum activities.[41] In contrast, homogenates of degenerating follicles, 3 days after blood feeding, display proteolytic activities without any acid treatment, suggesting that these proteases are already active in the mosquito ovaries at that time point and that they remained as inactive forms in the mosquito ovaries immediately after mosquito blood feeding. Chemical and immunohistochemical analyses also showed that, 3 days after blood feeding, more proteinases were located in the cytoplasm, rather than being associated with yolk granules.[41] In conclusion, these ovarian cysteine proteases from *C. pipiens*, apart from being activated at the onset of embryogenesis, are also activated during oogenesis, presumably to enhance oosorption, i.e., degeneration of some follicles in the developing ovaries. In this way, oosorption may enable females to recycle nutrients to developing oocytes when nutrient availability is poor.

Mosquito Caspases

The development and tissue homeostasis of *Aedes aegypti* is also safeguarded by the regulated activity of its caspases, key enzymes that serve as the initiators and executioners of apoptosis throughout the animal kingdom.[42] These enzymes fall into two categories depending on the size of their prodomain. Those caspases containing long prodomains (caspase recruitment or death effector domains) are called initiator caspases, because they activate the early apoptotic cascades. Those with short prodomains are called effector caspases, as they are involved in downstream apoptotic steps such as cleavage of the substrates/cellular proteins required to execute cell death (protein kinases, chromatin modifying enzymes and DNA repair proteins).[42] There is a key difference between *Drosophila* and vertebrates in the regulation of caspase activity; insect caspase activity is almost constitutive, abundant in all insect cells and inhibited by members of the inhibitors-of-apoptosis family (DIAP1: *Drosophila* inhibitor of apoptosis 1) in the cells that do not have to undergo apoptosis. In vertebrates, caspase activation requires the presence of death signals in the environment of the cells that will undergo apoptosis and members of the vertebrate inhibitors-of-apoptosis family contribute to apoptosis termination.[42] Although most information about insect caspases comes from studies in *Drosophila*, the cloning and functional characterisation of initiator caspases (named *Aedes* Dredd and *Aedes* Dronk, respectively) from the mosquito *Ae. aegypti* has been reported.[43,44] *Aedes* Dredd (*Ae*Dredd) contains two N-terminal death effector domains and the well-conserved caspase catalytic domain. Multiple sequence alignments, homology modeling and functional substrate assays of the recombinant protein suggest that *Ae*Dredd is an ortholog of *Drosophila* Dredd and human caspase-8, all key players in the death receptor-mediated apoptotic pathway.[43] qPCR analysis revealed low levels of *Ae*Dredd transcripts throughout the body of the mosquito and throughout its development. Transcript levels in callow pupae were five- to-six-fold higher than in other developmental stages. In adult mosquitoes, transcript levels were the highest in the fat body tissues, up to four-fold higher than the levels found in salivary glands and ovaries. *Ae*Dredd transcripts were also two-fold higher in midgut tissues when compared with those in salivary glands and ovaries.[43] Exposure of third instar larvae for 24 or 48 hrs to ecdysone did not significantly change *Ae*Dredd transcription, but 30 min exposure of adult midguts to ultraviolet light induced a six-fold higher *Ae*Dredd transcription level compared with the control.[43]

Aedes Dronc (*Ae*Dronc) is another initiator caspase in *Ae. aegypti* that is predicted to contain an N-terminal caspase recruitment domain and the conserved caspase subunit, which has a unique sequence, SIRCG, surrounding the catalytic Cys.[44] Molecular modeling, phylogenetics, sequence comparison analyses and its substrate specificity show that it is the mosquito homolog of *Drosophila* Dronc and human caspases-2 and -9. *Ae*Dronc appears to be involved principally in insect development; qPCR revealed that *Ae*Dronc transcript levels in third instar larvae were 26-fold higher than those in the early instar stages. Transcript levels in fourth instar larvae were 12-fold higher than those in the third instar larvae, while transcript levels in callow and black pupae were 126 and 53-fold higher, respectively, than those in the third instar larvae.[44] Thus, the highest levels of transcripts are detected in late instar larvae and early and late pupae, which corresponds to tissue reorganization events and pulses of the steroid hormone ecdysone. Indeed, a six-fold increase in *Ae*Dronc transcripts was observed after 24 hrs of exposure of third instart larvae to ecdysone and a 16-fold increase after 48 hrs of exposure. An analysis of total cellular protein from larvae treated with ecdysone showed an increase in both *Ae*Dronc-specific activity and the downstream, effector caspase-like activity. The *Ae*Dronc-specific activity increased at 24 hrs and remained relatively constant up to 48 hrs, whereas the more generic effector caspase-like activity steadily increased up to 48 h.[44] Overall these data demonstrate that exposure of third instar larvae to ecdysone results in a significant increase in both transcript levels and caspase activity, suggesting that *Ae*Dronc may play a role in the tissue reorganization that takes place during mosquito metamorphosis. *Ae*Dronc transcripts were also found in all adult tissues, with the highest levels detected in the fat body, a tissue with high cell turnover rates and the primary immune response tissue in insects.[44]

Recently, it was shown that *Ae*Dredd interacts with a mosquito FADD caspase adaptor, named *Aedes* FADD, that is required for antibacterial immunity in *Ae. aegypti*.[45] Moreover, caspase-like activity was detected in the invaded midgut cells of *Ae. aegypti* when infected with *Plasmodium gallinaceum*.[46] Currently, little is known about the biochemical pathways involved in controlling apoptosis in the malaria-infected mosquito midgut, but anacaspase-7 is one of the three apoptosis-related molecules that has been identified in cDNA libraries enriched in mosquito midgut sequences expressed immediately after midgut invasion by a malaria parasite.[47] Anacaspase-7 shares 40% identity and 60% homology with the *Drosophila* caspase DECAY and immunolocalization experiments revealed a putative protease present only in gut cells after invasion by large numbers of ookinetes. Increased apoptosis has also been detected in malaria-infected mosquito ovaries. Activation of a caspase-like molecule in *Anopheles gambiae* ovaries upon *Plasmodium yoelli nigeriensis* infection has been detected with FAM-VAD.fmk in approximately 25% of the ovarian follicles, but its molecular identity remains unknown.[48]

CYSTEINE PROTEASES IN OTHER BLOODFEEDING ARTHROPODS

Tsetse Fly Cathepsin B

A cathepsin B enzyme (GmCatB) is involved in protein digestion in the midgut of *Glossina morsitans morsitans* (tsetse fly), the vector of African trypanosomes.[49] The cDNA for GmCatB was shown as bloodmeal-induced in a subtractive suppression hybridization experiment and it encodes a protein of 340 amino acids with a predicted molecular mass of 38.2 kDa.[49] The first 19 amino acids of the protein are hydrophobic, suggesting a clear secretory signal peptide sequence at its N-terminus. Further structural analysis indicates that the following 65 amino acid residues (from position 20 to 84) correspond to the propeptide of the enzyme, demonstrating that the protease is expressed as an inactive zymogen.[49] This prodomain is likely proteolytically removed for the protease to be activated, resulting in a mature protein of 28.6 kDa. GmCatB transcription is constitutive and further induced throughout the digestion cycle within a few hours following ingestion of the first bloodmeal. It is also parasite-responsive, as its expression is two- to three-fold higher in trypanosome-infected flies, not only in their midgut but in their fat body as well.[49] Analysis of GmCatB transcripts showed high levels of expression in larvae and pupae, while in adult flies, although it was found to be preferentially expressed in the midgut, transcripts corresponding to GmCatB could be detected in the proventriculus and fat body tissues. No transcripts were detected in tsetse fly salivary glands.[49]

Cathepsin-Like Proteases of Triatomids

Using specific substrates and inhibitors, cathepsin B- and cathepsin L-like activities were also identified in the gut extracts of the bloodsucking bug *Triatoma infestans*, the vector of *Trypanosoma cruzi* (the etiologic agent of Chagas' disease).[50] More specifically, small intestine extracts from unfed bugs hydrolysed ZPhe-Arg-pNA; this activity was optimal at pH 5 and low but variable outside this pH range. It decreased during the first 2 days after feeding, but then increased to a maximum value at 5 and 10 days post feeding.[50] The activity was inhibited by both E-64 and CA-074, which are known chemical inhibitors of cysteine proteases, but inhibition was always less when using the specific cathepsin B inhibitor CA-074.[50] A cDNA encoding a cathepsin B-like proteinase (CatB1) was cloned, revealing an open reading frame of 996 nucleotides that encodes for a deduced protein of 332 amino acids with a predicted molecular weight of 36.3 kDa. The CatB1 protein contains a 16-residue secretory signal peptide, while a propeptide cleavage site was predicted between residues Thr69 and Leu70. The calculated mass of the mature protein is approximately 27 kDa, while the occluding loop, characteristic for cathepsin B-like enzymes, was found between the two Cys residues at positions 118 and 138. The deduced proenzyme sequence showed 48–60% identity to cathepsin-B enzymes from different arthropods and 60% identity to that from humans, while the lowest identity was observed with the fat body-cathepsin B of the dipteran *Ae. aegypti*.[50] The CatB1 gene was expressed at low, constitutive levels in unfed and fed *T. infestans* fifth instar larvae.[50] In addition, the cDNA encoding a cathepsin L-like proteinase from *T. infestans* (CatL1) was also cloned;[50] the open reading frame of 984 nucleotides encodes for a theoretical protein precursor with a molecular weight of 36.5 kDa. The CatL1 protein also possesses a 16-residue N-terminal signal peptide and its propetide contains 95 amino acids. As a result, the predicted mature enzyme consists of 217 amino acid residues and its estimated molecular weight is 23 kDa. The CatL1 precursor has 55–60% identity to cathepsin-L like enzymes from different insects and 68% identity to that of *Rhodnius prolixus*.[50] The three amino acid residues of the catalytic domain, CHN and the GCNGG motif were conserved in both cathepsins from *T. infestans*, but ERYNIN and KNFD motifs occurred only in the CatL1 sequence, defining it as a cathepsin L-like cysteine protease.[50]

Cathepsin B activity was present also in the midgut of *R. prolixus*, the main triatomine vector of Chagas' disease in Central America and the Andean region, not only before its feeding but also for 35 days after the blood meal. There was a 10-fold increase in this activity from unfed to 6 days post feeding of bugs and cathepsin B activity was higher in mated females than males.[51] Cathepsin B was localized to small, Golgi-derived vesicles in the intestinal cells of the small intestine and to lysosomes in cells of all midgut regions.[52] Apart from cathepsin B, cathepsin D activity has also been reported in the midgut of this insect based on the enzymatic activities of crude midgut extracts or their inhibition by specific inhibitors,[53,54] however, amino acid or nucleotide sequences of any of these proteases have yet to be identified. The cloning of a 1.2 kb cDNA encoding for a cysteine protease from *R. prolixus* (RpCat) that shows a high similarity to cathepsin L-like enzymes has also been reported.[55] The RpCat gene is expressed only in the midgut of adults, but not in the salivary glands or other tissues and in the 1st to 4th nymph instars, but not in 5th instar nymphal stages, indicating that its expression may be regulated by hormonal conditions that are modified before transformation into an adult.[55] The open reading frame of the cDNA encodes a zymogen of 316 amino acids that includes an N-terminal 99-residue propeptide. The enzyme has a C-terminus that could direct it to a secretory route, suggesting its participation in blood digestion.[55] The estimated molecular weights of the zymogen and the mature protease are 33.7 kDa and 23.8 kDa, respectively. The motif, ERFININ, that is characteristic for cathepsin-L like cysteine proteases was also identified in the prosegment region between amino acids 29 and 48.[55] Moreover, it has been demonstrated that insect cathepsin-like enzymes are involved in the proteolytic activation of canatoxin that results in production of entomotoxic peptide(s) involved in the deleterious effect of the protein in *R. prolixus*.[56] Canatoxin is a toxic protein isolated from the jackbean *Canavalia ensiformis*. The presence of canatoxin-like proteins in other leguminous seeds, including many edible ones, are suggestive of a physiological role, perhaps related to plant defense. Canatoxin toxicity has also been assessed in various insects, including *Manduca sexta*, *Schistocerca americana*, *Drosophila melanogaster*, *Ae. aegypti*, *R. prolixus* and *Callosobruchus maculatus*. Canatoxin was given in their diets and found to be lethal when fed to insects relying on cathepsins as their main digestive enzymes, such as *C. maculatus* and *R. prolixus*.[57] In contrast, insects with trypsin-based digestion (the other four tested) were not affected.[57]

Copepod Cysteine Proteases

Cysteine proteases are also present in the bloodfeeding parasitic copepod *Phrixocephalus cincinnatus*.[58] Substrate specificity, pH profile and inhibitor sensitivity indicate that the proteolytic activity found in these copepods could be partially attributed to cysteine proteases that they are mostly similar to mammalian cathepsins B, L and H.[58]

CONCLUSION

Cysteine peptidases play an important role in extracellular and intracellular protein degradation and processing in a wide range of organisms from bacteria to mammals. They have been studied in various species including bloodfeeding arthropods (Table 1). In ticks they have been shown to be the key digestive enzymes in the gut epithelium

Table 1. Cysteine proteases that have been identified and functionally characterized from bloodfeeding arthropods

Name	Species	Specificity	MW (kDa)	Tissue	Ref.
Ticks					
BmCL1	*Boophilus (Rhipicephalus) microplus*	Cathepsin L	42	Gut	12, 13
HlCPL-A	*Haemaphysalis longicornis*	Cathepsin L	29	Gut	16
IrCL1	*Ixodes ricinus*	Cathepsin L	30	Gut predominantly, also salivary glands, ovary, mal. glands	26, 32
IrCB1	*Ixodes ricinus*	Cathepsin B	32	Gut	26, 32
Longipain	*Haemaphysalis longicornis*	Cathepsin B	39	Gut	17
IrCC	*Ixodes ricinus*	Cathepsin C	23-25	Gut predominantly, also salivary glands, ovary, mal. glands	26, 32
IrAE	*Ixodes ricinus*	Legumain/AE	38-40	Gut	18, 32
HlLgm1	*Haemaphysalis longicornis*	Legumain/AE	38	Gut	21
HlLgm2	*Haemaphysalis longicornis*	Legumain/AE	36	Gut	22
Other arthropods					
Vitellogenic cathepsin B	*Aedes aegypti*	Cathepsin B	33	Adult ovaries, fat body	39
Aedes Dredd	*Aedes aegypti*	Initiator caspase	55 (Estimated, containing the death domains)	Abundant, higher expression in early pupae and adult fat body	43
Aedes Dronk	*Aedes aegypti*	Initiator caspase	52 (Estimated, containing the caspase recruitment domain)	Abundant, higher expression in late instar larvae, early and late pupae and adult fat body	44

continued on next page

Table 1. Continued

Name	Species	Specificity	MW (kDa)	Tissue	Ref.
GmCatB	*Glossina morsitans morsitans*	Cathepsin B	28.6 (Estimated)	Abundant, high levels in larvae and pupae, expressed in all adult tissues apart from the salivary glands	49
CatB1	*Triatoma infestans*	Cathepsin B	27 (Estimated)	Low, constitutive levels in unfed and fed fifth instar larvae	50
CatL1	*Triatoma infestans*	Cathepsin L	23 (Estimated)	Not known	50
RpCat	*Rhodnius prolixus*	Cathepsin L	23.8 (Estimated)	Only in adult midguts and in 1st to 4th nymph instars	55

The name, the species in which it was characterized, specificity, molecular weight of the mature enzyme (kDa) and tissue specificity are provided for each cysteine protease, as well as the corresponding reference in the chapter. 'Estimated' means that the gene sequence was used for molecular weight calculation, but there are no experimental data to support the calculation.

and to have specific roles in embryogenesis (i.e., degradation of yolk proteins). The evolutionarily older gut-associated cysteine and aspartic peptidase matrix, best described from bloodfeeding helminths, was conserved up to the insects and in Coleoptera, Diptera and Hemiptera, cysteine proteases have been considered as targets for pest control. However, these organisms have also evolved serine peptidase-based extracellular machinery not present in ticks. Cysteine proteases, such as cathepsin B and L, have also been shown to have important roles in embryogenesis and tissue remodelling during insect metamorphosis, while they can also participate in nutrient recycling and availability to the developing organism. Moreover, caspases have been proposed as important players in the apoptotic machinery of bloodfeeding arthropods. Finally, there is accumulating evidence that cysteine proteases may play a role in disease transmission as well, but this hypothesis remains to be tested as to whether it applies to pathogen transmission by different bloodfeeding arthropods.

ACKNOWLEDGEMENTS

This work was supported by grant No. Z60220518, grant No. IAA600960910, grant No. IAA600960811, and grant No. KJB600960911 from the Grant Agency of the Academy of Sciences of the Czech Republic and the Research Center No. LC06009 from the Ministry of Education, Youth and Sports of the Czech Republic. I.M.B.F., E.C. and M.K. were supported by the Intramural Research Program of the Division of Intramural Research, National Institute of Allergy and Infectious Diseases, National Institutes of Health. We thank NIAID intramural editor Brenda Rae Marshall for assistance and Dr Mark W. Robinson and Professor John P. Dalton for critical comments on the chapter.

CONFLICT OF INTEREST

The authors declare that there are no conflicts of interest regarding this manuscript.

REFERENCES

1. Nava S, Guglielmone AA, Mangold AJ. An overview of systematics and evolution of ticks. Front Biosci 2009; 14:2857-2877.
2. de la Fuente J, Estrada-Pena A, Venzal JM et al. Overview: Ticks as vectors of pathogens that cause disease in humans and animals. Front Biosci 2008; 13:6938-6946.
3. Grandjean O, Aeschlimann A. Contribution to the study of digestion in ticks: histology and fine structure of the midgut ephithelium of Ornithodorus moubata, Murray (Ixodoidea, Argasidae). Acta Trop 1973; 30:193-212.
4. Obenchain FD, Galun R. Physiology of Ticks, 1st ed. Oxford: Pergamon Press, 1982.
5. Harrison FW, Foelix RF. Chelicerate arthropoda. New York ; Chichester: Wiley; 1999.
6. Tarnowski BI, Coons LB. Ultrastructure of the midgut and blood meal digestion in the adult tick Dermacentor variabilis. Exp Appl Acarol 1989; 6:263-289.
7. Sauer JR, Hair JA. Morphology, Physiology and Behavioral Biology of Ticks. New York: Halsted Press, 1986.
8. Kaufman WR. Tick-host interaction: a synthesis of current concepts. Parasitol Today 1989; 5:47-56.
9. Sonenshine DE. Biology of Ticks. Oxford: Oxford University Press, 1991.
10. Reich CI, Zorzopulos J. Boophilus microplus: characterization of larval proteases. Exp Parasitol 1978; 44:1-6.
11. Zorzopulos J, Reich CI, Galassi N. Boophilus microplus: characterization of larval phosphomonoesterases and isolation of subcellular fractions with high phosphatase activity. Exp Parasitol 1978; 45:128-138.
12. Renard G, Garcia JF, Cardoso FC et al. Cloning and functional expression of a Boophilus microplus cathepsin L-like enzyme. Insect Biochem Mol Biol 2000; 30:1017 1026.
13. Renard G, Lara FA, de Cardoso FC et al. Expression and immunolocalization of a Boophilus microplus cathepsin L-like enzyme. Insect Mol Biol 2002; 11:325-328.
14. Mulenga A, Sugimoto C, Onuma M. Characterization of proteolytic enzymes expressed in the midgut of Haemaphysalis longicornis. Jpn J Vet Res 1999; 46:179-184.
15. Mulenga A, Sugimoto C, Ingram G et al. Molecular cloning of two Haemaphysalis longicornis cathepsin L-like cysteine proteinase genes. J Vet Med Sci 1999; 61:497-502.
16. Yamaji K, Tsuji N, Miyoshi T et al. Hemoglobinase activity of a cysteine protease from the ixodid tick Haemaphysalis longicornis. Parasitol Int 2009; 58:232-237.
17. Tsuji N, Miyoshi T, Battsetseg B et al. A cysteine protease is critical for Babesia spp. transmission in Haemaphysalis ticks. PLoS Pathog 2008; 4:e1000062.
18. Sojka D, Hajdusek O, Dvorak J et al. IrAE: an asparaginyl endopeptidase (legumain) in the gut of the hard tick Ixodes ricinus. Int J Parasitol 2007; 37:713-724.
19. Gotz MG, James KE, Hansell E et al. Aza-peptidyl Michael acceptors. A new class of potent and selective inhibitors of asparaginyl endopeptidases (legumains) from evolutionarily diverse pathogens. J Med Chem 2008; 51:2816-2832.
20. Ovat A, Muindi F, Fagan C et al. Aza-peptidyl Michael acceptor and epoxide inhibitors—potent and selective inhibitors of Schistosoma mansoni and Ixodes ricinus legumains (asparaginyl endopeptidases). J Med Chem 2009; 52:7192-7210.
21. Abdul Alim M, Tsuji N, Miyoshi T et al. Characterization of asparaginyl endopeptidase, legumain induced by blood feeding in the ixodid tick Haemaphysalis longicornis. Insect Biochem Mol Bio 2007; 37:911-922.
22. Alim MA, Tsuji N, Miyoshi T et al. HlLgm2, a member of asparaginyl endopeptidases/legumains in the midgut of the ixodid tick Haemaphysalis longicornis, is involved in blood-meal digestion. J Insect Physiol 2008; 54:573-585.

23. Alim MA, Tsuji N, Miyoshi T et al. Developmental stage- and organ-specific expression profiles of asparaginyl endopeptidases/legumains in the ixodid tick Haemaphysalis longicornis. J Vet Med Sci 2008; 70:1363-1366.
24. Alim MA, Tsuji N, Miyoshi T et al. Legumains from the hard tick Haemaphysalis longicornis play modulatory roles in blood feeding and gut cellular remodelling and impact on embryogenesis. Int J Parasitol 2009; 39:97-107.
25. Boldbaatar D, Sikalizyo Sikasunge C, Battsetseg B et al. Molecular cloning and functional characterization of an aspartic protease from the hard tick Haemaphysalis longicornis. Insect Biochem Mol Biol 2006; 36:25-36.
26. Sojka D, Franta Z, Horn M et al. Profiling of proteolytic enzymes in the gut of the tick Ixodes ricinus reveals an evolutionarily conserved network of aspartic and cysteine peptidases. Parasit Vectors 2008; 1:7.
27. Hatta T, Kazama K, Miyoshi T et al. Identification and characterisation of a leucine aminopeptidase from the hard tick Haemaphysalis longicornis. Int J Parasitol 2006; 36:1123-1132.
28. Caffrey CR, McKerrow JH, Salter JP et al. Blood 'n' guts: an update on schistosome digestive peptidases. Trends Parasitol 2004; 20:241-248.
29. Delcroix M, Sajid M, Caffrey CR et al. A multienzyme network functions in intestinal protein digestion by a platyhelminth parasite. J Biol Chem 2006; 281:39316-39329.
30. Williamson AL, Brindley PJ, Knox DP et al. Digestive proteases of blood-feeding nematodes. Trends Parasitol 2003; 19:417-423.
31. Anderson JM, Sonenshine DE, Valenzuela JG. Exploring the mialome of ticks: an annotated catalogue of midgut transcripts from the hard tick, Dermacentor variabilis (Acari: Ixodidae). BMC Genomics 2008; 9:552.
32. Horn M, Nussbaumerova M, Sanda M et al. Hemoglobin digestion in blood-feeding ticks: mapping a multipeptidase pathway by functional proteomics. Chem Biol 2009; 16:1053-1063.
33. Grunclova L, Horn M, Vancova M et al. Two secreted cystatins of the soft tick Ornithodoros moubata: differential expression pattern and inhibitory specificity. Biol Chem 2006; 387:1635-1644.
34. Fagotto F. Yolk degradation in tick eggs: I. Occurrence of a cathepsin L-like acid proteinase in yolk spheres. Arch Insect Biochem Physiol 1990; 14:217-235.
35. Fagotto F. Yolk degradation in tick eggs: II. Evidence that cathepsin L-like proteinase is stored as a latent, acid-activable proenzyme. Arch Insect Biochem Physiol 1990; 14:237-252.
36. Seixas A, Dos Santos PC, Velloso FF et al. A Boophilus microplus vitellin-degrading cysteine endopeptidase. Parasitology 2003; 126(Pt 2):155-163.
37. Seixas A, Leal AT, Nascimento-Silva MC et al. Vaccine potential of a tick vitellin-degrading enzyme (VTDCE). Vet Immunol Immunopathol 2008; 124:332-340.
38. Estrela A, Seixas A, Termignoni C. A cysteine endopeptidase from tick Rhipicephalus (Boophilus) microplus) larvae with vitellin digestion activity. Comp Biochem Physiol B Biochem Mol Biol 2007; 148:410-416.
39. Cho WL, Tsao SM, Hays AR et al. Mosquito cathepsin B-like protease involved in embryonic degradation of vitellin is produced as a latent extraovarian precursor. J Biol Chem 1999; 274:13311-13321.
40. Nirmala X, Marinotti O, James AA. The accumulation of specific mRNAs following multiple blood meals in Anopheles gambiae. Insect Mol Biol 2005; 14:95-103.
41. Uchida K, Ohmori D, Ueno T et al. Preoviposition activation of cathepsin-like proteinases in degenerating ovarian follicles of the mosquito Culex pipiens pallens. Dev Biol 2001; 237:68-78.
42. Cooper DM, Granville DJ, Lowenberger C. The insect caspases. Apoptosis 2009; 14:247-256.
43. Cooper DM, Pio F, Thi EP et al. Characterization of Aedes Dredd: a novel initiator caspase from the yellow fever mosquito, Aedes aegypti. Insect Biochem Mol Biol 2007; 37:559-569.
44. Cooper DM, Thi EP, Chamberlain CM et al. Aedes Dronc: a novel ecdysone-inducible caspase in the yellow fever mosquito, Aedes aegypti. Insect Mol Biol 2007; 16:563-572.
45. Cooper DM, Chamberlain CM, Lowenberger C. Aedes FADD: a novel death domain-containing protein required for antibacterial immunity in the yellow fever mosquito, Aedes aegypti. Insect Biochem Mol Biol 2009; 39:47-54.
46. Zieler H, Dvorak JA. Invasion in vitro of mosquito midgut cells by the malaria parasite proceeds by a conserved mechanism and results in death of the invaded midgut cells. Proc Natl Acad Sci USA 2000; 97:11516-11521.
47. Abraham EG, Islam S, Srinivasan P et al. Analysis of the Plasmodium and Anopheles transcriptional repertoire during ookinete development and midgut invasion. J Biol Chem 2004; 279:5573-5580.
48. Ahmed AM, Hurd H. Immune stimulation and malaria infection impose reproductive costs in Anopheles gambiae via follicular apoptosis. Microbes Infect 2006; 8:308-315.
49. Yan J, Cheng Q, Li CB et al. Molecular characterization of three gut genes from Glossina morsitans morsitans: cathepsin B, zinc-metalloprotease and zinc-carboxypeptidase. Insect Mol Biol 2002; 11:57-65.
50. Kollien AH, Waniek PJ, Nisbet AJ et al. Activity and sequence characterization of two cysteine proteases in the digestive tract of the reduviid bug Triatoma infestans. Insect Mol Biol 2004; 13:569-579.

51. Houseman JG, Downe AER. Activity cycles and the control of four digestive proteinases in the posterior midgut of Rhodnius prolixus Stål (Hemiptera: Reduviidae). J Insect Physiol 1983; 29:141-148.
52. Billingsley PF, Downe AER. Ultrastructural localisation of cathepsin B in the midgut of Rhodnius prolixus Stål (Hemiptera: Reduviidae) during blood digestion. Int J Insect Morph Embryol 1988; 17:295-302.
53. Terra WR, Ferreira C, Garcia ES. Origin, distribution, properties and functions of the major Rhodnius prolixus midgut hydrolases. Insect Biochem 1988; 18:423-434.
54. Houseman JG, Downe AER. Characterization of an acidic proteinase from the posterior midgut of Rhodnius prolixus Stål (Hemiptera: Reduviidae). Insect Biochem 1982; 12:651-655.
55. Lopez-Ordonez T, Rodriguez MH, Hernandez-Hernandez FD. Characterization of a cDNA encoding a cathepsin L-like protein of Rhodnius prolixus. Insect Mol Biol 2001; 10:505-511.
56. Ferreira-DaSilva CT, Gombarovits ME, Masuda H et al. Proteolytic activation of canatoxin, a plant toxic protein, by insect cathepsin-like enzymes. Arch Insect Biochem Physiol 2000; 44:162-171.
57. Carlini CR, Oliveira AE, Azambuja P et al. Biological effects of canatoxin in different insect models: evidence for a proteolytic activation of the toxin by insect cathepsinlike enzymes. J Econ Entomol 1997; 90:340-348.
58. Perkins PS, Haley D, Rosenblatt R. Proteolytic enzymes in the blood-feeding parasitic copepod, Phrixocephalus cincinnatus. J Parasitol 1997; 83:6-12.

CHAPTER 12

HOW PATHOGEN-DERIVED CYSTEINE PROTEASES MODULATE HOST IMMUNE RESPONSES

Sheila Donnelly,*[,1] John P. Dalton[2] and Mark W. Robinson[1]

[1] *i3 Institute (formerly IBID), University of Technology, Sydney, Ultimo, Australia;* [2]*Institute of Parasitology, McDonald Campus, McGill University, St. Anne de Bellevue, Quebec, Canada*

**Corresponding Author: Sheila Donnelly—Email: sheila.donnelly@uts.edu.au*

Abstract: In mammals, cysteine proteases are essential for the induction and development of both innate and adaptive immune responses. These proteases play a role in antigen- and pathogen-recognition and elimination, signal processing and cell homeostasis. Many pathogens also secrete cysteine proteases that often act on the same target proteins as the mammalian proteases and thereby can modulate host immunity from initial recognition to effector mechanisms. Pathogen-derived proteases range from nonspecific proteases that degrade multiple proteins involved in the immune response to enzymes that are very specific in their mode of action. Here, we overview current knowledge of pathogen-derived cysteine proteases that modulate immune responses by altering the normal function of key receptors or pathways in the mammalian immune system.

INTRODUCTION

It is the role of the immune system to synthesise and release effector molecules that detect, implement or orchestrate an appropriate response to a potential threat or danger signal. To regulate many aspects of this process, cells of the immune system utilise protease activity. One major group of cysteine proteases, termed cathepsins, are common constituents of the endolysosomal compartments of immune cells where they have critical roles in events such as antigen presentation and zymogen processing. Despite a widely held belief that the cathepsins were enzymes that function only at the acidic pH within the lysosomal compartment, it is now evident that they also operate in the cytosolic

Cysteine Proteases of Pathogenic Organisms, edited by Mark W. Robinson and John P. Dalton.

Table 1. Summary of the immuno-modulatory cysteine proteases of pathogenic organisms

Pathogen	Disease	Cysteine Protease	Observed Effect on Immune System
Helminth			
Fasciola hepatica	Fascioliasis	Cathepsin L1	Degrades endosomal TLR3 Cleaves hinge region of IgG
		Cathepsin L2	Cleaves hinge region of IgG
Necator americanus	Helminthiasis	Uncharacter-ised	IL-4 production by basophils
Paragonimus westermani	Paragonimiasis	Uncharacter-ised	Cleaves hinge region of IgG
Schistosoma mansoni	Schistosomiasis	Cathepsin B	Inhibits macrophage activation by Th1 stimulants Cleaves hinge region of IgG
Spirometra mansoni	Sparaganosis	Cathepsin S-like protease	Cleaves hinge region of IgG
Protozoan			
Entamoeba histolytica	Amebiasis	EhCp1,EhCp2, EhCp5	Converts Pro-IL-1ß to mature, active IL-1ß Cleaves Pro-IL-18 Degrades C5a and C3a of the complement pathway Cleaves hinge region of IgA1 and IgA2
Leishmania mexicana	Leishmaniasis	Cathepsin B	Induces Th2 cytokine responses Cleaves CD25 from T-cells Inhibition of NFκB binding to DNA
Plasmodium falciparum	Malaria	PfUCHL3	De-ubiquitinase activity—no specific target identified
		PfUCH54	
Toxoplasma gondii	Toxoplasmosis	TgUCHL3	De-ubiquitinase activity—no specific target identified
Trypanosoma cruzi	Trypanosomiasis	Cruzipain	Degradation of NFκB
			Cleaves hinge region of IgG
Bacteria			
Chlamydia trachomatis	Chlamydia	ChlaDub1	De-ubiquitinase activity—no specific target identified

continued on next page

Table 1. Continued

Pathogen	Disease	Cysteine Protease	Observed Effect on Immune System
		ChlaDub2	De-ubiquitinase activity—no specific target identified
Escherichia coli	Gastroenteritis	ElaD	De-ubiquitinase activity—no specific target identified
		StcE	Degrades the complement C1-inhibitor serpin
Porphyromonas gingivalis	Periodontal disease	Arg-specific gingipain	Cleaves CD14 from monocytes and fibroblasts
		Lys-specific gingipain	Cleaves IL-6, IL-1Ra, IL-12, TNFα, IFN-ϒ Cleaves IL-8 Activates complement C3, C4, C5
Prevotella intermedia	Periodontal disease	Interpain A	Activates the C1 complement complex
Streptococcus pyogenes	Pharyngitis	Streptopain (SpeB)	Converts Pro-IL-1β to mature, active IL-1β Degrades IgA, IgD, IgG, IgE and IgM Cleaves C3 complement protein
		IdeS	Cleaves hinge region of IgG
Salmonella sp	Salmonellosis	AvrA	De-ubiquitinates IκBα
Yersinia sp	Yersiniosis	YopJ	De-ubiquitinates TRAF2, TRAF3 and TRAF6 De-ubiquitinates IκBα
Virus			
Coronaviridae	Gastroenteritis, SARs, common cold	PLP domains	De-ubiquitinates IRF3
Herpes simplex virus 1	Oral and/or genital herpes	UL36	De-ubiquitinase activity—no specific target identified
adenovirus	Respiratory disease, Gastroenteritis	adenain	De-ubiquitinase activity—no specific target identified

and nuclear compartments of cells, as well as the extracellular space.[1] As a result of this widespread localisation, cathepsins can exert diverse effects on the development and regulation of immune responses by playing key roles in cytokine regulation, cell development, induction of apoptosis and influencing Toll-like receptor (TLR) signalling.[1,2]

Many pathogenic organisms synthesize proteases that resemble host cathepsins (Table 1), both functionally and structurally and have fundamental roles in the life-cycle and survival of the pathogen. For example, parasitic helminths (worms) secrete proteases in large quantities to degrade the extracellular matrix during host invasion,[3,4] intracellular protozoans require proteases for invasion, nutrition and exit from host cells,[5] bacterial cathepsins perform housekeeping roles like amino acid uptake and fimbriae maturation[6,7] and viral proteases are involved in processing of viral gene products.[8,9] A number of the target amino acid sequences for pathogen cathepsins are also present in key molecules involved in regulating the host immune response including immunoglobulins (Ig), cytokines, TLRs and ubiquitin (Ub). It has been suggested that adaptation of pathogen proteases to target these structures has occurred in an evolutionary process to increase pathogenicity.[2,10] This has been influenced by the unique biochemical challenges presented to each pathogenic organism and is linked, in part, to their localisation as extracellular or intracellular organisms and therefore their exposure to different cellular structures and macromolecules. Here, we review the functions that have been attributed to pathogen cysteine proteases in the evasion, suppression and modulation of the host immune response.

ADAPTIVE IMMUNE RESPONSES

The immunity and pathology occurring in response to infection with any pathogen is predominantly mediated by T-lymphocytes. In general, control of infection and healing is associated with a polarized Th1 type response whereas the induction of interleukin (IL)-4-dominated Th2 responses are largely suboptimal against a number of pathogens.

Induction of Th2 Immune Responses and/or Suppression of Th1 Responses

In certain protozoan infections, cysteine proteases are critical for the induction of Th2 type immune responses. Leishmania are obligate intracellular parasites that live as nonmotile amastigotes within cells of the mononuclear phagocyte lineage of their mammalian hosts. The outcome of infection in leishmaniasis is determined by the Th1 *versus* Th2 nature of the effector response with parasites successfully establishing infection by driving a Th2 immune response.[11] Inhibition of *Leishmania mexicana* cathepsin B, with a specific inhibitor, caused a switch in the polarisation of T-cell differentiation from a Th2 to a protective Th1 phenotype in mice.[12,13] In addition, mutants of *L. mexicana* lacking cysteine protease activity induced less IL-4 and IgE in BALB/c mice compared to wild type parasites.[14] Finally, the delivery of recombinant cysteine proteases derived from either *L. mexicana*, or *Trypanosoma cruzi*, elicited strong Th2-type responses typified by enhanced mRNA expression and production of Th2 cytokines (IL-4, IL-5 and IL-13) from the draining lymph nodes and also a polarised splenocyte response toward a Th2 bias in response to stimulation with anti-CD3.[15,16]

The ability of cysteine proteases to induce Th2-type immune responses is dependent on their enzymatic activity[12-17] and also on their substrate specificity as only cysteine proteases belonging to clan CA demonstrate an ability to promote the differentiation of

Th2.[18] Injection of the plant-derived cysteine protease papain to mice was shown to promote the activation of basophils, which then present antigen to CD4+ T-cells and induce Th2 cell responses through the secretion of IL-4 and thymic stromal lymphopoietin.[17] Therefore, it was suggested that basophils detect the presence of protease products which in turn activates these cells to induce a Th2 response.[18] This strategy of immune recognition is distinct from typical recognition by cells of the innate system since it is not dependent on the detection of pathogen molecular structures. Instead, it requires the detection of enzymatic activity associated with the presence of a pathogen or antigen. The immune outcome closely resembles the type of reaction that is triggered by helminths and has led to the suggestion that the innate immune system has evolved a detection mechanism based on sensing the abnormal protease activity associated with helminth infection.[18] This idea has support in studies showing that the secreted proteases of the hookworm *Necator americanus* induce Type 2 cytokine production by basophils.[19]

Another means by which cysteine proteases could modulate host immune responses towards a Th2 environment may be associated with their ability to inhibit Th1 immune responses. In support of this idea, it was shown that the predominant secreted product of the human and animal helminth pathogen *Fasciola hepatica*, a cathepsin L1 cysteine protease, suppressed the onset of protective Th1 immune responses in mice to infections with the respiratory microbe *Bordetella pertussis*, making them more susceptible to disease. In addition, injection of the cysteine protease immediately prior to immunization of mice with a *B. pertussis* whole-cell pertussis vaccine prevented the development of a Th1 response to the vaccine.[20-22] Studies with *L. mexicana* cathepsin B suggest that cleavage of CD25 from the surface of T-cells may prevent the development of Th1 responses.[14] Infection with the helminth *Schistosoma mansoni* may provide another example of how cysteine proteases suppress Th1 responses. The acute stages of schistosomiasis leads to the development of a weak Th1 response, but this switches to a potent Th2 response when female worms become fecund and release eggs that get trapped in host liver and intestinal tissues.[23] Secretions from the trapped eggs are responsible for the induction of this Th2 response. The major schistosome proteases, cathepsin B and cathepsin L, are expressed and secreted by the egg stage of the parasite[24] and, therefore, may facilitate this immune switching.

INNATE IMMUNE RESPONSES

Substantial evidence supports the view that cysteine proteases secreted by a range of pathogens specifically prevent cells of the innate immune response promoting Th1-adaptive immune responses.[10,25-27] The mechanisms are varied and depend on the substrate specificity of the protease and of the location of the pathogen within the host.

The innate immune system constitutes the first line of host defence during infection and plays a crucial role in the early recognition of invading pathogens. Unlike the adaptive immune response, the innate immune response is relatively nonspecific, relying on the recognition of evolutionarily conserved structures on pathogens, termed pathogen-associated molecular patterns (PAMPs), through a limited number of pattern recognition receptors (PRRs).[28] A number of different PRR families have been described but the best characterised is the family of TLRs. To date, 13 TLRs have been identified in humans and are distinguished by their recognition of distinct PAMPs derived from various pathogens.[29]

Interaction between a PAMP and its corresponding TLR, present either at the cell surface or intracellularly, leads to the recruitment of an adaptor molecule, followed by the activation of downstream signal transduction pathways.[28,29] These signaling pathways are classified on the basis of their utilization of different adaptor molecules, i.e., MyD88 or TIR domain-containing adaptor inducing IFNβ (TRIF) and, additionally, their respective activation of individual kinases and transcription factors.[30-31] Three major signaling pathways mediating TLR-induced responses have been described (i) NFκB, (ii) mitogen-activated protein kinases (MAPKs) and (iii) IFN regulatory factors (IRFs). NFκB and MAPKs are essential in the induction of a proinflammatory response whereas IRFs are required for stimulation of IFN production.[30-32] Ultimately, TLR-induced signal transduction pathways result in the activation of gene expression and synthesis of a range of molecules such as cytokines, chemokines and immunoreceptors.[28,30] Together, these coordinate the immediate host response to infection and provide an essential link to the adaptive immune response. Given that the containment and eradication of pathogens is dependent on efficient recognition and signaling through PRRs, it is perhaps not surprising that some pathogens have evolved strategies to interfere with this process.

Degradation of Pattern Recognition Receptors

Among the various bacterial species associated with the development of periodontitis, the Gram negative anaerobic bacterium *Porphyromonas gingivalis* is suspected to be one of the most important causative agents of the chronic form of this disease.[33] Arg- and Lys-gingipain cysteine proteases are the main endopeptidases produced by *P. gingivalis* and are considered to be important virulence factors[34] as both proteases inhibit CD14-dependent monocyte activation. CD14 is a 55-kDa glycosylphosphatidylinositol (GPI)-anchored membrane protein, which functions in the detection of many Gram-negative bacteria by cells of the innate immune response. Presentation of CD14-associated bacterial lipopolysaccharide (LPS) to TLR4 and the accessory molecule MD-2, leads to the release of protective inflammatory factors by macrophages and dendritic cells.[35] Both Arg- and Lys- specific gingipains from *P. gingivalis* preferentially cleave CD14, but not TLR4, from the surface of macrophages,[36,37] making the cells hyporesponsive to LPS stimulation. The gingipains display a preference for cleaving CD14 over other polypeptides on human monocytes which can be explained by the high frequency of Arg-X and Lys-X peptide bonds in the amino acid sequence of CD14.[38] This preference is further demonstrated by the observation that gingipains also specifically remove CD14 from the surface of gingival fibroblasts.[39] As a consequence of this proteolysis, macrophage recognition of the bacterium is attenuated thus neutrophil degranulation and respiratory burst mediated by fibroblast-secreted IL-8 is reduced. Therefore bacterial survival in the periodontal tissues is promoted.[40]

Like the gingipains, the major cysteine protease of *F. hepatica,* cathepsin L1, is secreted into the extracellular environment during infection and also inhibits macrophage recognition of bacterial products. However, this inactivation is not mediated by cleavage of cell surface CD14. Instead, cathepsin L1 is internalised by the host's phagocytic macrophages and trafficked to the endolysosomal compartments where it specifically degrades TLR3.[25] This variation in proteolytic activity reflects clear differences in substrate specificity between the Arg-/Lys-specific gingipains and cathepsin L1 which prefers hydrophobic residues such as Leu, Phe and Ala.[41,42] Considering both CD14 and members of the Toll receptor family contain multiple leucine-rich repeat motifs[43] it is perhaps surprising that cathepsin L does not cleave either CD14 or TLR4 from the surface of macrophages yet specifically degrades intracellular TLR3. Unlike the surface Toll receptors, TLR3 undergoes conformational changes in response to lysosomal acidification[44] which may make it susceptible to cleavage by cathepsin L, a protease that is stable and functional over a broad pH range.[41]

In general, macromolecules internalized by macrophages are degraded into antigenic peptides by the range of endogenous cathepsins resident in the endolysosomal compartments. However, *F. hepatica* cathepsin L1 is resistant to this endosomal degradation and data shows that the mature enzyme is highly resistant to proteolytic degradation by various endopeptidases.[42] It has also been reported that *F. hepatica* cathepsin L1 can degrade cystatins/serpins such as SCCA1 and SCCA2[45] which may protect the enzyme from inhibition by cystatins within the lysosome that are known to regulate the activity of resident endolysosomal cathepsins.[46]

During schistosomiasis, the eggs released by *S. mansoni* suppress the maturation of dendritic cells in response to the bacterial products poly-I:C and LPS.[47] Similar to *F. hepatica* cathepsin L1, immature dendritic cells rapidly internalized egg antigens and targeted them to endolysosomal compartments.[48] The egg antigens contain cysteine proteases cathepsin B and cathepsin L which, like *F. hepatica* cathepsin L1, may cleave TLRs within the endosome thus preventing the dendritic cells from maturing in response to activation signals.[25] This in turn would inhibit Th1 responses and allow the promotion of Th2 responses which correlates with the immune switching during egg deposition.

Inhibition of NFκB Signalling

Several studies have demonstrated that eradication of the intramacrophage-dwelling *Leishmania sp.* requires the induction of Th1 cells. Prolonged survival of these pathogens is associated with the parasites ability to regulate IL-12 production by macrophages and therefore control the production of protective IFN-γ.[50-54] Most mouse strains are resistant to infection with *L. major* but develop nonhealing lesions following infection with *L. mexicana*.[20] The inability of these mouse strains to heal following infection with *L. mexicana* is associated with a higher level of parasite cysteine protease B activity and subsequently lower induction of IL-12.[55] This role for the parasite protease was confirmed by the observation that amastigotes of protease deletion mutants of *L. mexicana* had limited ability to inhibit IL-12 production and unlike the wild type parasites, were unable to suppress a Type 1 adaptive immune response.[55] The mechanism of action appears to be the specific proteolytic degradation of NFκB, which did not affect the nuclear translocation of this transcription factor, but did prevent it binding to DNA and therefore inhibited its ability to induce IL-12 gene expression.[56]

The infection of cells with the obligate intracellular protozoan parasite, *Toxoplasma gondii* also results in the inhibition of IL-12 expression via interference with NFκB activation. In this case the termination of NF-κB activity was associated with a reduction of the phosphorylation of p65/RelA, an event required for the translocation of NF-κB to the nucleus.[57] While a number of cysteine proteases have been identified from *T. gondii*,[58] the possibility of their involvement in this immune-modulatory effect has not been investigated.

Prevention of Ubiquitination

Following interaction between a PAMP and its corresponding PRR, the activation of downstream signal transduction pathways is dependent on the ubiquitination of protein components of the signalling cascade.[30,31] Ubiquitination is an enzyme-mediated process by which ubiquitin (an 8 kDa protein) is covalently attached to lysine residues of target proteins. While this alteration to proteins does not mediate degradation it has functional consequences for the modified protein such as changes in their conformation, subcellular localization or catalytic activity. However, ubiquitination is a reversible modification and the rapid removal of ubiquitin from substrates is catalysed by de-ubiquitinating enzymes (DUBs) which are predominantly classed as cysteine proteases. The human genome is predicted to encode almost 500 proteins that are known to recognize ubiquitin and attach it to specific substrates and over 90 DUBs that reverse that reaction.[60]

During virus replication, the innate immune response is activated, resulting in the production of several hundred antiviral proteins converting the intracellular environment into a suboptimal context for replication.[60] In response to this selective environment, evidence suggests that a number of viral proteases have adapted a ubiquitin-removal specificity as a mechanism to disable the host immune system and optimize the intracellular environment for efficient virus replication and release.[10] For example, the nonstructural protein of severe acute respiratory syndrome (SARS) corona virus, nsp3, known to be involved in the processing of replicase polyproteins, has recently been shown to carry a conserved deubiquitinase (DUB) motif within its papain-like protease (PLpro) domain.[61,62] This protein efficiently inhibits IRF3 ubiquitination in a protease dependent mechanism.[63] IRF3 is a critical transcription factor for the activation of antiviral IFN and requires ubiquitination to translocate to the nucleus. Its inhibition by a viral DUB could explain why cultured cells infected with *Coronaviridae* characteristically produce very low levels of IFNs.[64-66] Indeed, most viruses, including all highly pathogenic human viruses, attempt to modulate this aspect of the innate immune response early in infection.[66,67]

Virus-encoded DUBs from a variety of virus families have been described,[10] including UL36 of herpesviruses such as herpes simplex virus Type 1, Epstein-Barr virus and mouse and human cytomegalovirus;[68-70] the adenain protease of adenovirus;[71] and the PLP domains from *Coronaviridae*.[72] These proteases are unique to the viral pathogens and quite distinct from host-encoded DUBs. While a pathogenic role has not been assigned to many of these viral DUBs, it is tempting to speculate that the prevention of ubiquitination of essential transcription factors is a common mechanism to suppress antiviral IFN production.

Despite having no intrinsic ubiquitin system, several bacterial strains have been found to contain ubiquitin-specific cysteine proteases. Like the viral DUBs, these proteases interfere with innate cell signalling, inhibiting activation of antibacterial responses by specifically targeting elements of the NFκB pathway. For example, YopJ secreted by *Yersinia sp.* was first described as a DUB, preventing ubiquitination of the TLR-adaptor proteins TRAF2, TRAF3 and TRAF6.[73,74] It has also been reported that YopJ can de-ubiquitinate the transcription factor IκBα, thereby preventing its degradation and the subsequent translocation of NF-κB to the nucleus.[75] Similarly, AvrA, a protease secreted by *Salmonella sp.* stabilises IκBα by preventing its ubiquitination and thereby inhibits NF-κB activated inflammatory responses of the host.[75] De-ubiquinitase activity has also been attributed to cysteine proteases secreted by *Chlamydia trachomatis* (ChlaDub1 and ChlaDub2)[77] and *Escherichia coli* (ElaD)[78] although a precise role has not yet been defined.

Bioinformatic analyses have predicted that a number of medically-relevant parasitic protozoa encode putative DUBs.[79] However, only three have been functionally assessed for their ability to bind to ubiquitin. *Plasmodium falciparum* (the causative agent of malaria) expresses two DUBs, namely PfUCHL3 and PfUCH54[80] and a third, TgUCHL3, a homologue of PfUCHL3, is expressed by *T. gondii*.[81] It has been hypothesized that these proteases may only function within the parasite itself and not target host proteins. There is currently no information on their physiological effects.

DEGRADATION OF SOLUBLE MEDIATORS OF IMMUNE RESPONSE

Degradation of Cytokines and Their Receptors

Cells of the innate and acquired immune systems communicate via the release of cytokines. These proteins are released in response to the presence of pathogens or to components of damaged tissue and are essential for the induction of an inflammatory response as well as its regulation and resolution. Interruption of this communication network profoundly impacts on the outcome of infectious disease. One method of disruption of signalling, performed promiscuously by cysteine proteases derived from *P. gingivalis*, occurs through the proteolytic modification of cytokines (IL-6, IL-1Ra, IL-12, TNF-α, IFN-γ) and their receptors (IL-6R).[27]

A second method of interference by pathogen proteases is mimicking the activity of host proteases used in the regulation of cytokine activity. Unlike most other cytokines, IL-18 and IL-1β lack a signal peptide and are first synthesized as biologically inactive precursors (proIL-18 and proIL-1β). These precursors are cleaved by caspase-1 (IL-1β-converting enzyme [ICE]), between Asp-116 and Ala-117[82,83] to produce the mature active cytokines. Cysteine proteases isolated from *Streptococcus pyogenes* (SpeB) and *Entamoeba histolytica* both convert Pro-IL-1β to a mature cytokine, targeting cleavage sites 1 and 5 amino acids from the caspase-1 site of action, respectively.[84,85] In both cases the resultant cytokine retains biological activity which leads to the augmentation of an inflammatory response. For *E. histolytica* this action may facilitate its spread beyond cells in direct contact with amoebic trophozoites.[84] In addition, the ICE-like activity of pathogen cysteine proteases suggests a mechanism which could activate caspases within infected cells and thus induce cell death by apoptosis.

Another cysteine protease from *E. histolytica* cleaves pro-IL-18.[86] However, the proteolytic action removes Glu-42, a key residue for biological activity of IL-18,[87] and therefore, the cleavage product is an inactive protein. It was suggested that the secondary structure of IL-18 may contribute to the choice of cleavage site, as the amino acid sequence does not correlate with known peptide substrate specificities.[87]

Effect on Chemokines

Chemokines are a distinct, large superfamily of cytokines encompassing small structurally-related proteins. Physiologically, chemokines have an important role in the recruitment of leukocytes during an acute inflammatory response. Host proteases tightly regulate this cellular recruitment by modulating the activity of chemokines via specific cleavage of their N- and C-termini.[88,89] The proteolytic cleavage of the N-terminus enhances activity for several CXC inflammatory chemokines, including the main neutrophil-attracting chemokine CXCL8 (IL-8). This function has been appropriated by the gingipains of *P. gingivalis* which efficiently cleave IL-8 at the N-terminus, enhancing its activity,[90] an action which correlates with the massive infiltration of neutrophils observed at sites of periodontitis. However, when associated with bacterial outer membrane vesicles the same proteases can also degrade IL-8. It has been suggested that this dual role of enhancing and inhibitory activity results in the association of pro- and anti-inflammatory reactions to distal and proximal positions of the bacterial plaque, respectively, and explains why there is no elimination of infection despite the accumulation of neutrophils.[90]

Degradation of Immunoglobulins

Immunoglobulins (Igs) play a key role in host immune defence mechanisms by specifically recognising invading organisms and mediating their killing by professional phagocytes or the complement system, or both. Igs are composed of antigen-recognising Fab regions, linked through a flexible hinge region to a constant Fc effector region. The Fc section interacts with Fc receptors on phagocytic cells and triggers the activation of the classical pathway of complement. The classical pathway is initiated by the binding of complement factor C1q to specific IgG or IgM. This interaction triggers a cascade of events resulting in the opsonisation of antigen with C3b forming an immune complex which is efficiently recognized by phagocytes.

Despite the fact that mammalian hosts infected with the helminth *F. hepatica* develop specific antibodies[91] and that the major protein isolated from eosinophils is highly toxic to newly excysted juvenile (NEJ) worms,[92] no evidence exists of antibody-mediated eosinophil damage to NEJs in nonpermissive bovine hosts.[93] While effector cells readily adhered to NEJs in the presence of immune sera, they failed to adhere if the parasite's excretory/secretory (ES) products were added which indicated that the contents of ES were preventing interaction between immune sera and eosinophils. In the presence of leupeptin, the effector cells remained attached to the NEJs, suggesting a role for cysteine proteases.[94] Subsequently, in vitro studies have demonstrated that the papain-like cathepsin L1 and L2 proteases secreted by *F. hepatica* cleaved all human IgG subclasses in the hinge region. Despite clear evidence that these proteases have distinct peptide bond preferences, both cathepsin L1 and L2 cleaved each of the IgG molecules at the same peptide bond.[95]

The ability to degrade human IgG subclasses is not exclusive to the cathepsin Ls of *Fasciola*. Similar proteolytic cleavage sites and specificity for human IgG subclasses have been shown for the major secreted cysteine proteases of other helminths, such as *Paragonimus westermani,*[96] *Spirometra mansoni*[97] and *Schistosoma mansoni*[98] which all degrade host IgG in vitro and prevent parasite specific antibody-mediated eosinophil activation.[99] Cruzipain, the cysteine protease secreted by the protozoan parasite *T. cruzi* also exhibits cathepsin-like activity, cleaving all human IgG subclasses at the hinge region in a similar but not identical region to cathepsins L1 and L2 of *F. hepatica.*[100] The fact that the main cleavage sites for this range of proteases all exist within the hinge regions of IgG, suggests that the conformation of the antibody molecules influences the accessibility of enzymes and therefore determines the specificity of cleavage, irrespective of the amino acid preference of each protease active site.

In all cases, the end result of proteolytic cleavage by these parasite proteases is the release of intact monomeric fragments of Fab.[95,98,100] The Fab fragments retain their capacity to bind to surface antigens, a term called 'fabulation'. Surface epitopes are therefore masked from intact and functional antibodies, while the loss of Fc fragments eliminates the interaction of IgG with effector cells and prevents resultant antibody-dependent cytotoxicity. In addition, the Fc fragments produced by initial proteolyic cleavage at the hinge region are further degraded by proteases from *F. hepatica* and *T. gondii* cruzipain in the CH2 region producing Fc-like-fragments of 14 kDa composed of the CH3 domain.[95,100] As the complement factor C1q specifically binds to the CH2 region of IgG, removal of this would prevent the initiation of the classical pathway.

Despite lacking sequence identity to the helminth papain-like proteases, the cysteine protease (IdeS) secreted by the pathogenic bacterium *S. pyogenes* adopts a canonical papain fold[101] and like papain, cleaves IgG in the flexible hinge region of the IgG heavy chain generating intact Fab and Fc fragments.[102] Similar to the outcome with helminth-specific immune responses, survival of opsonised bacteria against the phagocytic actions of macrophages or polymorphonuclear leukocytes is significantly enhanced in the presence of IdeS, due to the proteolytic interference of Fc-mediated killing.[103] Amongst the microbial secreted cathepsins, IdeS is unique in its specificity for IgG, with no additional substrates identified to date. The other major cysteine protease secreted by *S. pyogenes,* SpeB, is structurally homologous to papain and cleaves IgG in the flexible hinge region, although at a different cleavage site.[104] However, the proteolytic activity of SpeB towards immunoglobulins is not restricted to IgG. This protease also degrades the carboxy-terminal of the heavy chains of IgA, IgD, IgE and IgM.[103]

Specific-IgA protease activity is a well-established feature of many human infectious diseases that take place at, or originate from, mucosal surfaces.[105,106] For bacterial pathogens the IgA proteases have been primarily classified as serine or metallo-type proteases cleaving only the IgAl subclass of antibody, because the susceptible site is one of the Pro-Ser or Pro-Thr peptide bonds located within a 12-amino acid proline rich sequence in the hinge region of IgAl but absent from IgA2.[106,107] However, the cathepsin B-like cysteine protease (EhCp5) secreted by the protozoan *E. histolytica* with a preference for Arg-Arg residues displays an ability to cleave both IgA1 and IgA2 degrading the antibody structures at positions 245 and 250 of the hinge region.[108,109] Similar to the proteolytic cleavage of IgG described above, release of Fab fragments inhibits antigen disposal as immunogenic determinants are masked by fabulation. Moreover, removal of the Fc region eliminates the ability of IgA to agglutinate, preventing opsonophagocytosis thus facilitating survival of pathogenic organisms at the mucosal surface.

Inactivation of Complement Pathways

The complement system is composed of three distinct pathways: (1) classical pathway, activated by antigen-antibody complexes; (2) lectin pathway, activated by carbohydrate arrays found on microbial surfaces; and (3) alternative, activated by C3 binding to the surface of micro-organisms. Central to all pathways of activation is the formation of C3 and C5 convertase complexes. In particular, the C3 protein and its activated form, C3b, is an integral component of the C5 convertase complex whichever pathway of activation has been engaged. Cleavage of these molecules produces the C3a and C5a fragments which are responsible for attracting white blood cells to the site of infection and therefore form a link between the innate and adaptive immune systems.

Of the members of the complement cascade, cysteine proteases secreted by pathogens primarily target components of the C3 and C5 convertase complex. A number of bacterial proteases, including the SpeB cysteine protease of *S. pyogenes* cleave C3 and thus inactivate or prevent the formation of the C5 convertase complex.[110,111] Similarly, the cysteine protease from *E. histolytica* prevents the formation of the membrane attack complex (MAC) and the release of the pro-inflammatory mediator, C5a by degrading both C5a and C3a directly.[109,112]

The refractoriness of *P. gingivalis* to the immune response has been attributed to the biphasic effects of gingipains on the complement system. The lysine-specific gingipain is capable of degrading the C5a receptor, but more importantly the arginine-specific enzymes cleave C3, C4 and C5.[113] Sharing substrate specificity with the proteases required for complement activation, the gingipains activate C3, C4 and C5 by cleaving them at their activation sites. However, at high concentrations the Arg-gingipains completely degrade and thus inactivate the complement proteins.[113] This has led to the suggestion that early infections by *P. gingivalis* may actually stimulate complement activation to cause an inflammatory state that is advantageous to the bacterium. However, as infections become chronic the higher number of bacteria lead to the degradation of complement. This idea is supported by data showing that the cysteine protease, Interpain A, secreted by the periodontal bacterium, *Prevotella intermedia*, which co-aggregates with *P. gingivalis,* activates the C1 complex in serum, causing deposition of C1q on bacterial surfaces and resulting in a local inflammatory reaction during the initial stages of infection.[114]

Activation of the complement cascade involves a number of proteases which are tightly regulated by cystatins and serpins. The StcE cysteine protease of *E. coli* uniquely prevents complement activation via the classical pathway by specifically cleaving the serpin, C1-inhibitor, which regulates the proteases of the initiating C1 complex of the classical pathway.[115]

CONCLUSION

The immune system has evolved many different strategies to control many types of infections and thus prevent associated disease. Equally, pathogenic organisms appear to have developed processes to evade, suppress or subvert the immune response with pathogen-derived cysteine proteases emerging as key molecules. Whilst some of these proteases share a common origin with mammalian-encoded proteases, most of them have ancient intrinsic functions, such as processing pathogen protein components and may have acquired the specificity for host protein targets by interaction with their host's immune system over time. As many of these proteases have evolved distinct biochemical features from their mammalian counterparts they remain attractive as targets for new antimicrobial drugs. In addition, some of these proteases may be useful as novel therapeutics for the treatment of Th1 inflammatory disorders particularly in light of the current strategy of developing antagonists of innate immune responses as immunotherapeutics for sepsis and autoimmune disease.

REFERENCES

1. Colbert JD, Matthews SP, Miller G et al. Diverse regulatory roles for lysosomal proteases in the immune response. Eur J Immunol 2009; 39:2955-2965.
2. Bird PI, Trapani JA, Villadangos JA. Endolysosomal proteases and their inhibitors in immunity. Nat Rev Immunol 2009; 9:871-882.
3. Robinson MW, Menon R, Donnelly SM et al. An integrated transcriptomics and proteomics analysis of the secretome of the helminth pathogen Fasciola hepatica: proteins associated with invasion and infection of the mammalian host. Mol Cell Proteomics 2009; 8:1891-1907.
4. Robinson MW, Dalton JP, Donnelly S. Helminth pathogen cathepsin proteases: it's a family affair. Trends Biochem Sci 2008; 33:601-608.
5. McKerrow JH, Caffrey C, Kelly B et al. Proteases in parasitic diseases. Annu Rev Pathol 2006; 1:497-536.
6. Imamura T. The role of gingipains in the pathogenesis of periodontal disease. J Periodontol 2003; 74:111-118.
7. Onoe T, Hoover CI, Nakayama K et al. Identification of Porphyromonas gingivalis prefimbrilin possessing a long leader peptide: possible involvement of trypsin-like protease in fimbrilin maturation. Microb Pathog 1995; 19:351-364.
8. Maliar T, Baláz S, Tandlich R et al. Viral proteinases—possible targets of antiviral drugs. Acta Virol 2002; 46:131-140.
9. Harcourt BH, Jukneliene D, Kanjanahaluethai A et al. Identification of severe acute respiratory syndrome coronavirus replicase products and characterization of papain-like protease activity. Virol 2004; 78:13600-13612.
10. Edelmann MJ, Kessler BM. Ubiquitin and ubiquitin-like specific proteases targeted by Infectious pathogens: Emerging patterns and molecular principles. Biochim Biophys Acta 2008; 1782:809-816.
11. Alexander J, Bryson K. T helper (h) 1/Th2 and Leishmania: paradox rather than paradigm. Immunol Lett 2005; 99:17-23.
12. Maekawa Y, Himeno K, Katunuma N. Cathepsin B-inhibitor promotes the development of Th1 type protective T-cells in mice infected with Leishmania major. J Med Invest 1997; 44:33-39.
13. Maekawa Y, Himeno K, Ishikawa H et al. Switch of $CD4^+$ T-cell differentiation from Th2 to Th1 by treatment with cathepsin B inhibitor in experimental leishmaniasis. J Immunol 1998; 161:2120-2127.
14. Denise H, McNeil K, Brooks DR et al. Expression of multiple CPB genes encoding cysteine proteases is required for Leishmania mexicana virulence in vivo. Infect Immun 2003; 71:3190-3195.
15. Pollock KG, McNeil KS, Mottram JC et al. The Leishmania mexicana cysteine protease, CPB2.8, induces potent Th2 responses. J Immunol 2003; 170:1746-1753.
16. Giordanengo L, Guiñazú N, Stempin C et al. Cruzipain, a major Trypanosoma cruzi antigen, conditions the host immune response in favor of parasite. Eur J Immunol 2002; 32:1003-1011.
17. Sokol CL, Barton GM, Farr AG et al. A mechanism for the initiation of allergen-induced T helper type 2 responses. Nat Immunol 2008; 9:310-318.
18. Sokol CL, Chu NQ, Yu S et al. Basophils function as antigen-presenting cells for an allergen-induced T helper type 2 response. Nat Immunol 2009; 10:713-720.
19. Phillips C, Coward WR, Pritchard DI et al. Basophils express a type 2 cytokine profile on exposure to proteases from helminths and house dust mites. J Leukoc Biol 2003; 73:165-171.
20. Brady MT, O'Neill SM, Dalton JP et al. Fasciola hepatica suppresses a protective Th1 response against Bordetella pertussis. Infect Immun 1999; 67:5372-5378.
21. O'Neill SM, Brady MT, Callanan JJ et al. Fasciola hepatica infection downregulates Th1 responses in mice. Parasite Immunol 2000; 22:147-155.
22. O'Neill SM, Mills KH, Dalton JP. Fasciola hepatica cathepsin L cysteine proteinase suppresses Bordetella pertussis-specific interferon-gamma production in vivo. Parasite Immunol 2001; 23:541-547.
23. Pearce EJ, MacDonald AS. The immunobiology of schistosomiasis. Nat Rev Immunol. 2002; 2:499-511.
24. Sung CK, Dresden MH. Cysteinyl proteinases of Schistosoma mansoni eggs: purification and partial characterization. J Parasitol. 1986; 72: 891-900.
25. Donnelly S, O'Neill SM, Stack CM et al. Helminth cysteine proteases inhibit TRIF-dependent activation of macrophages via degradation of TLR3. J Biol Chem. 2010; 285:3383-3392.
26. Dowling DJ, Hamilton CM, Donnelly S et al. Major secretory antigens of the helminth Fasciola hepatica activate a suppressive dendritic cell phenotype that attenuates Th17 cells but fails to activate Th2 immune responses. Infect Immun 2010; 78:793-801.
27. Potempa J, Pike RN. Corruption of innate immunity by bacterial proteases. J Innate Immun 2009; 1:70-87.
28. Akira S, Uematsu S, Takeuchi O. Pathogen recognition and innate immunity. Cell 2006; 124:783-801.
29. Kawai T, Akira S. The role of pattern-recognition receptors in innate immunity: update on Toll-like receptors. Nat Immunol 2010; 11:373-384.
30. Akira S, Takeda K. Toll-like receptor signalling. Nat Rev Immunol 4:499-511.

31. O'Neill LA, Bowie AG. The family of five: TIR-domain-containing adaptors in Toll-like receptor signalling. Nat Rev Immunol 2007; 7:353-364.
32. Kawai T, Akira S. Signaling to NF-kappaB by Toll-like receptors. Trends Mol Med 2007; 13:460-469.
33. Slots J, Ting M. Actinobacillus actinomycetemcomitans and Porphyromonas gingivalis in human periodontal disease: occurrence and treatment. Periodontol 1999; 20:82-121.
34. Holt SC, Kesavalu L, Walker S et al. Virulence factors of Porphyromonas gingivalis. Periodontol 1999; 20:168-238.
35. Gioannini TL, Weiss JP. Regulation of interactions of Gram-negative bacterial endotoxins with mammalian cells. Immunol Res 2007; 39:249-260.
36. Sugawara S, Nemoto E, Tada H et al. Proteolysis of human monocyte CD14 by cysteine proteinases (gingipains) from Porphyromonas gingivalis leading to lipopolysaccharide hyporesponsiveness J Immunol 2000; 165:411-418.
37. Duncan L, Yoshioka M, Chandad F et al. Loss of lipopolysaccharide receptor CD14 from the surface of human macrophage-like cells mediated by Porphyromonas gingivalis outer membrane vesicles. Microb Pathog 2004; 36:319-325.
38. Kim JI, Lee CJ, Jin MS et al. Crystal structure of CD14 and its implications for lipopolysaccharide signaling. J Biol Chem 2005; 280:11347-11351.
39. Tada H, Sugawara S, Nemoto E et al. Proteolysis of CD14 on human gingival fibroblasts by arginine-specific cysteine proteinases from Porphyromonas gingivalis leading to down-regulation of lipopolysaccharide-induced interleukin-8 production. Infect Immun 2002; 70:3304-3307.
40. Imamura T. The role of gingipains in the pathogenesis of periodontal disease. J Periodontol. 2003; 74:111-8.
41. Lowther J, Robinson MW, Donnelly SM et al. The importance of pH in regulating the function of the Fasciola hepatica cathepsin L1 cysteine protease. PLoS Negl Trop Dis 2009; 3:e369.
42. Stack CM, Donnelly S, Lowther J et al. The major secreted cathepsin L1 protease of the liver fluke, Fasciola hepatica: a Leu-12 to Pro-12 replacement in the nonconserved C-terminal region of the prosegment prevents complete enzyme autoactivation and allows definition of the molecular events in prosegment removal. J Biol Chem 2007; 282:16532-16543.
43. Ferrero E, Hsieh CL, Francke U et al. CD14 is a member of the family of leucine-rich proteins and is encoded by a gene syntenic with multiple receptor genes. J Immunol 1990; 145:331-336.
44. Ranjith-Kumar CT, Miller W, Xiong J et al. Biochemical and Functional Analyses of the Human Toll-like Receptor 3 Ectodomain. J Biol Chem 2007; 282:7668-7678.
45. Kanaji S, Tanaka Y, Sakata Y et al. Squamous cell carcinoma antigen 1 is an inhibitor of parasite-derived cysteine proteases. FEBS Lett 2007; 581:4260-4264.
46. Kopitar-Jerala N. The role of cystatins in cells of the immune system. FEBS Lett 2006; 580:295-301.
47. Kane CM, Cervi L, Sun J et al. Helminth antigens modulate TLR-initiated dendritic cell activation. J Immunol 2004; 173:7454-7461.
48. van Liempt E, van Vliet SJ, Engering A et al. Schistosoma mansoni soluble egg antigens are internalized by human dendritic cells through multiple C-type lectins and suppress TLR-induced dendritic cell activation. Mol Immunol 2007; 44:2605-2615.
49. Alexander J, Satoskar AR, Russell DG. Leishmania species: models of intracellular parasitism. J Cell Sci 1999; 112:2993-2998.
50. Reiner SL, Zheng S, WangAE et al. Leishmania promastigotes evade interleukin-12 (IL-12) induction by macrophages and stimulate a broad range of cytokines from CD4+ T-cells during initiation of infection. J Exp Med 1994; 179:447-252.
51. Carrera L, Gazzinelli RT, Badolato R et al. Leishmania promastigotes selectively inhibit interleukin-12 induction in bone marrow-derived macrophages from susceptible and resistant mice. J Exp Med 1996; 183:515-520.
52. Heinzel FP, Schoenhaut DS, Rerko RM et al. Recombinant interleukin 12 cures mice infected with Leishmania major. J Exp Med 1993; 177:1505-1510.
53. Sypek JP,Chung CL, Mayor SHE et al. Resolution of cutaneous leishmaniasis: interleukin-12 initiates a protective T helper type 1 immune response. J Exp Med 1993; 177:1797-1802.
54. Afonso LCC, Scharton TM, Vieira, LQ et al. The adjuvant effect of interleukin-12 in a vaccine against Leishmania major. Science 1994; 263:235-240.
55. Buxbaum LU, Hubert D, Coombs GH et al. CPB cysteine proteases of Leishmania mexicana inhibit host Th1 responses and protective immunity. J Immunol 2003; 171:3711-3718.
56. Cameron P, McGachy A, Anderson M et al. Inhibition of lipopolysaccharide-induced macrophage IL-12 production by Leishmania mexicana amastigotes: the role of cysteine peptidases and the NF-kappaB signaling pathway. J Immunol 2004; 173:3297-3304.
57. Shapira S, Harb OS, Margarit J et al. Initiation and termination of NF-kappaB signaling by the intracellular protozoan parasite Toxoplasma gondii. J Cell Sci 2005; 118:3501-3508.

58. Roiko MS, Carruthers VB. New roles for perforins and proteases in apicomplexan egress. Cell Microbiol 2009; 11:1444-1452.
59. Nijman SM, Luna-Vargas MP, Velds A. A genomic and functional inventory of deubiquitinating enzymes. Cell 2005; 123:773-786.
60. Levy DE, Garcia-Sastre A. The virus battles: IFN induction of the antiviral state and mechanisms of viral evasion. Cytokine Growth Factor Rev 2001; 12:143-156.
61. Barretto N, Jukneliene D, Ratia K et al. The Papain like protease of severe acute respiratory syndrome coronavirus has deubiquitinating activity, J Virol 2005; 79:15189-15198.
62. Ratia K, Pegan S, Takayama J et al. A noncovalent class of papain-like protease/deubiquitinase inhibitors blocks SARS virus replication. Proc Natl Acad Sci USA 2008; 105:16119-16124.
63. Frieman M, Ratia K, Johnston RE et al. Severe acute respiratory syndrome coronavirus papain-like protease ubiquitin-like domain and catalytic domain regulate antagonism of IRF3 and NF-kappaB signaling. J Virol 2009; 83:6689-6705.
64. Frieman M, Heise M, Baric R. SARS coronavirus and innate immunity. Virus Res 2007; 133:101-112.
65. Zhou H, Perlman S. Mouse hepatitis virus does not induce Beta interferon synthesis and does not inhibit its induction by doublestranded RNA. J Virol 2007; 81:568-574.
66. Versteeg GA, Bredenbeek PJ, van den Worm SH et al. Group 2 coronaviruses prevent immediate early interferon induction by protection of viral RNA from host cell recognition. Virology 2007; 361:18-26.
67. Hengel H, Koszinowski UH, Conzelmann KK. Viruses know it all: new insights into IFN networks. Trends Immunol 2005; 26:396-401.
68. Kattenhorn LM, Korbel GA, Kessler BM et al. A deubiquitinating enzyme encoded by HSV-1 belongs to a family of cysteine proteases that is conserved across the family Herpesviridae. Mol. Cell 2005; 19:547-557.
69. Schlieker C, Korbel GA, Kattenhorn LM et al. A deubiquitinating activity is conserved in the large tegument protein of the Herpesviridae. J Virol 2005; 79:15582-15585.
70. Wang J, Loveland AN, Kattenhorn LM et al. High-molecular-weight protein (pUL48) of human cytomegalovirus is a competent deubiquitinating protease: mutant viruses altered in its activesite cysteine or histidine are viable. J Virol 2006; 80:6003-6012.
71. Balakirev MY, Jaquinod M, Haas AL et al. Deubiquitinating function of adenovirus proteinase. J Virol 2002; 76:6323-6331.
72. Barretto N, Jukneliene D, Ratia K et al. The papain-like protease of severe acute respiratory syndrome coronavirus has deubiquitinating activity. J Virol 2005; 79:15189-15198.
73. Sweet CR, Conlon J, Golenbock DT et al. YopJ targets TRAF proteins to inhibit TLR-mediated NF-kappaB, MAPK and IRF3 signal transduction. Cell Microbiol 2007; 9:2700-2715.
74. Zhou H, Monack DM, Kayagaki N et al. Yersinia virulence factor YopJ acts as a deubiquitinase to inhibit NF-kappa B activation. J Exp Med 2005; 202:1327-1332.
75. Orth K, Xu Z, Mudgett MB et al. Disruption of signaling by Yersinia effector YopJ, a ubiquitin-like protein protease. Science 2000; 290:1594-1597.
76. Ye Z, Petrof EO, Boone D et al. Salmonella effector AvrA regulation of colonic epithelial cell inflammation by deubiquitination, Am J Pathol 2007; 171:882-892.
77. Misaghi S, Balsara ZR, Catic A et al. Chlamydia trachomatis-derived deubiquitinating enzymes in mammalian cells during infection. Mol Microbiol 2006; 61:142-150.
78. Catic A, Misaghi A, Korbel GA et al. ElaD, a deubiquitinating protease expressed by E. coli. PLoS ONE 2007; 2:e381.
79. Ponder EL, Bogyo M. Ubiquitin-like modifiers and their deconjugating enzymes in medically important parasitic protozoa, Eukaryot. Cell 2007; 6:1943-1952.
80. Artavanis-Tsakonas K, Misaghi S, Comeaux CA et al. Identification by functional proteomics of a deubiquitinating/deNeddylating enzyme in Plasmodium falciparum. Mol Microbiol 2006; 61:1187-1195.
81. Frickel EM, Quesada V, Muething L et al. Apicomplexan UCHL3 retains dual specificity for ubiquitin and Nedd8 throughout evolution. Cell Microbiol 2007; 9:1601-1610.
82. Kostura MJ, Tocci MJ, Limjuco G et al. Identification of a monocyte specific pre-interleukin 1 beta convertase activity. Proc Natl Acad Sci USA 1989; 86:5227-5231.
83. Black RA, Kronheim SR, Merriam JE et al. A pre-aspartate-specific protease from human leukocytes that cleaves pro-interleukin-1 beta. J Biol Chem 1989; 264:5323-5326.
84. Kapur V, Majesky MW, Li LL et al Cleavage of interleukin 1 beta (IL-1 beta) precursor to produce active IL-1 beta by a conserved extracellular cysteine protease from Streptococcus pyogenes.Proc Natl Acad Sci USA 1993; 90:7676-7680.
85. Zhang Z, Wang L, Seydel KB et al. Entamoeba histolytica cysteine proteinases with interleukin-1 beta converting enzyme (ICE) activity cause intestinal inflammation and tissue damage in amoebiasis. Mol Microbiol 2000; 37:542-548.
86. Que X, Kim SH, Sajid M et al. A surface amebic cysteine proteinase inactivates interleukin-18. Infect Immun 2003; 71:1274-80.

87. Kim SH, Azam T, Novick D et al. Identification of amino acid residues critical for biological activity in human interleukin-18. J Biol Chem 2002; 277:10998-11003.
88. Wolf M, Albrecht S, Märki C. Proteolytic processing of chemokines: implications in physiological and pathological conditions. Int J Biochem Cell Biol 2008; 40:1185-1198.
89. Dean RA, Cox JH, Bellac CL et al. Macrophage-specific metalloelastase (MMP-12) truncates and inactivates ELR+ CXC chemokines and generates CCL2, 7, 8 and 13 antagonists: potential role of the macrophage in terminating PMN influx. Blood 2008;112:3455-3464.
90. Mikolajczyk-Pawlinska J, Travis J, Potempa J. Modulation of interleukin-8 activity by gingipains from Porphyromonas gingivalis: implications for pathogenicity of periodontal disease. FEBS Lett 1998; 440:282-286.
91. O'Neill SM, Parkinson M, Strauss W et al. Immunodiagnosis of Fasciola hepatica infection (fascioliasis) in a human population in the Bolivian Altiplano using purified cathepsin L cysteine proteinase. Am J Trop Med Hyg 1998; 58:417-423.
92. Duffus WP, Thorne K, Oliver R. Killing of juvenile Fasciola hepatica by purified bovine eosinophil proteins. Clin Exp Immunol 1980; 40:336-344.
93. Duffus WP, Franks D. In vitro effect of immune serum and bovine granulocytes on juvenile Fasciola hepatica. Clin Exp Immunol 1980; 41:430-440.
94. Carmona C, Dowd AJ, Smith AM et al. Cathepsin L proteinase secreted by Fasciola hepatica in vitro prevents antibody-mediated eosinophil attachment to newly excysted juveniles. Mol Biochem Parasitol 1993; 62:9-17.
95. Berasain P, Carmona C, Frangione B et al. Fasciola hepatica: parasite-secreted proteinases degrade all human IgG subclasses: determination of the specific cleavage sites and identification of the immunoglobulin fragments produced. Exp Parasitol 2000; 94:99-110.
96. Chung YB, Yang HJ, Kang SY et al. Activities of different cysteine proteases of Paragonimus westermani in cleaving human IgG. Korean J Parasitol 1997; 35:139-142.
97. Kong Y, Chung YB, Cho SY et al. Cleavage of immunoglobulin G by excretory-secretory cathepsin S-like protease of Spirometra mansoni plerocercoid. Parasitology 1994;109:611-621.
98. Auriault C, Ouaissi MA, Torpier G et al. Proteolytic cleavage of IgG bound to the Fc receptor of Schistosoma mansoni schistosomula. Parasite Immunol 1981; 3:33-44.
99. Shin MH, Kita H, Park HY et al. Cysteine protease secreted by Paragonimus westermani attenuates effector functions of human eosinophils stimulated with immunoglobulin. G Infect Immun 2001; 69:1599-1604.
100. Berasain P, Carmona C, Frangione B et al. Specific cleavage sites on human IgG subclasses by cruzipain, the major cysteine proteinase from Trypanosoma cruzi. Mol Biochem Parasitol 2003; 130:23-29.
101. Wenig K, Chatwell L, von Pawel-Rammingen U et al. Structure of the streptococcal endopeptidase IdeS, a cysteine proteinase with strict specificity for IgG. Proc Natl Acad Sci USA. 2004; 101: 17371-17376.
102. von Pawel-Rammingen U, Johansson BP, Björck L. IdeS, a novel streptococcal cysteine proteinase with unique specificity for immunoglobulin. G EMBO J 2002; 21:1607-1615.
103. von Pawel-Rammingen U, Björck L. IdeS and SpeB: immunoglobulin-degrading cysteine proteinases of Streptococcus pyogenes. Curr Opin Microbiol 2003; 6:50-55.
104. Collin M, Olsén A. Effect of SpeB and EndoS from Streptococcus pyogenes on human immunoglobulins. Infect Immun 2001; 69:7187-7189.
105. Plaut AG. The IgAl proteases of pathogenic bacteria. Annu Rev Microbiol 1983; 37:603-622.
106. Kilian M, Mestecky J, Russell MW. Defense mechanisms involving Fc-dependent functions of immunoglobulin A and their subversion by bacterial immunoglobulin A proteases. Microbiol Rev 1988; 52:296-303.
107. Mistry D, Stockley RA. IgA1 protease. Int J Biochem Cell Biol 2006; 8:1244-1248.
108. Garcia-Nieto RM, RicoMata R, Arias-Negrete S et al. Degradation of human secretory IgA1 and IgA2 by Entamoeba histolytica surface-associated proteolytic activity. Parasitol Int 2008; 57:417-423.
109. Que X, Reed SL. Cysteine proteinases and the pathogenesis of amebiasis. Clin Microbiol Rev 2000; 13:196-206.
110. Kuo CF, Lin YS, Chuang WJ et al. Degradation of complement 3 by streptococcal pyrogenic exotoxin B inhibits complement activation and neutrophil opsonophagocytosis. Infect Immun 2008; 76:1163-1169.
111. Terao Y, Mori Y, Yamaguchi M et al. Group A streptococcal cysteine protease degrades C3 (C3b) and contributes to evasion of innate immunity. J Biol Chem 2008; 283:6253-6260.
112. Reed SL, Ember JA, Herdman DS et al. The extracellular neutral cysteine proteinase of Entamoeba histolytica degrades anaphylatoxins C3a and C5a. J Immunol 1995; 155:266-274.
113. Popadiak K, Potempa J, Riesbeck K et al. Biphasic effect of gingipains from Porphyromonas gingivalis on the human complement system. J Immunol 2007; 178:7242-7250.
114. Potempa M, Potempa J, Kantyka T et al. Interpain A, a cysteine proteinase from Prevotella intermedia, inhibits complement by degrading complement factor C3. PLoS Pathog 2009; 5:e1000316.
115. Lathem WW, Bergsbaken T, Welch RA. Potentiation of C1 esterase inhibitor by StcE, a metalloprotease secreted by Escherichia coli O157:H7. J Exp Med 2004; 199:1077-1087.

CHAPTER 13

CYSTATINS OF PARASITIC ORGANISMS

Christian Klotz,* Thomas Ziegler, Emilia Daniłowicz-Luebert and Susanne Hartmann

Department of Molecular Parasitology, Humboldt University Berlin, Germany
**Corresponding Author: Christian Klotz—Email: christian.klotz@hu-berlin.de*

Abstract: The cystatin superfamily comprises several groups of protease inhibitors. In this chapter we will focus on I25 family members, which consist predominantly of the Type 2 cystatins. Recently, a wealth of information on these molecules and their activities has been described. Parasite cystatins are shown to have dual functions via interaction with both parasite and host proteases. Thereby, parasite cystatins are not only essentially involved in the regulation of physiological processes during parasite development, but also represent important pathogenicity factors. Interestingly, some studies indicate that parasite cystatins evolved exceptional immuno-modulatory properties. These capacities could be exploited to interfere with unwanted immune responses in unrelated human inflammatory diseases. We highlight the different biological roles of parasite cystatins and the anticipated future developments.

INTRODUCTION: CYSTATIN SUPERFAMILY-CLASSIFICATION AND STRUCTURAL CHARACTERISTICS

In this chapter we will discuss recent understanding of the biological role of parasite cystatins belonging to the MEROPS family I25, which consists predominantly of Type 2 cystatins (http://merops.sanger.ac.uk). Cystatins are inhibitors of cysteine proteases of the papain-like family (C1) in animal and plant taxa. The endogenous role of cystatins in multicellular organisms is mainly to protect cells from proteolytic damage. The general function of the cystatin superfamily and their phylogenomic analysis has previously been extensively reviewed.[1-5] Currently, two main classification methods for cystatins are used. The older classification divides cystatins into different types according to sequence similarities and on the basis of typical cystatin-like features such as the occurrence of

Cysteine Proteases of Pathogenic Organisms, edited by Mark W. Robinson and John P. Dalton.

disulfide bonds and the number of cystatin-like elements.[6] A more recent classification system is provided by the MEROPS database for peptidases and their inhibitors that groups molecules hierarchically into families based on amino acid sequence and structure (http://merops.sanger.ac.uk).[7] According to this classification cystatins belong to family I25 that has been further subdivided into 3 subfamilies. Type 1 cystatins (also named stefins) are grouped in I25 subfamily A and consist of low molecular weight cytoplasmic proteins of about 11 kDa that possess no disulfide bonds or glycosylation sites. Human cystatin A is a representative prototype for this subfamily. Type 1 cystatins exclusively inhibit papain-like peptidases of family C1 and have related structural features to Type 2 cystatins.[1]

Most cystatins, including those expressed by parasites, are classified as Type 2 cystatins that group into I25 subfamily B. The representative molecule for this subfamily is chicken cystatin (ovocystatin) for which the first crystallographic cystatin structure was determined.[8] Unlike Type 1 cystatins, these widely distributed inhibitors are secretory proteins of about 13-15 kDa. They typically possess an N-terminal signal peptide and 2 conserved disulfide bonds (Fig. 1). A further structural characteristic is the existence of a conserved glycine residue within the N-terminal region of the protein that may function as a hinge between the flexible N-terminal segment and the rest of the molecule.[9] The N-terminal region containing the conserved glycine residue, together with a central Q-X-V-X-G motif and a C-terminal PW hairpin loop forms the cysteine protease interaction site. These three elements direct the cystatin molecule into the active site cleft of the cysteine protease. All three regions form a hydrophobic wedge-like structure, which is highly complementary in shape to the active site of the protease, which mediates protease inhibition.

Cystatins of the I25B subfamily inhibit family C1 peptidases. However, some also possess a distinct conserved S-N-D/S motif, between the first conserved glycine and the central cystatin motif, that is responsible for the inhibition of legumain-like peptidases (C13 family).[10] Of note, an unusual characteristic of some cystatins is the existence of dimeric or oligomeric structures that reduce the inhibitory function of the protein.[11] A naturally-occurring point mutation (L68Q) in human cystatin C supports protein aggregation and leads to a severe phenotype of amyloidosis.[12] The point mutation favours the natural feature of cystatins to dimerize through three-dimensional domain swapping leading to oligomerization and/or amyloid formation through conformational changes of the β-hairpin loop L1.[12]

The Type 3 cystatins are divergent and include kininogens, fetuins and histidine rich proteins.[13] The only true inhibitors of cysteine proteases within this group are the kininogens, which consist of multi-domain proteins and group primarily into the MEROPS I25B subfamily. These high molecular weight proteins (>60 kDa) possess tandem repeats of cystatin Type 2-like domains and an extended number of disulfide bridges. They are found in the extracellular fluid and are typically glycosylated. Fetuins and histidine-rich proteins are also secreted proteins but lack cystatin activity and group into the MEROPS I25C subfamily. They are sometimes categorized as 'unclassified cystatins'.

PARASITE CYSTATINS

Parasite cystatins of the family I25 are predominantly found in three distinct parasitic groups: parasitic nematodes (phylum Nematoda), trematoda (phylum Platyhelminthes) and ticks (phylum Arthropoda). Table 1 summarizes the functions of characterized

```
Ce-CPI-1       -----------MRFILLIALVFAVLDGINC----------------------QIAGGLSD--VNAS--EYTGAAWNSVPEINSK-NNGQNYMVP 56
Hc-cystatin    ----------------------------------------------------MLAGGLTD--QSTDDPEFMEQAWKAATKVNEEANDGDYYMIP 40
Av-cystatin    --MMLSIKEDGLLVVLLLSFGVTTVLVRCEEPANMESE-------VQAPN---LLGGWQE--RNPEEKEIQDLLPKVLIKLNQLSNV-EYHLMP 79
Bm-CPI-2       MMSTMSIKEGLLVILLSLFLFDTTALIHRREIPHMESK-------GQMQRGQVLLGGWQE--RSPEDNEILELLPSVLTKVNQQSND-EYHLMP 84
Onchocystatin  ---MLTIKDGTLLIHLLLFSVVALVQLQGAKSARAKNPSKMESKTGENQDRPVLLGGWED--RDPKDEEILELLPSILMKVNEQSND-EYHLMP 88
Sialostatin    -------MTSTFALVLLLGGMAVCVATG-------------------------VFGGYSERANHQANPEFLNLAHYATSTWSAQQPGKTHFDTV 62
CCC            -------MAGARGCVVLLAAALMLVGAVLGSEDRS-----------------RLLGAPVP--VDENDEGLQRALQFAMAEYNRASND-KYSSRV 67
HCC            -------MAGPLRAPLLLLAILAVALAVSPAAGSSPGK------------PPRLVGGPMD--ASVEEEGVRRALDFAVGEYNKASND-MYHSRA 72
                                                                     : *.                           .

Ce-CPI-1       IKVVKAQVQVVAGTNTVLEVLVGESTCPRQGSVQASQVTAANCPLKSGGKRELYKVSIWEKPWENFKQTKAEKIRGVKPDEKI 139
Hc-cystatin    TKVLSAKTQVVSGVMFTSKVLFEESFCKK-GDVPVDQLKASNCAPKEGGKRVIYEISVLLQPWVKSEQVGVKVLRVFDPGEQV 122
Av-cystatin    IKLLKVSSQVVAGLRYKMEIQVAQSECKK---SSGEEVNLKTCKRLEGHPDQIITLEAWEKSWENFLQVKILEKKEVLSSV-- 157
Bm-CPI-2       IKLLKVSSQVVAGVKYKMEVQVARSECKK---SASEQVNLKTCKKLEGHPDQVMTLEVWEKPWEDFLQVNILETKVLSSV--- 161
Onchocystatin  IKLLKVSSQVVAGVKYKMDVQVARSQCKK---SSNEKVDLTKCKKLEGHPEKVMTLEVWEKPWENFMRVEILGTKEV------ 162
Sialostatin    AEVVKVETQVVAGTNYRLTLKVAESTCEL-----TSTYNKDTCLPKADAAHRTCTTVVFEN-LQGDKSVSPFECEAA------ 133
CCC            VRVISAKRQLVSGIKYILQVEIGRTTCPK---SSGDLQSCEFHDEPEMAKYTTCTFVVYSIPWLNQIKLLESKCQ-------- 139
HCC            LQVVRARKQIVAGVNYFLDVELGRTTCTK---TQPNLDNCPFHDQPHLKRKAFCSFQIYAVPWQGTMTLSKSTCQDA------ 146
                .:: .   *:*:*        : . .: *        .
```

Figure 1. A multiple-sequence alignment of nematode, arthropod and vertebrate cystatin sequences. Signal peptide (grey letters), conserved amino acids and regions important for inhibition of cysteine proteases, respectively (bold letters and grey box). Conserved residues for inhibition of legumain-like proteases (bold letters, grey background). Conserved cysteine residues forming disulphide bonds are indicated by brackets. Identical residues are indicated by (*), conserved amino acid substitutions with (:) and semi-conserved substitutions with (.). Accession numbers of protein sequences: *C. elegans* (*Ce*-CPI-1, AF100663); *H. contortus* (*Hc*-cystatin, AAB95324); *A. viteae* (*Av*-cystatin (Av17), AAA87228), *B. malayi* (*Bm*-CPI-2, AAB69857); *O. volvulus* (onchocystatin, P22085); chicken cystatin C (CCC, P01038); human cystatin C (HCC, CAA29096); *I. scapularis* (sialostatin L, 22164282).

Table 1. Parasite cystatins and their biological function

Inhibitor	Classification MEROPS (Superfamily)	Accession Number	Organism	Function	Refs.
NEMATODES					
Hc-cystatin	I25B (Type 2)	AF035945	*Haemonchus contortus*	• Inhibition of host cathepsin B and H • Control of cathepsin L-like protease activity	30
Onchocystatin *Ov*-CPI-1	I25B (Type 2) I25 (n/a)	M37105 AF177194	*Onchocerca volvulus*	• Inhibition of human cathepsin L and S • Modulation of molting from L3 to L4 • Inhibition of human PBMC T-cell proliferation • Down-regulation of MHC-II and CD86 • Induction of TNF-α, NO and Il-10	22, 23, 27, 45, 53, 54
Av-cystatin	I25B (Type 2)	L43053	*Acanthocheilo-nema viteae*	• Inhibition of host cathepsin L and S • Down-regulation of T-cell proliferation • Up-regulation of IL-10 production in mice • Induction of NO in INF-γ primed macrophages Ameliorating effect on asthma and colitis	25, 45, 52, 53, 56
Bm-CPI-1 *Bm*-CPI-2 *Bm*-CPI-3	I25B (Type 2) I25B (Type 2) n/a	U80972 AF015263 DS238934	*Brugia malayi*	• Important for parasite transmission from the mosquito to the mammalian host • Inhibition of host cathepsin S, L, B and AEP • Inhibition of antigen presentation	14, 26, 46
Ls-cystatin	I25B (Type 2)	AF229173	*Litosomoides sigmodontis*	• Enhancement of mRNA transcription for TNF-α • Down-regulation of antigen-specific splenocyte proliferation and NO production	47

continued on next page

Table 1. Continued

Inhibitor	Classification MEROPS (Superfamily)	Accession Number	Organism	Function	Refs.
Nippocystatin	I25B (Type 2)	AB050883	*Nippostrongylus brasiliensis*	• Inhibition of host cathepsin B and L • Inhibition of spleen cell proliferation • Reduction of allergic parameters in OVA-immunized mice • Increased host resistance to larval stages	28, 29
PLATHELMINTES					
Fh-cystatin	I25A (Type 2)	AJ312374	*Fasciola hepatica*	• Inhibition of parasite cathepsin L1	31
Fg-stefin1	n/a (Type 1)	FJ827152	*Fasciola gigantica*	• Inhibition of mammalian cathepsin B, L, S • Protection of parasite intestine/tegument surface	33
Sm-cystatin	I25A (Type 1)	AY334553	*Schistosoma mansoni*	• Regulation of haemoglobin degradation	34,35
ARTHROPODS					
Sialostatin L Sialostatin L2	I25B (Type 2) n/a (Type 2)	AAM93646 AAY66685	*Ixodes scapularis*	• Inhibition of host cathepsin L and S in dendritic cell cultures • Down-regulation of CD80/CD86 expression on dendritic cells • Inhibition of T-cell proliferation • Suppression of EAE	42, 43, 48
Hl-cystatin-1 *Hl*-cystatin-2 *Hl*-cystatin-3	I25A (Type 1) I25B (Type 2) I25B (Type 2)	EU426544 DQ364159 EU426545	*Haemaphysalis longicornis*	• Inhibition of cathepsin L and B • Involved in blood feeding, midgut physiology and tick innate immunity	37-39

continued on next page

Table 1. Continued

Inhibitor	Classification MEROPS (Superfamily)	Accession Number	Organism	Function	Refs.
Hl-SC-1	n/a (Type 2)	AB510962	*Haemaphysalis longicornis*	• Inhibition of host cathepsin L	36
Om-cystatin-1 *Om*-cystatin-2	I25B (Type 2) I25B (Type 2)	AY521024 AY547735	*Ornithodorus moubata*	• Inhibition of host cathepsin B and H • Regulation of blood digestion and heme detoxification	40
Bm-cystatin	I25A (Type 1)	DQ646915	*Boophilus microplus*	• Inhibition of host cathepsin L • Inhibition of tick vitellin degrading cysteine endopeptidase (VTDCE) • Protection from ingested harmful host factors	41

n/a: not assigned

parasitic cystatins listed in the database of family I25. Undoubtedly, this list will expand in the future as detailed phylogenomic *in silico* analysis of the cystatin superfamily from a larger number of genomes will be possible.[3] However, it is interesting that all parasite family I25 cystatins so far listed derive from parasites that include phases of tissue migration (nematodes and trematodes) or prolonged contact with host tissue during blood feeding (ticks). Parasite cystatins are mainly shown to regulate the activities of histolytic proteases or to contribute to the immunomodulatory mechanisms of parasites (Table 1 and discussed below in detail). The three dimensional structure of parasite cystatins and their molecular interactions with cysteine proteases has yet to be determined. However, based on their homology with vertebrate cystatins, it is likely that parasite cystatins will represent similar structures and mechanisms of inhibition as shown for other cytstatins.[14]

Migration of nematodes and trematodes within tissue or through tissue barriers requires a high energy expenditure. However, this migration enables parasites to escape from harmful immune responses at the site of mucosal surfaces or within tissues and thus facilitates parasite survival.[15,16] Larval stages of liver flukes (e.g., *Fasciola hepatica* and *Fasciola gigantica*) migrate through the parenchyma of the liver, while adult worms are located in the bile ducts. Trematode cystatins from *Fasciola spp.* and *Schistosoma spp.* are members of the I25 family, which group either into Type 1 or Type 2 cystatins (Table 1). In contrast, filarial nematodes are transmitted via arthropod vectors and are either found in lymphatic/subcutaneous tissue (adult stages) or in the peripheral blood and cutaneous tissue (microfilariae). Filarial cystatins are the best characterized parasite cystatins within the I25 family and all belong to subfamily B (Table 1). Among other molecules, they take part in the evasion strategies of the worm by conferring immune modulatory functions (see below). Although cystatins from nematodes lack the second disulfide bond that is characteristic of other Type 2 cystatins, it has been proposed that the overall structural features are similar (Fig. 1).[14] Ticks are ectoparasites and comprise important vectors of human and animal diseases, such as lyme disease and borreliosis. Some tick cystatins are transmitted to the host as part of the saliva and protect ticks from proteases and harmful immune reactions.[17] Cystatins from ticks are heterogenous and are found in I25 subfamilies A and B. In addition to classical I25 family cystatins, parasites express other unrelated cysteine protease inhibitors. These belong mainly to protozoan parasites. Examples are the homologues to chagasin and amoebiasin from *Trypanosoma spp.* and *Entamoeba spp.* that are members of the MEROPS family I42.[18] Other examples of protozoan parasites that form their own families are falstatin (*Plasmodium spp.*)[19,20] and toxostatin[21] (*Toxoplasma gondii*) from MEROPS family I72 and I81, respectively. However, these will not be discussed in detail in this chapter.

BIOLOGICAL FUNCTION OF PARASITE CYSTATINS

Protease Inhibition in Physiological Processes

One major function of parasite cystatins is the regulation of endogenous physiological processes like oogenesis, moulting during larval development and migration through host tissues. Thereby, parasite cystatins inhibit parasite and host proteases to promote parasite development.

One of the first parasite cystatins identified was onchocystatin, a highly-antigenic protein from the human pathogenic filaria *Onchocerca volvulus*. Onchocystatin is expressed in the cuticle of L3, L4, male and female adult worms and within the eggshell of microfilariae. It was suggested that onchocystatin might interfere with physiological processes during moulting from the larval stage L3 to L4 and the development of microfilariae in the uterus.[22,23] Studies with *Caenorhabditis elegans* revealed an essential role of cystatin during oogenesis and fertilization, indicating a similar role for cystatins of parasitic nematodes.[24] In further studies, homologous cystatins from various parasitic nematodes were identified and their inhibitory capacity to block host proteases has been characterized. Protease inhibition assays and Ki-value determinations are mainly performed with host proteases as there is often limited access to purified parasite proteases. Analysis with recombinant filarial cystatins such as onchocystatin (*Ov*17, *Ov*-CPI-2), *Av*-cystatin (Av17, from *Acanthocheilonema viteae*) and *Bm*-CPI-2 (from *Brugia malayi*) revealed that papain-like host proteases are efficiently inhibited.[25-27] For example, onchocystatin inhibited host cathepsin L (Ki = 0.038 nM), cathepsin S (Ki = 0.033 nM) and cathepsin B (Ki = 0.494 μM).

The release of biologically active cystatins is not restricted to filarial nematodes, as recombinant cystatins from the gastrointestinal nematodes *Nippostrongylus brasiliensis* (nippocystatin) and *Hemonchus contortus* revealed similar functions by inhibiting host cathepsins.[28-30] In *B. malayi* three distinct cystatins (*Bm*-CPI-1, -2, -3) have been identified and the analysis of their expression patterns revealed stage specific differences. *Bm*-CPI-2 is expressed throughout the life-cycle while *Bm*-CPI-1 and *Bm*-CPI-3 are expressed at later stages in the mosquito vector but not in the mammalian host. This indicates a specific role for different cystatins in the host or during parasite transmission.[14]

In parasitic trematodes a multi-domain cystatin of *F. hepatica* efficiently inhibits parasite cathepsin L1 activity, which is important for successful penetration of host intestine.[31,32] The protein showed structural features familiar to Type 3 cystatins but phylogenic analysis revealed that it was more closely related to Type 2 cystatins.[31] *F. gigantica* expresses a heat stable Type 1 cystatin designated as *Fg*-Stefin-1, that was proposed to protect the intestine and tegumental surface of the parasite from extracellular proteolytic damage.[33] The cystatin in the blood fluke *Schistosoma mansoni* (*Sm*-cys), the causative agent of schistosomiasis, belongs to the Type 1 cystatins and is equally expressed by male and female worms. Studies suggest an involvement of *Sm*-cys in the regulation of haemoglobin degradation in the schistosomula, which use red blood cells as a main source of amino acids. This suggests a possible function for *Sm*-cys in the gut of the parasite.[34,35]

In disease-transmitting ectoparasites, most of the cystatins have been identified in ticks. They are expressed mainly in the saliva or midgut in order to regulate proteolytic processes or help to evade harmful immune responses that occur during the relatively long feeding process on the host. For example, knock-down of cystatin expression by siRNA in salivary glands and midgut of *Amblyomma americanum* led to reduced feeding abilities, which suggests that cystatin protects ticks from harmful ingested host factors.[17] Further, the Type 2 cystatins *from Haemaphysalis longicornis* are potent cathepsin L inhibitors (*Hl*-cyst-1: IC_{50} = 0.202 pM; *Hl*-cyst-2: IC_{50} = 1.68 μM) and are differentially expressed in the salivary glands (*Hl*-cyst-1) or the midgut (*Hl*-cyst-2, -3). The expression levels of these cystatins are markedly induced during blood feeding suggesting an important role in the regulation of midgut cysteine proteases and/or in immune evasion.[36-39] Cystatins from *Ornithodorus moubata* (*Om*-cystatin-1 and -2) prevent endogenous peptidases from digesting stored blood meals and possibly play a role in heme detoxification.[40] OmCystatins are also differentially expressed in the midgut (*Om*-cystatin-1, -2) and the salivary glands, ovaries and malpighian tubulus (*Om*-cystatin-2), highlighting the various functions of cystatins in ticks. In addition, cystatin from *Boophilus microplus* inhibits the vitellin-degrading cysteine endopeptidase and therefore might affect embryogenesis by regulating endogenous proteolysis.[41] Sialostatin L and L2 from the saliva of *Ixodes scapularis* inhibit various mammalian cathepsins such as cathepsin L (Ki = 4.7 nM) and cathepsin V (Ki = 57 nM) and contribute to the blood-feeding success of the ticks.[42,43] In conclusion, parasite cystatins perform key roles in the inhibition of parasite and host proteases that are involved in essential physiological processes during the parasite life-cycle.

INTERFERENCE WITH ANTIGEN PROCESSING AND PRESENTATION

Many immunological processes rely on extracellular and intracellular activities of proteases and their respective inhibitors. In particular, cysteine proteases are involved in processes such as antigen processing and presentation and regulation of pattern recognition receptor (PRR) signalling.[44] These proteases are mainly localized within the endolysosomal compartment and are regulated by a variety of endogenous inhibitors, including cystatins.[1,2] Most parasite cystatins are potent inhibitors of host cysteine proteases and thus they are involved in immune modulation of host cells.[45]

The first evidence that parasite cystatins directly interfere with antigen presentation came from studies with filarial cystatins. *Bm*-CPI2, from *B. malayi,* inhibits the cysteine proteases cathepsin B/L, S and asparginyl endopeptidase (AEP) in lysosomal extracts of human EBV-transformed B-cell lines. These proteases are important enzymes for antigen processing and presentation; thus, their inhibition probably leads to a reduced MHC class-II-mediated antigen presentation in antigen presenting cells (APCs) (Fig. 2). Indeed, *Bm*-CPI-2 inhibited in vitro processing of tetanus toxin antigen by these lysosome fractions and also inhibited the processing of the invariant chain (Ii) of class-II molecules, which together mediated a reduced presentation of selected tetanus toxin epitopes by APC.[26] AEP inhibitory domains are common in mammalian cystatins and were identified for various filarial cystatins, but not for the free-living nematode *C. elegans* raising the possibility that AEP inhibitory function was acquired during co-evolution of the host and the parasite.[46] Further evidence, for the manipulation of parasite specific antigen presentation by filarial cystatins, was provided by an experiment with constant in vivo application of *Ls*-cystatin, from *Litomosoides sigmodontis*, by micro-osmotic pumps implanted into the peritoneum of mice. Subsequent challenge with microfilariae revealed a reduced antigen-specific response in spleen cells of *Ls*-cystatin-treated animals.[47] Similarly, nippocystatin treatment of ovalbumin-immunized mice inhibited antigen-specific restimulation of spleen cells, indicating that other nematode cystatins mediate similar effects.[29] These studies indicated that not only unrelated antigen presentation is inhibited but also parasite-specific responses are blocked by parasite cystatins.

In human PBMC cultures, onchocystatin reduced the purified protein derivate (PPD)-induced proliferation, giving further evidence for a direct modulation of antigen presentation in APCs by parasite cystatins.[27] Interestingly, these experiments also show that HLA-DR and CD86 surface expression on human monocytes was reduced in PBMC cultures treated with onchocystatin, an effect that was dependent on a direct stimulation of IL-10 and other cytokines. This suggests a second, receptor-mediated effect of cystatins on their target cells. Comparable effects were mediated by the tick cystatin sialostatin L, which was shown to down-modulate MHC-II and costimulatory molecule CD86 of lipopolysaccharide-primed murine dendritic cells and to suppress antigen-specific

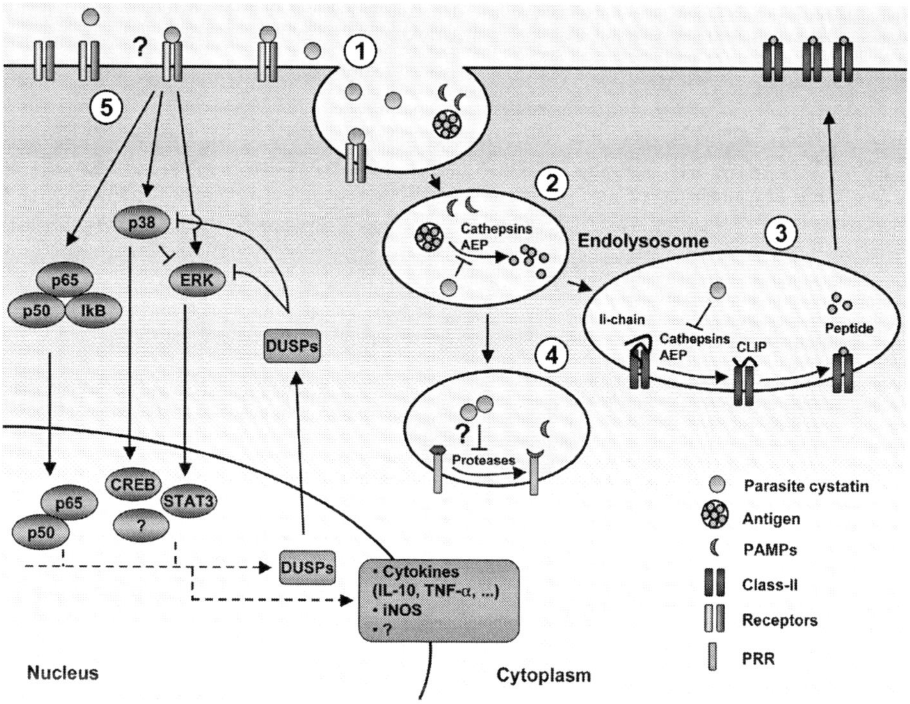

Figure 2. Proposed mechanisms of host cell modulation by parasite cystatins. Parasite cystatins can manipulate host antigen presenting cells at different sites. Parasite cystatins are actively taken up via endocytosis (1). Within the endolysosomal compartment parasite cystatins prevent antigen degradation by inhibiting proteases such as cathepsins and AEP (2). The inhibition of these target enzymes also blocks the processing of the MHCII invariant chain (Ii) peptide to CLIP, thus leading to reduced antigen presentation (3). Theoretically, parasite cystatins may also interfere with the processing of pattern recognition receptors, which might render APCs insensitive for activation via specific pathogen-associated molecular patterns (PAMPs) (4). In addition, parasite cystatins induce activation of MAPK and NF-κB pathways to induce IL-10 and low levels of pro-inflammatory cytokines such as TNF-α and IL-6. The release of TNF-α subsequently induces the expression of iNOS. Stimulation with parasite cystatin leads also to the expression of dual specificity phosphatases (DUSPs) that negatively regulate MAPK pathways (5).

proliferation in vitro and in vivo.[48] Thus, interference with antigen processing and presentation is a major feature of parasite cystatins (Fig. 2).

MODULATION OF CYTOKINE RESPONSES AND NITRIC OXIDE PRODUCTION

An intrinsic feature of endogenous and parasite cystatins is the modification of the cytokine and nitric oxide (NO) production of APCs, such as macrophages.[5] Thereby, the presence or absence of pro-inflammatory stimuli such as IFN-γ and TNF-α or an anti-inflammatory mediator like IL-10 define the modulatory features of cystatins. Stimulation of IFN-γ-primed peritoneal macrophages with chicken cystatin drastically increases NO production by a mechanism that is independent of the protease inhibitory activity.[49] The NO production is an indirect effect of the release of TNF-α and IL-10 and could be blocked by neutralizing anti-TNF-α antibodies.[50] Interestingly, endogenous murine cystatin C also induces the release of TNF-α and IL-10 in IFN-γ primed mouse macrophages.[51] Similarly, our laboratory showed that cystatins of filaria, onchocystatin and *Av*-cystatin induced NO production in IFN-γ primed macrophages. This influence was dependent on TNF-α and IL-10, suggesting a conserved, yet unknown mechanism by which cystatins activate macrophages to produce cytokines.[52] The increased IL-10 production in mouse spleen cells and human PBMC was responsible for the reduced proliferative T-cell responses and reduced HLA-DR and CD86 surface expression on APC in the presence of *Av*-cystatin and onchocystatin respectively.[25,27,53] Comparison of the effects of filarial cystatins and cystatin from the free-living nematode *C. elegans* revealed similarities such as NO production in IFN-γ primed macrophages. However, filarial cystatin was a more potent suppressor of immune responses including T-cell proliferation. This was accompanied by the release of higher IL-10 levels.[54] Recent data indicated that filarial cystatin exploits signalling events via mitogen-activated protein kinases (MAPK) to induce the production of the anti-inflammatory cytokine IL-10 in macrophages. In this setting the MAPK activation by cystatin was regulated via induction of dual specificity phosphatases, which are known feedback regulators of MAPK. Preliminary data further suggests that MAPK activation induced, in parallel, low level expression of pro-inflammatory cytokines such as TNF-α. Cystatin-induced cytokine production was associated with the activation of transcription factors such as CREB, STAT3 and NF-κB (Klotz et al, unpublished data).[51,55] On the contrary, studies with sialostatin L from *I. scapularis* indicate that parasite cystatins interfere with toll like receptor-mediated release of IL-12 and TNF-α but not IL-10 by dendritic cells.[48] Therefore cystatins, in addition to their function as protease inhibitors, exhibit intrinsic structural features that enable them to modulate the host immune system, possibly by receptor engagement. Parasite cystatins have further evolved immune regulatory capacities and probably comprise additional sequence characteristics that enable them to manipulate and reprogram host immune cells. Which common receptors (if any) and pathways are manipulated by various parasite cystatins or whether different cystatins ligate specific receptors and effect different downstream signalling pathways remains to be investigated in future studies.

ANTICIPATED DEVELOPMENTS

Interestingly, the intrinsic immunomodulatory activities of cystatins, in particular those expressed by parasites, might be utilized to modulate unrelated inflammatory immune responses. In this respect, we showed that a cystatin from a parasitic nematode significantly reduced allergic and inflammatory responses in two murine disease models. In a mouse model of allergic airway hyperreactivity, we showed that treatment with filarial cystatin during sensitization with a model allergen, or before challenge, led to a significant reduction of several disease-related parameters. Filarial cystatin induced a significant reduction in influx of inflammatory cells and suppressed the production of IgE, IL-4 and bronchiole mucus in the lung. These effects were dependent on macrophages and IL-10, because the cystatin effect could be abrogated by applying IL-10 receptor antibodies or by depletion of macrophages.[56] Similarly, *Escherichia coli*-expressed cystatin from *N. brasiliensis* suppressed antigen–specific, but not total serum, IgE in ovalbumin-immunized mice and further reduced levels of IL-4 and IFN-γ in response to stimulation with antigen.[29]

Another cystatin from the tick *Ixodes scapularis*, sialostatin L, has been shown to exhibit anti-inflammatory and immunosuppressive activities.[43] Target cells of sialostatin are myeloid-derived cells and its immunomodulatory activities were mediated via inhibition of cathepsin S. In addition, administration of sialostatin in experimental murine autoimmune encephalomyelitis significantly prevented disease symptoms.[48] The applicability of cystatins to interfere with disease progression has also been shown in a completely different approach with endogenous cystatin. Chicken cystatin, like other cystatins, was shown to induce NO production in IFN-γ primed macrophages and this property was successfully exploited to cure lethal murine visceral leishmaniasis.[57] Further studies on chicken cystatin identified the specific immuno-modulatory peptide within cystatin that ameliorates disease progression[58] via protein kinase-dependent signalling pathway in macrophages.[55] Taken together, the data on parasite as well as host cystatins clearly indicate the strong immuno-modulatory capacity of cystatins. It is likely that cystatins from parasitic organisms have adapted during co-evolution to achieve stronger modulatory effects.

CONCLUSION

An inevitable feature of the parasitic life style is the close contact with host tissue. In particular, helminths reside in the tissues and are thereby directly exposed to the immune system. In chronic infections like filariasis, those interactions last for decades. The necessity to modulate the host immune system might be one fundamental reason why modulatory cystatins have evolved in parasites that are exposed for a prolonged time to the host immune system. However, it is not surprising that parasites adopted the general immune-modulating features of cystatins during evolution in order to ensure their parasitic life style. Future studies have to draw particular attention to the mechanisms manipulated by parasite cystatins in host cells that could potentially lead to sophisticated applications against overshooting inflammation either by applying parasite cystatins directly or by targeting particular pathways addressed by these molecules. In this respect, future studies will pinpoint human diseases where modulatory mechanisms exhibited by parasite cystatins could be beneficial. Hence, it may be speculated that the regulatory influence of parasite cystatins could lead to the development of new drugs that interfere with inflammatory-driven diseases.

REFERENCES

1. Abrahamson M, Alvarez-Fernandez M, Nathanson CM. Cystatins. Biochem Soc Symp 2003; 70:179-199.
2. Kopitar-Jerala N. The role of cystatins in cells of the immune system. FEBS Lett 2006; 580:6295-6301.
3. Kordis D, Turk V. Phylogenomic analysis of the cystatin superfamily in eukaryotes and prokaryotes. BMC Evol Biol 2009; 9:266.
4. Turk V, Stoka V, Turk D. Cystatins: biochemical and structural properties and medical relevance. Front Biosci 2008; 13:5406-5420.
5. Zavasnik-Bergant T. Cystatin protease inhibitors and immune functions. Front Biosci 2008; 13:4625-4637.
6. Rawlings ND, Barrett AJ. Evolution of proteins of the cystatin superfamily. J Mol Evol 1990; 30:60-71.
7. Rawlings ND, Barrett AJ, Bateman A. MEROPS: the peptidase database. Nucleic Acids Res 2010; 38:D227-233.

8. Bode W, Engh R, Musil D et al. The 2.0 A X-ray crystal structure of chicken egg white cystatin and its possible mode of interaction with cysteine proteinases. Embo J 1988; 7:2593-2599.
9. Hall A, Ekiel I, Mason RW et al. Structural basis for different inhibitory specificities of human cystatins C and D. Biochemistry 1998; 37:4071-4079.
10. Alvarez-Fernandez M, Barrett AJ, Gerhartz B et al. Inhibition of mammalian legumain by some cystatins is due to a novel second reactive site. J Biol Chem 1999; 274:19195-19203.
11. Staniforth RA, Giannini S, Higgins LD et al. Three-dimensional domain swapping in the folded and molten-globule states of cystatins, an amyloid-forming structural superfamily. Embo J 2001; 20:4774-4781.
12. Janowski R, Kozak M, Jankowska E et al. Human cystatin C, an amyloidogenic protein, dimerizes through three-dimensional domain swapping. Nat Struct Biol 2001; 8:316-320.
13. Lee C, Bongcam-Rudloff E, Sollner C et al. Type 3 cystatins; fetuins, kininogen and histidine-rich glycoprotein. Front Biosci 2009; 14:2911-2922.
14. Gregory WF, Maizels RM. Cystatins from filarial parasites: evolution, adaptation and function in the host-parasite relationship. Int J Biochem Cell Biol 2008; 40:1389-1398.
15. Mulcahy G, O'Neill S, Fanning J et al. Tissue migration by parasitic helminths—an immunoevasive strategy? Trends Parasitol 2005; 21:273-277.
16. Read AF, Skorping A. The evolution of tissue migration by parasitic nematode larvae. Parasitology 1995; 111:359-371.
17. Karim S, Miller NJ, Valenzuela J et al. RNAi-mediated gene silencing to assess the role of synaptobrevin and cystatin in tick blood feeding. Biochem Biophys Res Commun 2005; 334:1336-1342.
18. Wang SX, Pandey KC, Scharfstein J et al. The structure of chagasin in complex with a cysteine protease clarifies the binding mode and evolution of an inhibitor family. Structure 2007; 15:535-543.
19. Pandey KC, Singh N, Arastu-Kapur S et al. Falstatin, a cysteine protease inhibitor of Plasmodium falciparum, facilitates erythrocyte invasion. PLoS Pathog 2006; 2:e117.
20. Rennenberg A, Lehmann C, Heitmann A et al. Exoerythrocytic Plasmodium parasites secrete a cysteine protease inhibitor involved in sporozoite invasion and capable of blocking cell death of host hepatocytes. PLoS Pathog 2010; 6:e1000825.
21. Huang R, Que X, Hirata K et al. The cathepsin L of Toxoplasma gondii (TgCPL) and its endogenous macromolecular inhibitor, toxostatin. Mol Biochem Parasitol 2009; 164:86-94.
22. Lustigman S, Brotman B, Huima T et al. Characterization of an Onchocerca volvulus cDNA clone encoding a genus specific antigen present in infective larvae and adult worms. Mol Biochem Parasitol 1991; 45:65-75.
23. Lustigman S, Brotman B, Huima T et al. Molecular cloning and characterization of onchocystatin, a cysteine proteinase inhibitor of Onchocerca volvulus. J Biol Chem 1992; 267:17339-17346.
24. Hashmi S, Zhang J, Oksov Y et al. The Caenorhabditis elegans CPI-2a cystatin-like inhibitor has an essential regulatory role during oogenesis and fertilization. J Biol Chem 2006; 281:28415-28429.
25. Hartmann S, Kyewski B, Sonnenburg B et al. A filarial cysteine protease inhibitor down-regulates T-cell proliferation and enhances interleukin-10 production. Eur J Immunol 1997; 27:2253-2260.
26. Manoury B, Gregory WF, Maizels RM et al. Bm-CPI-2, a cystatin homolog secreted by the filarial parasite Brugia malayi, inhibits class II MHC-restricted antigen processing. Curr Biol 2001; 11:447-451.
27. Schonemeyer A, Lucius R, Sonnenburg B et al. Modulation of human T-cell responses and macrophage functions by onchocystatin, a secreted protein of the filarial nematode Onchocerca volvulus. J Immunol 2001; 167:3207-3215.
28. Dainichi T, Maekawa Y, Ishii K et al. Molecular cloning of a cystatin from parasitic intestinal nematode, Nippostrongylus brasiliensis. J Med Invest 2001; 48:81-87.
29. Dainichi T, Maekawa Y, Ishii K et al. Nippocystatin, a cysteine protease inhibitor from Nippostrongylus brasiliensis, inhibits antigen processing and modulates antigen-specific immune response. Infect Immun 2001; 69:7380-7386.
30. Newlands GF, Skuce PJ, Knox DP et al. Cloning and expression of cystatin, a potent cysteine protease inhibitor from the gut of Haemonchus contortus. Parasitology 2001; 122:371-378.
31. Khaznadji E, Collins P, Dalton JP et al. A new multi-domain member of the cystatin superfamily expressed by Fasciola hepatica. Int J Parasitol 2005; 35:1115-1125.
32. Mulcahy G, O'Connor F, Clery D et al. Immune responses of cattle to experimental anti-Fasciola hepatica vaccines. Res Vet Sci 1999; 67:27-33.
33. Tarasuk M, Vichasri Grams S, Viyanant V et al. Type I cystatin (stefin) is a major component of Fasciola gigantica excretion/secretion product. Mol Biochem Parasitol 2009; 167:60-71.
34. Morales FC, Furtado DR, Rumjanek FD. The N-terminus moiety of the cystatin SmCys from Schistosoma mansoni regulates its inhibitory activity in vitro and in vivo. Mol Biochem Parasitol 2004; 134:65-73.
35. Wasilewski MM, Lim KC, Phillips J et al. Cysteine protease inhibitors block schistosome hemoglobin degradation in vitro and decrease worm burden and egg production in vivo. Mol Biochem Parasitol 1996; 81:179-189.

36. Yamaji K, Tsuji N, Miyoshi T et al. A salivary cystatin, HISC-1, from the ixodid tick Haemaphysalis longicornis play roles in the blood-feeding processes. Parasitol Res 2009; 106:61-68.
37. Zhou J, Liao M, Gong H et al. Characterization of Hlcyst-3 as a member of cystatins from the tick Haemaphysalis longicornis. Exp Appl Acarol 2010; Epub ahead of print.
38. Zhou J, Liao M, Ueda M et al. Characterization of an intracellular cystatin homolog from the tick Haemaphysalis longicornis. Vet Parasitol 2009; 160:180-183.
39. Zhou J, Ueda M, Umemiya R et al. A secreted cystatin from the tick Haemaphysalis longicornis and its distinct expression patterns in relation to innate immunity. Insect Biochem Mol Biol 2006; 36:527-535.
40. Grunclova L, Horn M, Vancova M et al. Two secreted cystatins of the soft tick Ornithodoros moubata: differential expression pattern and inhibitory specificity. Biol Chem 2006; 387:1635-1644.
41. Lima CA, Sasaki SD, Tanaka AS. Bmcystatin, a cysteine proteinase inhibitor characterized from the tick Boophilus microplus. Biochem Biophys Res Commun 2006; 347:44-50.
42. Kotsyfakis M, Karim S, Andersen JF et al. Selective cysteine protease inhibition contributes to blood-feeding success of the tick Ixodes scapularis. J Biol Chem 2007; 282:29256-29263.
43. Kotsyfakis M, Sa-Nunes A, Francischetti IM et al. Antiinflammatory and immunosuppressive activity of sialostatin L, a salivary cystatin from the tick Ixodes scapularis. J Biol Chem 2006; 281:26298-26307.
44. Bird PI, Trapani JA, Villadangos JA. Endolysosomal proteases and their inhibitors in immunity. Nat Rev Immunol 2009; 9:871-882.
45. Vray B, Hartmann S, Hoebeke J. Immunomodulatory properties of cystatins. Cell Mol Life Sci 2002; 59:1503-1512.
46. Murray J, Manoury B, Balic A et al. Bm-CPI-2, a cystatin from Brugia malayi nematode parasites, differs from Caenorhabditis elegans cystatins in a specific site mediating inhibition of the antigen-processing enzyme AEP. Mol Biochem Parasitol 2005; 139:197-203.
47. Pfaff AW, Schulz-Key H, Soboslay PT et al. Litomosoides sigmodontis cystatin acts as an immunomodulator during experimental filariasis. Int J Parasitol 2002; 32:171-178.
48. Sa-Nunes A, Bafica A, Antonelli LR et al. The immunomodulatory action of sialostatin L on dendritic cells reveals its potential to interfere with autoimmunity. J Immunol 2009; 182:7422-7429.
49. Verdot L, Lalmanach G, Vercruysse V et al. Cystatins up-regulate nitric oxide release from interferon-gamma-activated mouse peritoneal macrophages. J Biol Chem 1996; 271:28077-28081.
50. Verdot L, Lalmanach G, Vercruysse V et al. Chicken cystatin stimulates nitric oxide release from interferon-gamma-activated mouse peritoneal macrophages via cytokine synthesis. Eur J Biochem 1999; 266:1111-1117.
51. Frendeus KH, Wallin H, Janciauskiene S et al. Macrophage responses to interferon-gamma are dependent on cystatin C levels. Int J Biochem Cell Biol 2009; 41:2262-2269.
52. Hartmann S, Schonemeyer A, Sonnenburg B et al. Cystatins of filarial nematodes up-regulate the nitric oxide production of interferon-gamma-activated murine macrophages. Parasite Immunol 2002; 24:253-262.
53. Hartmann S, Lucius R. Modulation of host immune responses by nematode cystatins. Int J Parasitol 2003; 33:1291-1302.
54. Schierack P, Lucius R, Sonnenburg B et al. Parasite-specific immunomodulatory functions of filarial cystatin. Infect Immun 2003; 71:2422-2429.
55. Kar S, Ukil A, Das PK. Signaling events leading to the curative effect of cystatin on experimental visceral leishmaniasis: involvement of ERK1/2, NF-kappaB and JAK/STAT pathways. Eur J Immunol 2009; 39:741-751.
56. Schnoeller C, Rausch S, Pillai S et al. A helminth immunomodulator reduces allergic and inflammatory responses by induction of IL-10-producing macrophages. J Immunol 2008; 180:4265-4272.
57. Das L, Datta N, Bandyopadhyay S et al. Successful therapy of lethal murine visceral leishmaniasis with cystatin involves up-regulation of nitric oxide and a favorable T-cell response. J Immunol 2001; 166:4020-4028.
58. Mukherjee S, Ukil A, Das PK. Immunomodulatory peptide from cystatin, a natural cysteine protease inhibitor, against leishmaniasis as a model macrophage disease. Antimicrob Agents Chemother 2007; 51:1700-1707.

INDEX

K

L

M

N

O

P

Q

R

S

T

V

Y